Lecture Notes in Electrical Engineering

Volume 394

About this Series

"Lecture Notes in Electrical Engineering (LNEE)" is a book series which reports the latest research and developments in Electrical Engineering, namely:

- Communication, Networks, and Information Theory
- Computer Engineering
- Signal, Image, Speech and Information Processing
- Circuits and Systems
- Bioengineering

LNEE publishes authored monographs and contributed volumes which present cutting edge research information as well as new perspectives on classical fields, while maintaining Springer's high standards of academic excellence. Also considered for publication are lecture materials, proceedings, and other related materials of exceptionally high quality and interest. The subject matter should be original and timely, reporting the latest research and developments in all areas of electrical engineering.

The audience for the books in LNEE consists of advanced level students, researchers, and industry professionals working at the forefront of their fields. Much like Springer's other Lecture Notes series, LNEE will be distributed through Springer's print and electronic publishing channels.

More information about this series at http://www.springer.com/series/7818

Kapila Rohan Attele · Amit Kumar
V. Sankar · N.V. Rao · T. Hitendra Sarma
Editors

Emerging Trends in Electrical, Communications and Information Technologies

Proceedings of ICECIT-2015

Springer

Editors
Kapila Rohan Attele
Mathematics and Computer Science
Chicago State University
Chicago, IL
USA

N.V. Rao
Department of Computer Science
CVR College of Engineering
Hyderabad
India

Amit Kumar
BioAxis DNA Research Centre
(P) Hyderabad
IEEE India Council
Hyderabad
India

T. Hitendra Sarma
Department of Computer Science and Engineering
Srinivasa Ramanujan Institute of Technology
Ananthapur
India

V. Sankar
Department of Electrical Engineering
JNTU College of Engineering
Ananthapur
India

ISSN 1876-1100 ISSN 1876-1119 (electronic)
Lecture Notes in Electrical Engineering
ISBN 978-981-10-1538-0 ISBN 978-981-10-1540-3 (eBook)
DOI 10.1007/978-981-10-1540-3

Library of Congress Control Number: 2016946938

Printed on acid-free paper

This Springer imprint is published by Springer Nature
The registered company is Springer Nature Singapore Pte Ltd.
The registered company address is: 152 Beach Road, #22-06/08 Gateway East, Singapore 189721, Singapore

Organizing Committee

General Chair

Prof. L. M. Patnaik, INSA Senior Scientist and Adjunct Faculty, National Institute of Advanced Studies, IISc. Campus, Bangalore

Chief Patron

Prof. M.M.M. Sarcar, Vice-Chancellor, JNTUA Ananthapuramu

Patron

Smt. J. Padmavathy, Chairperson, Srinivasa Ramanujan Institute of Technology, Ananthapuramu
Sri. A. Sambasiva Reddy, Correspondent & Secretary,
Srinivasa Ramanujan Institute of Technology, Ananthapuramu

Co-patron

Sri. K. Jagan Mohan Reddy, CEO, Srinivasa Ramanujan Institute of Technology, Ananthapuramu
Dr. K. Subba Reddy, Principal, Srinivasa Ramanujan Institute of Technology, Ananthapuramu

Organizing Chairs

Dr. P. Viswanath, Professor, Indian Institute of Information Technology–Sri City
Dr. T. Hitendra Sarma, Srinivasa Ramanujan Institute of Technology, Ananthapuramu
Dr. Amit Kumar, Vice Chair—IEEE India Council (Technical and Professional Activities)

Program Chairs

Prof. M. Ramasubba Reddy, Indian Institute of Technology, Madras
Prof. V. Sankar, JNTUA, Ananthapuramu
Prof. K. G Srinivasa, MS Ramaiah Institute of Technology, Bangalore

International Program Committee

Dr. N.D.R. Sarma, Electricity Reliability Council of Texas (ERCOT), United States of America
Dr. Kapila Rohan Attele, Chicago State University, United States of America
Dr. Pascal Paschos, Chicago State University, United States of America
Dr. V. Suresh Babu, United Technologies Research Center, Ireland
Dr. Chitti Babu Baladhandautham, VSB—Technical University of Ostrava, Ostrava—Poruba, Czech Republic

Technical Program Committee

Prof. R. Mahanthy, Indian Institute of Technology, Bhubaneswar
Prof. M. Veerachary, Indian Institute of Technology, Delhi
Prof. Srirama Srinivas, Indian Institute of Technology, Madras
Dr. K. Siva Kumar, Indian Institute of Technology, Hyderabad
Prof. T.S. Ramu, Indian Institute of Sciences, Bangalore
Dr. G. Narayanan, Indian Institute of Sciences, Bangalore
Prof. C. Nagamani, National Institute of Technology, Trichi
Prof. V.T. Somasekhar, National Institute of Technology, Warangal
Dr. D.M. Vinod Kumar, National Institute of Technology, Warangal
Dr. Srinivasa Rao Sandepudi, National Institute of Technology, Warangal
Dr. G. Siva Kumar, National Institute of Technology, Warangal
Dr. Akhilesh Swarup, National Institute of Technology, Kurukshetra
Dr. Sheron Figarado, National Institute of Technology, Karnataka
Dr. B. Venugopal Reddy, National Institute of Technology, Goa
Dr. Narayan Manjrekar, BITS Pilani, Goa
Prof. K. Vaisakh, Andhra University

Prof. M.L.S. Deva Kumar, JNTUA, Ananthapuramu
Prof. M. Vijaya Kumar, JNTUA, Ananthapuramu
Dr. B. Durga Prasad, JNTUA, Ananthapuramu
Prof. Yadaiah, JNTU, Hyderabad
Dr. Ch. Sai Babu, JNT University, Kakinada
Dr. S. Sivanagaraju, JNT University, Kakinada
Prof. G.V. Marutheswar, S K. University, Ananthapuramu
Prof. P.S. Raju, SVU College of Engineering, Tirupati
Dr. T. Gouri Manohar, SVU College of Engineering, Tirupati
Prof. K. Nagabhushan Raju, Sri Krishnadevaraya University, Ananthapuramu
Prof. R. Bala Subramanian, G. Narayanamma Institue of Technology and Science, Hyderabad
Prof. N. Moorthy Muthukrishnan, G. Narayanamma Institue of Technology and Science, Hyderabad
Prof. N. Malla Reddy, G. Narayanamma Institue of Technology and Science, Hyderabad
Prof. Atul Negi, Hyderabad Central University, Hyderabad
Prof. P. Viswanath, Indian Institute of Information Technology, Sri City
Dr. U.S.N. Raju, National Institute of Technology, Warangal
Dr. R.B.V. Subramanyam, National Institute of Technology, Warangal
Dr. R. Vijayaarjunan, Manipal University, Dubai, UAE
Dr. M. Lakshminarayana, Chairman, IEEE Hyderabad Section
Dr. Jyothi Bellari, Aditya College of Engineering, Madanapalle
Dr. Uday Bhaskar, Governmet Arts & Science College, Ananthapuramu
Prof. N. Rama Krishna, G. Narayanamma Institute of Technology & Science, Hyderabad
Dr. L.B. Raj Kumar, G. Narayanamma Institute of Technology & Science, Hyderabad
Dr. K. Rama Linga Reddy, G. Narayanamma Institute of Technology & Science, Hyderabad
Dr. N. Kalyani, G. Narayanamma Institue of Technology and Science, Hyderabad
Dr. V. Vijay Kumar, Anurag Group of Institutions, Hyderabad
Dr. V.V. Krishna, Anurag Group of Institutions, Hyderabad
Dr. Srinivas Rao Kothapalli, Anurag Group of Institutions, Hyderabad
Dr. B. Eswara Reddy, JNTUA, Ananthapuramu
Dr. C. Chandra Mohan, JNTUH, Hyderabad
Dr. C. Shobha Bindu, JNTUA, Ananthapuramu
Dr. K. Madhavi, JNTUA, Ananthapuramu
Dr. J. Sasi Kiran, Vidya Vikas Institute of Technology, Hyderabad
Dr. G. Satyanarayana Murthy, Aditya Institute of Technology and Management, Tekkali
Dr. Pullela S.V.V.S.R. Kumar, JNTU, Kakinada
Dr. L. Sumalatha, University College of Engineering Kakinada, JNTUK
Dr. C. Chandra Mohan, JNTUH, Hyderabad
Dr. A. Sri Krishna, RVR & JC of Engineering, Guntur

Prof. N.V. Rao, CVR College of Engineering, Hyderabad
Dr. Esther Rani, CVR College of Engineering, Hyderabad
Dr. Nookala Srinivasa Rao, Matrusri Engineering College, Hyderabad
Dr. N. Kasiviswanath, G. Pulla Reddy Engineering College, Kurnool
Dr. T. Kishore Kumar, National Institute of Technology, Warangal
Dr. Bindu A. Thomas, Vidya Vikas Institute of Engineering and Technology, Mysuru
Dr. A. Srikrishna, RVR & JC College of Engineering, Chowdavaram, Guntur
Mr. N. Venkatesh, Redpine Signals, Hyderabad
Dr. B. Bhaskar, SJB Institute of Technology, Bangalore
Dr. P. Dinesha, Dayananda Sagar College of Engineering, Bengaluru
Dr K.V. Mahendra Prashanth, SJB Institute of Technology, Bengaluru
Mr. K. Raghavendra, Scientist 'SF', Department of Space, ISRO
Dr. Rajarshi Mahapatra, CVR College of Engineering, Hyderabad
Dr. G.N. Kodandaramaiah, Kuppam Engineering College, Kuppam
Dr. A.R. Reddy, Madanapalle Institute of Technology and Science
Dr. Abdul Jilani, Madanapalle Institute of Technology and Science

Contents

Part I
Trends in Knowledge and Data Engineering, Big Data and Advanced Computing Techniques, Computer Networks, Software Engineering and Artificial Intelligence

An Enhanced Mechanism for Balanced Job Scheduling Based on Deadline Control in Computational Grid

K. Jairam Naik, A. Jagan and N. Satyanarayana

Abstract Grid can be thought of as a network of heterogeneous interactive computational resources from multiple administrative domains that collectively works towards achieving a common goal. Inefficient scheduling and work load distribution among the various computational resources in a network is one of the major issues that affect grid performance. Some resources may tend to be heavily loaded while some are kept idle, thus affecting the overall performance of the grid. Balanced load scheduling is thus a serious issue which needs to be properly addressed in the grid. Balancing the load affects some factors like job execution and service selection, thus making it all the more necessary to be well implemented. In this paper we propose a distributed, dynamic and balanced load scheduling scheme on grids which considers deadline of jobs. Our approach for solving the problem goes as follows: The resources first check their state and make a request to the Grid Broker based on the change in state of their load. Then, the Grid Broker assigns Jobs (Gridlets) among resources, provides schedules for load balancing and selecting best node of a resource for execution under the given deadline. We applied our balanced job scheduling mechanisms into a popular simulation platform called GridSim Tool kit. Experimental results prove that our balanced job scheduling mechanism can reduces the make span, failure tendency, and resubmitted time by maximizing the throughput.

K. Jairam Naik (✉)
Department of CSE, Vasavi College of Engineering, Hyderabad, Telangana, India
e-mail: jairam.524@gmail.com

A. Jagan
Department of CSE, Dr. BV Raju Institute of Technology, Narsapur, Telangana, India
e-mail: jagan.amgoth@bvrit.ac.in

N. Satyanarayana
Department of CSE, Nagole Institute of Technology and Science, Hyderabad, Telangana, India
e-mail: nsn1208@gmail.com

K.R. Attele et al. (eds.), *Emerging Trends in Electrical, Communications and Information Technologies*, Lecture Notes in Electrical Engineering 394,
DOI 10.1007/978-981-10-1540-3_1

1 Introduction

Balanced job scheduling is one of the old issues of the present days. Many algorithms and strategies have been already introduced and using to solve the issues in a grid. The most important task accomplished by those algorithms is to improve the response time of a user's application while guaranteeing that the resources are utilized to their greatest extent. However the important aim of our algorithm is to make sure that all the efficient resources are loaded equally with their set of tasks, which results in enhancing the attainment. By the phrase "equally loaded" what we mean is that, no processor in the grid should be heavily loaded while others are idle or even lightly loaded [1]. By efficient resources, we mean that the resources with higher speed and lower possibility to fail while executing job approaching deadline.

Balance job scheduling can be defined by the 6 policies they are, firstly the information policy gives information on what work load should be collected, from where it should be collected and when it should be collected. Secondly, the triggering policy determines the time at which load balancing operation should initiate. Thirdly, the resource type policy classifies a resource as a server/receiver of job based on its availability status. Fourthly, the job selection policy defines tasks that should be shift from overloaded source (resources) to the efficient idle receiver (resources). Fifthly, the resource selection policy determines the successive resource for job approaching the deadline. Policy SIX, the location policy uses the results of the resource policy type to find a most suitable partner for a resource provider/receiver.

Every distributed computing consists of policies like load balancing, scheduling and fault tolerance etc., for the efficient utilization of resources. However these policies cannot be applied directly to the grid even though it belongs to the class of distributed systems [2, 3]. In addition to this, the load balancing techniques cannot be applied to grid architectures despite their use in conventional, parallel and other distributed computing because the classes have architectures that are specific. Also, the scheduling of tasks on multiprocessors/multiple computers assume that the processors are not heterogeneous but homogeneous and linked with homogeneous and fast networks which may not possible all the times. The motive behind our approach is as follows [2, 3].

(1) All the resources have the same speed. (2) High bandwidth interconnection network among PE's. (3) Input data is readily available at the processing site. (4) The complete time for transferring I/O data is negligible.

There are several challenges to design an efficient and effective load balancing system for the grid environments. Few of them include heterogeneity, scalability, adaptability, selection of resources and computation data separation. In addition to this, the challenge of large size data that is to be moved and the improper distribution of the resources in the Grid environment. Few problems resulting from the above have not yet solved and till today remained as open research issues. Hence, designing an effective balance job scheduling system for the grid is a major challenging task.

2 Related Work

Till date many load balancing strategies have been proposed [4–11]. Cao [4] use an ant like self organizing policy to achieve the system wide load balancing by collection of simple local interactions between Grid elements. In this exemplary, numerous resource management agents coordinate to achieve automatic load balancing of job queues which are distributed. Each individual ant holds tow of m steps in succeeding to decide the least and the most loaded nodes, Loads on these tow nodes will be then redistributed between themselves. After a series of successive redistributions, uniform system wide load balancing can be attained. To attain dynamic load balancing in Grid, Yagoubi [8] proposed a tree model based layered algorithms. Their algorithm presents the following important aspects: (1) it is layered one; (2) it supports scalability, heterogeneity; and (3) it is fully not dependent of any physical architecture of a Grid. To balance the load, Lenders [9] uses a field based routing and service discovery scheme for sensor networks. Novel anycast routing and density based anycast methods they used to show that the number of group members can be effective in the routing decision. This is contrast to routing based on proximity techniques, which consider distance details only. Information dissemination protocols introduced by Erdil [10], can distribute the load in a way without use of load rebalancing through job migration. This is more costly and difficult in large scale grids. Two novel distributed swarm intelligence inspired load balance algorithms were introduced by Ludwig and Maollem [11]. Those are based on ant colony, particle swarm optimization.

In Naik et al. [6] tried to balance the load in the grid by grouping the resources into levels according to their speed. Computed fault rate of each resource to determine the suitable resources for the jobs, and thereafter to take scheduling decision. As per resource speed requirement, job is scheduled to the resource of a desired level. Naik et al. [7] considered fault rate of each resource along with other QOS parameters in calculating the Performance Indicator (PI). Value of PI decides suitable resources for job (lower the PI gives the resource with less failure tendency). EGDC of Hao [5] gives deadline control based load balancing strategy in GridSim, where they worked on deadline. Success and failure rate of the resources are also need to be considered along with job deadline to make scheduling decision. As stated earlier, this paper pays attention on success rate of PE's also and presents a balanced job scheduling mechanism. Our approach to balanced job scheduling goes as follows: First, load of the resource is calculated and then classified into groups according to load. Namely over loaded, normally loaded and under loaded. Second, failure and success rate of every resource in the group is determined; this rate is used to schedule the jobs in the queue which approaching its deadline on to the resource having more cost rate. Finally, our proposed balanced job scheduling mechanism is applied in GridSim, to schedule on and to change the resource state. Our experimental results used to demonstrate proposed mechanism not only reduces the make span, but also reduces failure tendency, resubmitted times, and improves the throughput.

In the Sect. 3, we introduce GridSim and summarize some proposed load balancing schemes. Section 4 presents our balanced scheduling mechanism based on success rate and deadline control. Simulation, comparison among without load balancing (WLB) [12], load balancing on enhanced GridSim (LBEGS) [13] and EGDC [5] presented in Sect. 5. Finally, Sect. 6 concludes the paper.

3 Resource Load Balancing Schemes on Grid

A. Simulation of Balanced Job scheduling on GridSim

GridSim [12] is a very popular Grid simulation tool and has been used widely in the studies of Grids [14]. GridSim supports entities for simulation of various types, like single processors, multiprocessors, and for heterogeneous resources that are configured as time shared or space shared systems in the actual time. All along the system of simulation, GridSim creates a number of multi threaded entities for each and everything that use it i.e. the grid users, grid brokers, grid resources, grid information service, statistics, and network-based I/O, each of which runs in parallel in its own process. An entity's behavior needs to be simulated within its method body(), as suggested in SimJava [15]. The layers of GridSim are shown in Fig. 1. Grid Broker occupies the highest position in the Grid environment which is responsible for maintaining the scheduling and rescheduling activities of the Grid resources. The Grid Broker basically gets information of a load from a grid resource, converts into a gridlet and sends this gridlet to resources for further optimization of scheduling. During this, it also checks the state of the resource and performs any scheduling if needed. In the GridSim hierarchy, Resource is next to the Grid broker. Grid Resource coordinates activities like scheduling; load balancing on its machines and sending an event to the Grid Broker about its load status.

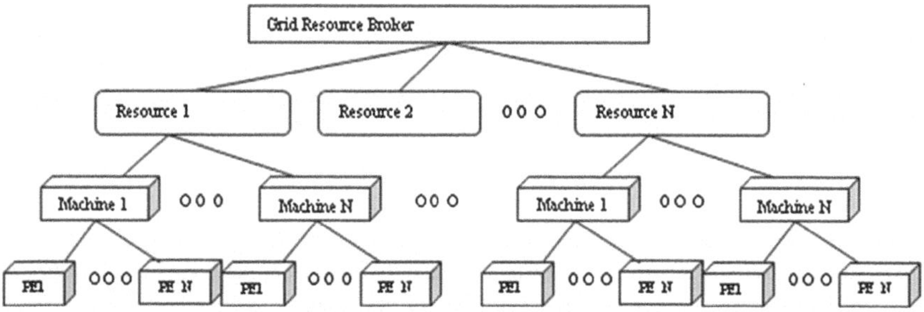

Fig. 1 The grid structure in GridSim

Machine is a processing entity (PE) owns one or more PE's and manages them. It performs task scheduling and load balancing on its PE's. Resource simulator of GridSim uses internal events to simulate the execution and allocation of PE's to Gridlet jobs. When job arrives at and if there is a free PE, space-shared approach starts its execution immediately, jobs queued otherwise. During assignment of Gridlet, its execution time is determined and the event is scheduled for delivery at the end of execution time. Whenever a job completes its execution, an internal event is issued to indicate the completion of the scheduled job. Resource simulator then release the PE allocated to it, and checks other jobs waiting in the queue if any. If jobs are waiting in the queue, then the scheduler in resource simulator selects a suitable (approaching deadline) job depending on the policy and allocates it to the successive free PE. The completed Gridlet is sent back to its originator (broker/user) and then removed from execution set. GridSim schedules new internal event that is to be delivered at the completion time of the scheduled job.

A Gridlet/job is an entity which contains information of its execution management details such as job length expressed in MIPS, disk I/O operations, the size of input/output files, deadline, job originator and other execution management details if any. Using these basic parameters, GridSim determine the job execution time, the time required to transfer input and output files between users/remote resources, and sending the processed jobs back to the originator along with the results. A resource entity represents a Grid resource may having different characteristics and hence may be different. Resource speed and job execution time can be defined in terms of ratings the standard benchmarks. Upon obtaining the resource contact details from the Grid Resource information service (GRIS), Grid Brokers may query resources directly for their static and dynamic properties.

Throughout our paper, we considered three parameters they are, (1) resource success rate s, (2) deadline t of job and (3) the resource load level l, the details can be found in Sect. 5.

B. ***Balanced scheduling schemes in GridSim***

Most of the scheduling approaches in Grid belong to no load balancing approaches such as FPLTF (fastest processor to largest task first), min–min, and max–min. There are many centralized load balancing techniques [13, 16, 17] but all of them cannot be suitable in all the scenarios. We just introduce WLB (without load balancing) [12], LBEGS [13] and EGDC [5] in this paper. To simulate the execution and allocation of PEs to Gridlet jobs, the GridSim resource simulator adopts internal events. LBEGS [13] gives the details of algorithms for load calculation of the PE, Machine, and GridResource. This also provides load balancing algorithms among PEs, machines, and resources. The advantages of this scheme include less communication overhead and reduced idle of Grid resources.

EGDC [5] gives load balancing strategy into a GridSim based on deadline control, where they concentrated on deadline. But to meet jobs deadline, it is also required to consider success rate of the resource assigning.

4 Enhanced Gridsim for Balanced Job Scheduling Based on Deadline Control

Every resource must manage itself and hence all the processing capabilities cannot be given to the gridlet. Suppose that, the load of the resource is denoted by *l. current-load*, and that Gridlet *g* (denotes the job) is assigned to Resource *r* with a deadline in *t* seconds. *Current-load, l* is the percentage of the resource calculation ability that is given to Gridlets. The Method for the calculation of the new load of a resource is as follows.

```
double forecastnewload (Gridlet g, Resource r, Deadline t)
 Begin
 l.currentload = l.currentload + g.gridletlength
/t/resource.speed;            Return l.currentload;
End
```

In fact, *l.currentload* is the present load on a node; if *g* can be finished quicker, the load of *l* might be less than the result of above calculation. If no more Gridlets are present for load calculation, the Gridlet will be completed sooner as. Under this condition, the load is lesser than the result that we obtain by above method, which only provides methods for scheduling. Considering deadline *t*, we elaborate the Gridlets as follows.

Gridlet (gridletID, gridletLength, gridletFileSize, gridletOutputSize, record, deadline);

Let us presume that the resource *l* has a machine list as follows: [machine0, machine1…machinem, $machine0_{temp}$, $machine1_{temp}$ …, $machinem_{temp}$]. Here $machine_{temp}$ can be explained as $Machine_{temp}$ (int $Machineid_{temp}$, int $numPE_{temp}$, int $ratingPE_{temp}$). $Machineid_{temp}$, $numPE_{temp}$, $ratingPE_{temp}$ denote the id of the resource machine, the number of PE's, and the rating of every PE's, respectively. Here $ratingPE_{temp}$ can be expressed in MIPS. The resource's speed can be calculated as follows:

$$Resource_i.Speed = \sum_{temp=1}^{m} NumPE_{temp} * RatingPE_{temp}$$

We must check the state of a resource r time to time because some machines or PE's might go online or offline from the resource and also new machines or PE's can be joined to the resource.

```
Algorithm1: Finding Resource State
static double A=loadtemp<rb, B=loadtemp>rt, C=!(loadtemp<rb)&&!(loadtemp>rt) ;
 Checkstate(Resource r)
Begin
  loadtemp=Obtain the load of Resource r
  statetemp= r.state
  // Now I set my statetemp to have A=true, B=true, C=true, effectively
  double statetemp = A | B | C ;
  switch( statetemp )
    Begin
    case (A): newstate= 0;     break ;
    case (B):  newstate= 1;    break ;
    default: newstate= 2;
    Endswitch
  If(statetemp!=newstate)
  Begin
      r.state=newstate; If(newstate==0)
      Begin Trigger its Grid Broker stating that the resource r was ''under loaded''; Endif.
      If(newstate==1)
   Begin Trigger its Grid Broker telling  r is '' normally loaded'' resources; Endif
      If(newstate == 2)
      Begin Trigger its Grid Broker that the resource r is ''over loaded'';   Endif Endif
EndCheckstate
```

Algorithm 1 gives the specifications for finding state of resource, which can either be classified as underloaded, normallyloaded, and overloaded and can be denoted by 0, 1, and 2 respectively. We know that the CPU load on Windows/Unix OS can be scheduled even on a Grid, so we can forecast the load of the resource by itself. Then, by summing the load of other Gridlets allocated to the resource, the resource load is obtained as follows. When a resource finds its state change, request was made to the Brokers for changing the state in its Grid information. Suppose $1 > = rt > rb > = 0$, if the resource load is more than *rt*, we say resource is *over loaded*; if the load of resource is less than *rb*, the state of the resource is called as *under loaded*. If the resourceload is in the range *rb* to *rt*, resource state is called *normally loaded*. When the resource understands that its state was transformed, it transmits a request to the Brokers and adjusts the state to new state of load in its Grid information. The algorithm for checking the state is as mentioned. The state of every resource is checked by the grid broker and it inserts it into one of the three states based on the load. In our approach, Grid Broker puts the resource into "*overloaded list*" if the state was over-loaded, spot into "*normally loaded list*" when the state was normally-loaded and inserted into "*underload list*" if the state was lightly-loaded. Assume $0 < rb < rt < 1$, if the load of resource is greater than *rt*, then the state of the resource is *over loaded*, and the resource belongs to the *Overloaded list*; if the load at resource is lesser than *rb*, then the state of the resource is called *underloaded*, and this resource is from the *Underload list*, otherwise, the resource belongs to *Normallyloaded list*. The detail of system initialization is given in Algorithm 2.

```
Algorithm 2: Update Resource State based on load
Begin
    For resource=1 to n do Begin
      Insert into Overloadlist whose loadtemp>rt;
      Insert into Underloadlist whose loadtemp<rb;
      Insert into Normalyloadllist whose !(loadtemp<rb)&&!(loadtemp>rt);
      Endfor; End
```

The Grid Broker updates the record of the resource information whenever any trigger is received from the resource (see Algorithm 1). If the state of the resource is overloaded, some of the Gridlets of the current resource are assigned to the unassigned gridlet list and if the state of the resource is underloaded, more number of Gridlet are assigned to it. The details are illustrated in Algorithm 3.

```
Algorithm 3: Balance Resource load
Begin Algo
 If Grid Broker receives ''under loaded OR normally loaded OR over loaded'' event from a
Resource then   Begin
   Delete the resource r from the list which it belongs to;
   Insert the resource r into under loaded OR normally loaded OR over loaded list
    Endif
While (Underloadelist & Overloadlist != empty) Begin
   get gridletlist which we want to transform from Overloadlist;
   For each gridlet gtemp concern to gridletlist does fill gtemp in to unassignedgridletlist   Begin
Endfor Endwhile   End Algo
```

In Algorithm 4, Unassigned Gridlet list *l* queues the newly arriving jobs and the unfinished jobs coming from the resource as the execution fails. The Grid Broker schedules jobs time to time and simultaneously, the resource also checks the load and supervises the state. If a resource can give more accounting ability to a Gridlet and can change of state from under-loaded to over-loaded state, we select the scheduling with the maximum supluscapacity1 of all Gridlets. The supluscapacity1 maintains the leaving processing capacity to the state of over loaded. Otherwise, we select the scheduling whose state is under loaded with the maximum supluscapacity2 of all Gridlets. In particular, supluscapacity2 stores the leaving computing capacity to the state of normally-loaded when no scheduling gets resource state changing into over loaded. Supluscapacity1 and supluscapacity2 provide a flexible mechanism for scheduling if some resources/machines/PEs go offline occasionally. Supluscapacity1 and supluscapacity2 transform the resource state approach a ''normally load'' one, so these approaches give more free resource than other and maintains an accepted level of load of the scheduled resource. Algorithm 4 gives details for scheduling Gridlets.

```
Algorithm 4: Algorithm for Scheduling the waiting Gridletlist
Begin Algo
Maxvalue1 = −∞; Maxvalue2 = −∞;
While (Underloadelist is not empty && unassignedgridletlist l is not
empty)        Begin
   For every resource r in Underloadelist          Begin
     For every Gridlet g in unassignedgridletlist l    Begin
       temp = r.load × r.speed + g.gridletlength/g.deadline;
       surpluscapacity1 = r.r t × r.speed - temp;
       surpluscapacity2 = temp- r.rt × r.capacity;
       If((r . rb × r . capacity < temp < r . rt × r . capacity) AND
(surpluscapacity1 > Maxvalue1))
       Begin
            Maxvalue1=surpluscapacity1; Assignedresource1=r;
Assignedgridlet1=g;
       Endif
       If((temp< r.rb × r.capacity) && (surpluscapacity2 > Maxvalue2))
       Begin Maxvalue2=surpluscapacity2; Assignedresource2=r;
Assignedgridlet2 = g;  Endif
   EndFor    EndFor
   If (Minvalue1! = −∞) Begin
     Assigning Assignedgridlet1 to Assignedresource1; Checkstate(r);
     Else
     Assigning Assignedgridlet2 to Assignedresource2; Endif
Endwhile
End Algo
```

Algorithm 5 calculates the success rate of resources, to be used by Algorithm 4. Keeping in mind the success rate of a resource helps one to recognize an efficient resource. Fault handler agent is a static entity that stays in the grid and works in collaboration with scheduler to keep a record of the success and fault rate of grid resources. It performs on the basis of notifications received, and updating list obtained from the Gridlet Agent. A mark is assigned to any resource$_{i,}$ which are employed to complete the nominated Gridlets on time, are termed or marked as success otherwise marked as failed.

```
Algorithm 5: Algorithm for calculating resource successRate
double faultRate = 0;
  for(BalancedTaskGridResourceWithFailure resource : resourcesi) {
  faultRate+=resourcei.getFaultRate();
  Print Failure tendency:     resource.getResourceCharacteristics().getResourceName() +
"=" +resourcei.getFaultRate();}
  CostCalculatorWithFault{
double executionTime = ((job.getGridletLength() / element.getMIPSRating()));
    double faultRate = resourcei.getFaultRate(); successRatei = 1 – faultRatei;
    double cost = executionTime;
    costi = executionTimej * (1 + successRatei);        }
    return cost;  return successRate;  } }
```

The history of all resources in the list is updated by the agents and value of the resource's fault rate is calculated. If the PE executes the job in the given deadline, its success rate is incremented by 1, otherwise decreased by –1. PE's speed and its success rate are used to calculate the cost acquired for it. The available PE with higher cost will be the best and most suitable PE for executing the job with a near deadline and the job will be scheduled on it.

5 Simulation

Makespan is the main parameter that determines the performance of a grid. The makespan is defined as the ''total time for executing tasks of a Gridlet'', which is period of time between the moment at which the first job is sent to the grid and the moment at which the last job exits out of the grid. In our simulation, Gridlets with variable sizes are assigned. In order to consider realistic conditions, we made use of "Makespan" as the criterion for determining the performance of the grid. Other criteria used to test the performance include: failure tendency, throughput, total time and resource unavailability. The proposed algorithms are simulated on GridSim 5.0 [12], using same configuration of resources in our simulations. We use Windows 8.1 on an Intel Core (TM) i-5 CPU (1.70 GHz and 2.40 GHz), with 4 GB of RAM and 1000 GB of hard disk.

Simulation with constant Resources and (variable) Gridlets
There are 4 resource providers (from R0 to R3), every resource provider has five machines associated with it, every machine has five PE's, thus the Grid system has 100 PE's in total. The load factor is initially configured to 0. The configuration of Gridlet parameters, resource load and threshold parameters and are listed in Table 1. Every PE is given a grade between 1 and 5 (lowest cost is assigned grade 1,..., higher is assigned grade 5). The cost of each PE is calculated on its speed (MIPS) and success rate. Every job arrives randomly between 1 and 10 time units(s) and each has a deadline with a range of time units [1...6]. Each given Gridlet length is between grade 10,000 and grade 50,000 (10,000 million instructions is assigned grade 1 ,..., 50,000 million instructions is assigned grade 5). So, the PE with highest cost assigned to the job with nearest deadline. If this PE is above normal load then the PE which has next higher cost will be provided. This process will continue till the best PE is identified and assigned for the job. The threshold assumed for resource, machine, PE is 0.6, 0.75, 0.8 respectively. We set $rb = 0.75$, $rt = 0.80$. The readings are taken by varying number of Jobs 200, 500, 1000 and 1500 and the number of PEs is kept constant at 100. The comparisons of finished jobs, unfinished jobs, throughput, failure tendency and makespan between OurAppr, EGDC, LBEGS, and WLB are shown in Figs. 2, 3, 4, 5, 6 and 7.

Figures 2 and 3 show the number of finished jobs of OurAppr is always higher in number than that of EGDC, LBEGS and WLB. With more number of jobs submitted, the number of jobs finished increases. Unfinished Gridlets are directly proportional to the jobs submitted for constant PE's and for changing PE's. The reason is that EGDC will not consider success rates of resources. EGDC selects a PE possessing highest speed and is at normal load for execution, without taking into

Table 1 Simulation parameters with constant gridlets

Parameters of Gridlet			Load-level		
Jobs, arrive time	Gridlet length	Deadline	Resource	Machine	PE
500, [1, 10] s	[1, 5] PE's	[1, 6] s	0.6	0.75	0.8

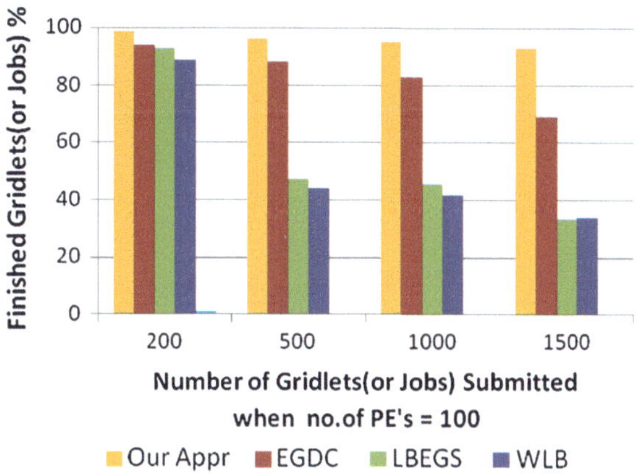

Fig. 2 Throughput—no. of gridlets submitted versus finished Gridlets unfinished

Fig. 3 Efficiency—no. of gridlets submitted versus Gridlets

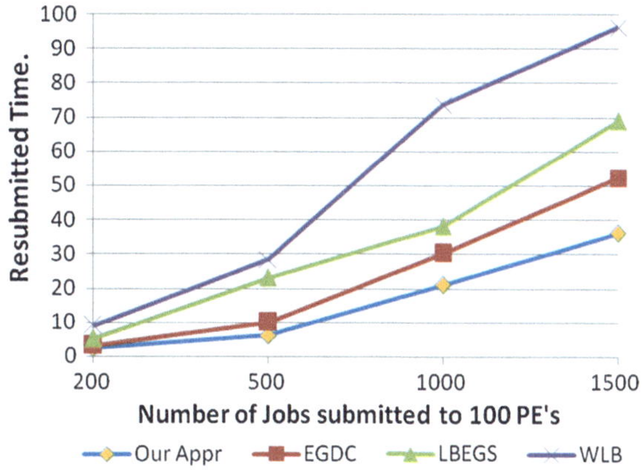

Fig. 4 Resubmission time—no. of gridlets submitted versus resubmission time

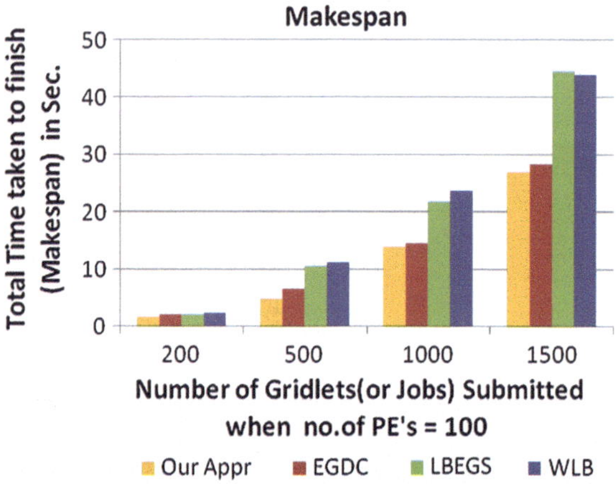

Fig. 5 Makespan—no. of gridlets submitted versus total time to finish

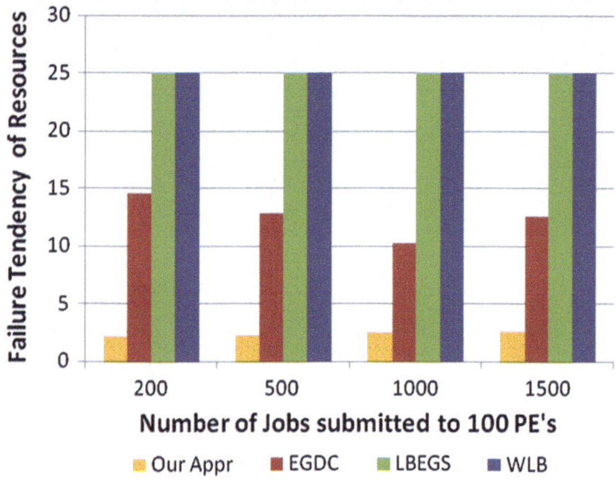

Fig. 6 Failure Tendency—no. of jobs submitted versus jobs failed

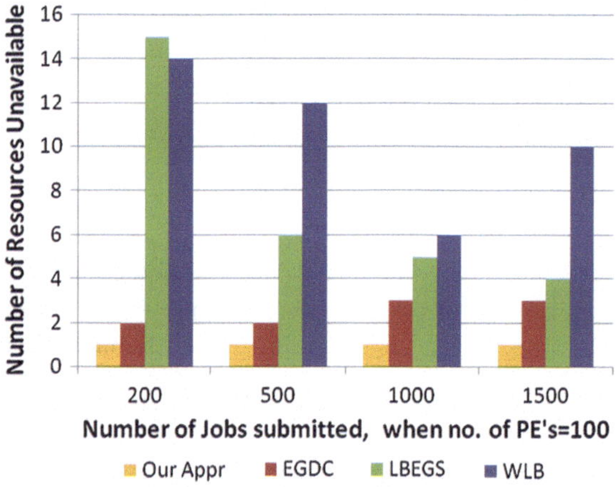

Fig. 7 Unavailability—no. of gridlets submitted versus resources unavailable

account PE failure history, The PE may have worst success rate history. So, the probability that job will be executed in given deadline is less. Thus, the PE/machine/resource threshold produces time waste and approaching job deadline. Our approach acknowledges PE/resource history as well as PE/resource speed and calculates its cost. The PE which has the highest cost is selected as the best PE hence it has more speed and good success rate.

An example of a Grid which has only one resource is taken. The resource has only one machine, and the machine possesses 100 PEs; considering OurAppr, the number of PEs used is between 75 and 80. Some resources go online or offline randomly, posing more problems to job scheduling. With EGDC resubmission time will increase as it may schedule the job on PE with more failure with rise of gridlets.WLB has no flexibility rules for a Grid and is considered to be the worst of the three schemes. Once the Gridlet grow to 1500, the finished Gridlets become less and even remain a constant after a while. Hence the Grid system has no more power to schedule Gridlets as much as that of Gridlets for deadline and processing capacity.

Figure 4 depicts that, the resubmitted time of OurAppr is less than that of the others with the increasing number of Gridlets. The reason behind this is that OurAppr provides lot of flexibility in areas of resource selection and scheduling. If some machines/PE's are offline, execution won't be hampered and can be completed by scheduling between successive resources. For EGDC, LBEGS and WLB, if a resource/machine/PE had less success rate or they go offline, the scheduling is inefficient/fails.

Figure 5 shows that the makespan of OurAppr (1.5–25), EGDC (2–28), LBEGS (2–42) and WLB (2–44) for 100 PE's and Gridlets 200, 500, 1000, 1500. We executed our test many number of times and obtained this average range. The makespan is the "total application execution time". It is less for OurAppr when compared to EGDC, LBEGS and WLB, because we schedule the jobs on most successful PE's for continuous and uninterrupted execution. From the above experimental results, it can be concluded that OurAppr is more efficient and has better performance than EGDS, LBEGS and WLB on constant PE's and with Gridlets 200, 500, 1000, 1500.

Figure 6 display another metric which has been introduced in this work to scrutinize the efficiency of OurAppr, known as Failure tendency. It gives the number of jobs that failed to complete execution in the defined deadline. The reason for failing may be shorter deadline periods assigned to low speed PE's, few machines/PE's may be offline or the week success history of that resource. The success or failure history of the grid resources are predicted when using a certain scheduling system. The Failure Tendency of OurAppr is compared to EGDC, LBEGS and WLB for 100 PE's and submitted Gridlets 200, 500, 1000, 1500. The Tendency of failure for OurAppr is 2–4 resources. Thus lower number of resource failure with OurAppr depicts that the scheduling and load balancing strategies used in OurAppr are the best. We state that OurAppr is more efficient than any other mechanisms. The experimental result prove the failure tendency of resources with EGDC (12–14 resources), LBEGS (25 resources) and WLB (25 resources) is more.

We have averaged this test to obtain an accurate result. In Fig. 7, unavailability of resources for execution Gridlets is seen. This situation may result into Gridlet resubmission and lead to crossing deadline. If the numbers of resubmitted job increase, resubmission time, turnaround time also increase and throughput decreases, thus reducing the efficiency. Unavailability for OurAppr is just 1–2 resources. Thus smaller number of resource unavailability with OurAppr proves that the success rate occupying scheduling and load balancing way used in OurAppr is most efficient when compared to any other. Our experimental result says the unavailability of resources with EGDC (2–5 resources), LBEGS (4–15 resources) and WLB (6–14 resources) is noted. We executed our test several times and obtained this average range.

6 Conclusion

In this paper, a dynamic, distributed balanced job scheduling approach for a computational Grid is proposed, the simulations of which prove that the performance of a grid are optimized and improved to a large extent while decreasing the failure tendency and resubmitted time. This approach defines a perfectly elastic and efficient mechanism for balanced job scheduling. The resources as well as the Grid Broker are a part of the balanced job scheduling operations of the grid. In our future work, more features and characteristics of the grid, such as bandwidth, processing ability, Gridlet requirement etc. will be taken into account and further research will be carried out.

References

1. Xu C (1997) Load Balancing in parallel comp: theory and practice. Boston
2. Yagoubi B, Slimani Y (2007) Task load balancing strategy in grid environment. J Compt Sci 3 (3):186–194
3. Yagoubi B (2007) Load balancing strategy in grid environment. J IT App 4:285–296
4. Cao J (2004) Self-organizing agents for grid load balancing. In: 5th IEEE/ACM international workshop on grid computing
5. Hao Y (2012) Enhanced load balancing mechanism based on deadline control on GridSim. FGCS 28:657–665
6. Naik KJ (2012) Scheduling tasks on most suitable fault tolerant resource for execution in computational grid. IJGDC 5(3)
7. Naik KJ (2013) A novel fault-tolerant task scheduling algorithm for computational grid (15th ICACT-978-1-4673-2818-0/13 ©2013 IEEE)
8. Yagoubi B, Slimani Y (2006) Dynamic load balancing strategy for grid computing. Eng Technol 90–95
9. http://www.lenders.ch/publications/books/thesis.pdf [visit:2011–04-01]]
10. Erdil D, Lewis M (2010) Dynamic grid load sharing with adaptive dissemination protocols. J Supercomput 1–28

11. Ludwig S, Moallem A (2011) Swarm intelligence approaches for grid load balancing. J Grid Comp 1–23
12. Buyya R, Murshed M (2012) GridSim: practice and experience 14:13–15
13. Qureshi K, Rehman A, Manuel P (2010) Enhanced GridSim architecture with load balancing. J Supercomput 1–11
14. http://www.buyya.com/GridSim/ [visit:2011-1-27]
15. Howell F (1998) SimJava: a discrete Java event simulation package with applications in computer systems modelling. In: 1st international conference on web-based modelling and simulation, Society for Computer Simulation, San Diego, CA
16. Li Y (2009) A hybrid load balancing strategy of sequential tasks for grid computing environment. FGCS (ISSN: 0167-739X) 25(8):819–828
17. Subrata R (2008) Game-theoretic approach for load balancing in computational grids. IEEE Trans Parallel Distrib Sys 19(1):66–76

A Secure Location-Based Coupon Redeeming System

J. Maruthi Nagendra Prasad and A. Subramanyam

Abstract With the rapid evolution of mobile computing technologies, Mobile location based services are identified as one of the most promising target application. Mobile location based services have lot of limiting factors. In this paper we propose a new rewarding system based on location of the Mobile User where Mobile Units will collect Coupons from the Coupon Distribution Center and then redeem their Coupon's at the Coupon Collection Center. Coupons acts as virtual currency Coupon's Distributers and Collectors can be any entity or retailer. This rewarding system based on location of the Mobile Unit is a secure one and preserves privacy of the Mobile user.

Keywords Mobile location based services · Rewards · Coupons · Privacy · Security

1 Introduction

With the proliferation of mobile devices, mobile location-based services (MLBSs) have emerged as a new type of mobile marketing. Mobile commerce is poised to make a qualitative leap. Knowledge of the end user's location will be used to deliver relevant, timely, and engaging content and information. For mobile network operators, location-based services represent an additional stream of revenue that can be generated from their investments in fixed infrastructure. For the end user, these services can help reduce confusion, improve the consumption experience, and deliver high-quality service options. As per a report 1 % of Americans used MLBSs [1]. Research conducted by Juniper predict that revenue generated by MLBS's will be more than $12.7 billion by 2014 [2].

J. Maruthi Nagendra Prasad (✉) · A. Subramanyam
AITS, Rajampet, India
e-mail: maruthiprasad1986@gmail.com

A. Subramanyam
e-mail: smarige@gmail.com

K.R. Attele et al. (eds.), *Emerging Trends in Electrical, Communications and Information Technologies*, Lecture Notes in Electrical Engineering 394,
DOI 10.1007/978-981-10-1540-3_2

There are various kinds of Mobile Location Based Services. First one is social networking based on Location [3]. Second one require users to provide current or historical location proofs to fulfill some purposes [4]. Third one is Mobile Commerce [5] and Fourth one is check-in games based on location [6].

Location-based check in games are having 3 restrictions first mobile users can get benefits from the same store, second security is not guaranteed in this system [7, 8] and third is preserving users privacy is difficult.

Here, we propose a rewarding system based on mobile location. The proposed system consists of a Mobile Unit, Coupon Distribution Center, Coupon Collection Center, Data Center and Authentication Center. Mobile Unit will collect Coupon's from the Coupon Distribution Center and Mobile Unit can redeem the Coupons for rewards at the Coupon Collection Center. Data Center stores the Mobile Unit and Coupon validation details, used to validate the Coupon before rewarding the Mobile User. Data Center is used when Coupon Distribution Center and Coupon Collection Center happens to be of the same retailer.

If the Coupon Distribution Center and Coupon Collection Center are from different retailer instead of Data Center Authentication Center is used which validate the Mobile User and Coupon. Communication between MU and CDC/CCC carried out via WiFi interface. CDC/CCC are connected to the DC through wired network.

2 Related Work

In spite of the fact that there are many flavors of Mobile Location based systems they cannot ensure security of the system and privacy for the users.

First users can fake their location to get more benefits, this problem can be addressed by sun et al. [9] uses signal patterns to position users. Anisetti et al. [10] explore geographical information and can achieve location accuracy and using Bluetooth for generating location proofs [11]. Lenders et al. [12] uses geo-tags. [13, 8] propose to use Wi-Fi.

Second User's Privacy can be compromised. [14–20] Propose schemes to achieve communication anonymity and data privacy k-anonymity Clocking scheme [15–19], propose to hide real location of the user and others include obfuscation of location [20], and using pseudonyms.

3 System Design

This system consists of Mobile User's (MU), Coupon Distribution Center (CDC), Coupon Collection Center (CCC) and Data Center (DC) or Authentication Center (AC) as shown in Figs. 1 and 2.

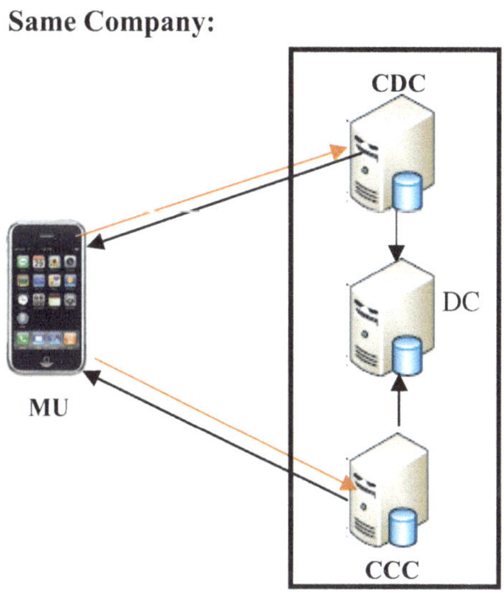

Fig. 1 Same Company Scenario where CDC and CCC belongs to same company or retailer

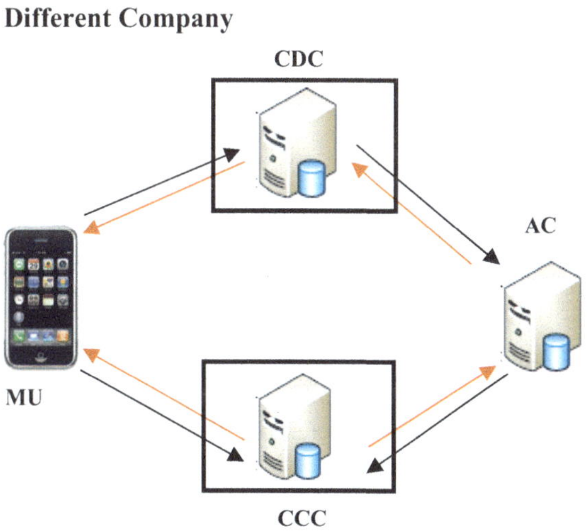

Fig. 2 Different Company scenario where CDC and CCC belongs to the different retailers

Mobile Users (MUs): A Mobile Unit collects the coupons and redeem for rewards. For collecting coupons Mobile User communicates with Coupon Distribution Center and redeems the Coupon when it communicates with Coupon Collection Center.

Coupon Distribution Center (CDC): Coupon Distribution Center generates the Coupon for the Mobile Unit and Stores the Mobile User validation information and coupon information in the Data Center or in Authentication Center.

Data Center or Authentication Center: it validates the Mobile User and coupon.

Table 1 Shared secret keys in the Rewarding system based on the location of the mobile user

Shared secret key	Purpose
K d, c	Shared secret key between CDC and CCC
K i, d	Shared secret key between MU and CDC
K i, c	Shared secret key between MU and CCC
K d, dc	Shared secret key between CDC and DC
K c, dc	Shared secret key between CCC and DC
K d, a	Shared secret key between CDC and AC
K c, a	Shared secret key between CCC and AC

Coupon Collection Center (CCC): This entity verify the Mobile User' coupon and rewards them with benefits.

Shared Secret Key Generation:
As Assume two mobile users user1 and user2 either one of them pick large prime numbers n and g. User1 picks a large number x and keep it secret and User2 picks a large number y and keep it secret User1 initiates the key exchange protocol by sending User2 a message containing (n, g, g^x mod n).

User2 responds by sending User1 a message containing g^ymodn. Now User1 computes (g^ymod n)^{x}mod n = g ^{xy}mod n. Similarly User2 computes (g^x mod n)^y mod n = g ^{xy}mod n, in the similar fashion shared secret keys shown in Table 1 can be generated.

4 Rewarding System

As Here the system consists of following processes:

Coupon Distribution:
Whenever a Mobile User visits a CDC it requests a Coupon. To protect identity of the Mobile User and preserve the location details MU randomly generates a pseudonym based on real identity. CDC needs to check MU's identity before allocating a coupon.

Thus CDC consists of two phases

- MU's Identity authentication:
  Purpose of validating an MU's identity is defend against misbehaving users who use fake ids.
- Coupon Distribution:
  If the MU completes the identity authentication the CDC will process the MU's token.

Coupon Redemption:
Whenever an MU communicates with CCC it initiates a token redemption process.

Thus CCC consists of three phases:

- MU's Identity Authentication at CCC:
  CCC first checks the MU's identity to make sure it's an authorized user and this phase is similar to MU's Identity authentication in CDC.
- Coupon validation:
  This phase is used to validate the coupon submitted.
- Reward Distribution:
  After the MU and coupon validation completes then benefits will be rewarded to the MU.

Rewarding system based on Location of the Mobile User includes the following steps:

Sample Scenario:
See (Figs. 3, 4, and 5)

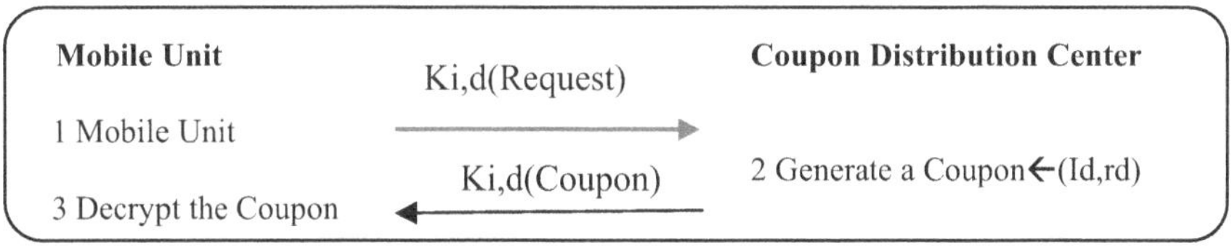

Fig. 3 Shows interaction between Mobile Unit and CDC using shared secret key Ki, d where generated coupon will be consisting of Id of the MU and a random number

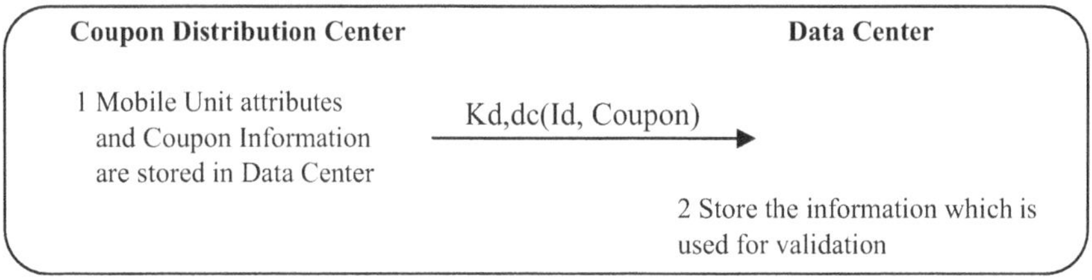

Fig. 4 Shows interaction between CDC and DC using shared secret key Kd, dc

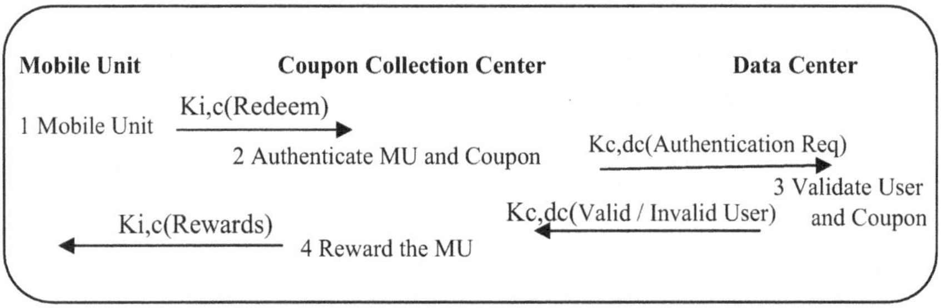

Fig. 5 Shows interaction between MU and CCC and DC using the shared secret keys Ki, c and Kc, dc

5 Conclusion

Here we proposed a secure and privacy preserving rewarding system based on location of the mobile user. We designed secured and privacy aware system for redeeming the rewards. We find the system is resilient to many attacks and privacy of the mobile unit will be protected well.

References

1. http://pewinternet.org/~/media/Files/Reports/2010/PIP-location%20based%20services.pdf (2010)
2. Juniper Research, Mobile Location Based Services Applications, Forecasts and Opportunities 2010–2014
3. http://www.facebook.com/about/location
4. Lenders V, Koukoumidis E, Zhang P, Martonosi M (2008) Location-based trust for mobile user-generated content: applications, challenges and implementations. In: Proceedings of the ninth workshop mobile computing systems applications (HotMobile'08)
5. Loreto S, Mecklin T, Opsenica M, Rissanen H-M (2009) Service broker architecture: location business case and mashups. IEEE Comm Mag 47(4):97–103
6. https://foursquare.com/
7. Sastry N, Shankar U, Wagner D (2003) Secure verification of location claims. In: Proceedings of the second ACM workshop wireless security (WiSe'03)
8. Luo W, Hengartner U (2010) Veriplace: a privacy-aware location proof architecture. In: Proceedings of the 18th SIGSPATIAL international conference advances geographic information systems (GIS'10)
9. Sun G, Chen J, Guo W, Liu KR (2005) Signal processing techniques in network-aided positioning. IEEE Signal Process Mag 22(4):12–23
10. Anisetti M, Ardagna CA, Bellandi V, Damiani E, Reale S (2011) Map-based location and tracking in multipath outdoor mobile networks. IEEE Trans Wireless Commun 10(3):814–824
11. Zhu Z, Cao G (2011)Towards privacy preserving and collusion resistance in location proof updating system. IEEE Trans Mobile Comput 99
12. Lenders V, Koukoumidis E, Zhang P, Martonosi M (2008) Location-based trust for mobile user-generated content: applications, challenges and implementations. In: ACM HotMobile, Napa Valley, California

13. Saroiu S, Wolman A (2009) Enabling new mobile applications with location proofs. In: ACM HotMobile, Santa Cruz, California
14. Kong J, Hong X (2003) Anodr: Anonymous on demand routing with untraceable routes for mobile ad-hoc networks. In: Proceeding of ACM MobiHoc, Annapolis, Maryland
15. Gruteser M, Grunwald D (2003) anonymous usage of location-based services through spatial and temporal cloaking. In: proceedings of the first international conference mobile systems, applications services (Mobisys'03)
16. Gedik B, Liu L (2008) Protecting location privacy with personalized K-anonymity: architecture and algorithms. IEEE Trans Mobile Comput 7(1):1–18
17. Kalnis P, Ghinita G, Mouratidis K, Papadias D (2007) Preventing location-based identity inference in anonymous spatial queries. IEEE Trans Knowl Data Eng 19(12):1719–1733
18. Gedik B, Liu L (2005) Location privacy in mobile systems: a personalized anonymization model. In: Proceedings of the IEEE 25th international conference distributed computing systems (ICDCS)
19. Kido H, Yanagisawa Y, Satoh T (2006) An anonymous communication technique using dummies for location-based services. In: Proceedings of the IEEE 25th international conference distributed computing systems (ICDCS)
20. Ardagna C, Jajodia S, Samarati P, Stavrou A (2010) Providing mobile users' anonymity in hybrid networks. In: Proceedings of the 15th European symposium on research in computer security (ESORICS), Athens, Greece

Sanskrit as Inter-Lingua Language in Machine Translation

Sunita Chand

Abstract This paper gives an insight into the role of Sanskrit as inter-lingua language in Multi-language machine translation. Inter-lingua and direct transformation based approaches have been used for a long period complementing each other while sometimes competing with each other. Inter-lingua based approach is efficient when used for multi-lingual machine translation e.g. Angla-Bharati system uses pseudo lingua for Indian language (PLIL) as inter-lingua language for translation from Hindi to other Indian regional language. It is proposed to use Sanskrit as an inter-lingua in Multi-language machine translation.

Keywords Inter-lingua · Machine translation · Natural language · Corpus based · Sanskrit · Hindi

1 Introduction

Natural language processing has evolved from artificial intelligence field in order to provide linguistic intelligence to machines so that it can unambiguously interpret the sentences given as input in regional language and do the desired processing. Machine translation (MT) is a task that requires knowledge of various other disciplines such as computational linguistics, cognitive science, computer science etc. MT has been made possible by various approaches classified as follows:

S. Chand (✉)
University of Delhi, New Delhi, India
e-mail: sunitamk@gmail.com

K.R. Attele et al. (eds.), *Emerging Trends in Electrical, Communications and Information Technologies*, Lecture Notes in Electrical Engineering 394,
DOI 10.1007/978-981-10-1540-3_3

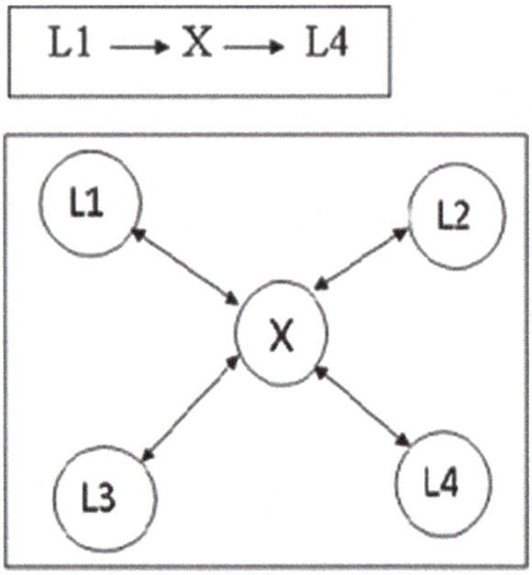

Fig. 1 Translation using single inter-lingua language

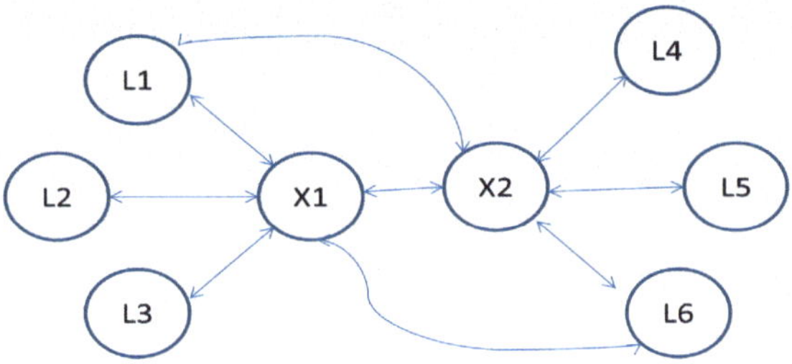

Fig. 2 Translation using two intermediate languages

1.1 Direct Substitution

In this approach the words or phrases in the source language are translated as they are to a target language by using a dictionary. It needs a comprehensive dictionary of all the words and their phrases. Obviously It seems very unrealistic that this method can cope-up with complexity and ambiguous nature of a natural language e.g. Anusaaraka.

1.2 Rule Based/Knowledge Based/Transfer Based Approach (RBMT)

This approach involves generating the database of rules used by the source language as well as target language and obtaining a parse of the source language for mapping to target language structure using these rules. There have been many systems developed using this approach e.g. Angla-Bharati, Matra, Anubaad etc. This approach has the limitations that we can't incorporate all the rules in a system due to which such system suffers from being inadequate, providing limited coverage and also sometimes producing incorrect translations.

1.3 Corpus Based Approach (CBMT)

Corpus based system are further divide as example based MT system e.g. ANUBHARATI and Statistical MT system (SBMT) e.g. GOOGLE Translator, Bing Translator etc. These systems learn how to translate by analyzing existing human translations (known as bilingual text corpora). The success of these systems is obviously dependent upon availability of representative parallel corpora with wide and adequate coverage in the domain of application.

1.4 Inter-Lingua Machine Translation

In this approach, the translation between source language and the target language is accomplished by using some intermediate language which is capable of presenting whole information contained in source language sentences in unambiguous form. Further the intermediate language is translated to target language text.

This approach is very efficient for designing of multilingual translation systems with minimum additional effort. The quality and success of this approach depends on the 'virtues' of the intermediate language and the intermediate structure obtained from the source text. PLIL is one such intermediate structure used by Angla-Bharati system [1]. Other languages used as inter-lingua are UNL (Universal networking language) [2], KANT system [3].

1.4.1 Types of Inter-Lingua Machine Translation Systems

A. ***Translation using single intermediate language.***

In this approach any source language is first translated into in intermediate language (X) and vice versa [4]. Now to translate a language L1–L4 the translations required are (Fig. 1):

B. ***Translation using two intermediate languages***

There might be cases that a single intermediate language is not sufficient for translation from language L1 to L2 i.e. there does not exist a common language which can be conveniently used as intermediate language for L1 and L2. Instead L1 can be conveniently converted to X1 whereas X2 can be used as intermediate language for L2. Also X1 and X2 are commonly used languages for which X1 to X2 conversion packages are already available. Then some least used languages can take the benefit of conversion through two intermediate language instead of no conversion solution.

Let L2 and L5 are least used languages having that does not have a common intermediate language, then L2 → L5 conversion can take place as L2 → X1 → X2 → L5 (Fig. 2).

The remaining paper is organized as follows: Sect. 2 gives an insight to the features required by any language to serve as inter-lingua. Section 3 describes the features of Sanskrit language that make it a strong candidate to be used as inter-mediated language for machine translation. The paper ends with the conclusion and future scope in Sect. 4.

2 Features of an Inter-Lingua

Inter-lingua based approach is advantageous over other machine translation approaches. The first advantage is that multilingual translation need not define explicit rules for each language pair in translation direction. Rather each of the N languages first need N mappings to be translated to intermediate language and then N mappings to translate each language from inter-lingua. So a total of 2 N mappings are required to translate among N natural languages. Whereas, in transfer based approach, a separate mapping is required to translate in each direction for every pair of languages resulting in a total of N (N–1) mappings [5].

For a language to be capable of being an inter-lingua language, it should possess some important characteristics. First of all, it should be unambiguous. Each word in the inter-lingua should be explicit in representation. Secondly, an inter-lingua should be universal, i.e., it should be capable of representing the abstract meaning of any text belonging to any language or domain. Third characteristic that an inter-lingua should possess is that it should be capable of presenting the wholesomeness of the input text i.e., it should be able to represent morphological, syntactic, semantic and even pragmatic meaning of the input text. Fourth, an inter-lingua should not get influenced by the formal representation of the content in source language. Rather it should represent the content of the input language only. Fifth, an inter-lingua should be independent of both, the source language and the target language. The analysis part of the Source-to-target language translation should be based on source language, whereas the generation part should be target specific [6].

The next section describes the features of sanskrit language that make it suitable to be selected as inter-lingua language.

3 Sanskrit as an Inter-Lingua Language

The entire Sanskrit grammar, known as Ashtadhyayi was created by sage Panini with the help of fourteen distinctive sounds that he conceived from God Shiva's damru (small hand-drum which God Shiva holds in His hand).

The perfection of Sanskrit grammar can be proved very easily by the extensiveness of its grammatical tenses, one form for the present tense, three forms for the past tense and two forms for the future tense. There is an exclusive representation for, potential mood, imperative mood, benedictive mood (called asheerling, which is used for indicating blessing), and conditional. It has three separate words for each of the three grammatical persons (first, second and third person), and it further distinguishes among ekvachan, dvi-vachan and bahu-vachan i.e., if it is referring to one, two or more than two people. Also the three categories of the verbs, known as atmanepadi, parasmaipadi and ubhaipadi. signifies that the outcome of the action is related to the doer or the other person or both respectively.

In this way there are ninety forms of one single verb.

For example: 'kri' root word (known as dhatu) means 'to do'. Sanskrit has ninety forms of verbs like this e.g., karoti, kurutah, kurvanti, etc. whereas in English language, there are only a few forms of each word e.g., do, doing, and done in the below figure. Additional words e.g. is, was, will, has been, have been, had, had been etc. are added to these forms of verb to distinguish the tenses. But in Sanskrit language there are distinct single words for all kinds of uses and situations. There are words for all the three genders for the nouns and pronouns and each word has twenty-one forms of its own to cover all situations.

कृ = करना (परस्मैपदी) kri = to do

	एकवचन (one person)	द्विवचन (two people)	बहुवचन (more than two)	
लट्	करोति (karoti)	कुरुतः (kurutah)	कुर्वन्ति (kurvanti)	प्र. (first person)
(Present)	करोषि	कुरुथः	कुरुथ	म. (second person)
	करोमि	कुर्वः	कुर्मः	उ. (third person)
लिट्	चकार	चक्रतुः	चक्रुः	प्र.
(Past Perfect)	चकर्थ	चक्रथुः	चक्र	म.
	चकार, चकर	चकृव	चकृम	उ.
वि. लिं.	कुर्यात्	कुर्याताम्	कुर्युः	प्र.
(Potential mood)	कुर्याः	कुर्यातम्	कुर्यात	म.
	कुर्याम्	कुर्याव	कुर्याम	उ.
आ. लिं.	क्रियात्	क्रियास्ताम्	क्रियासुः	प्र.
(Benedictive)	क्रियाः	क्रियास्तम्	क्रियास्त	म.
	क्रियासम्	क्रियास्व	क्रियास्म	उ.
लुङ्	अकार्षीत्	अकार्ष्टाम्	अकार्षुः	प्र.
(Aorist)	अकार्षीः	अकार्ष्टम्	अकार्ष्ट	म.
	अकार्षम्	अकार्ष्व	अलार्ष्म	उ.
लृङ्	अकरिष्यत्	अकरिष्यताम्	अकरिष्यन्	प्र.
(Conditional)	अकरिष्यः	अकरिष्यतम्	अकरिष्यत	म.
	अकरिष्यम्	अकरिष्याव	अकरिष्याम	उ.
लुट्	कर्ता	कर्तारौ	कर्तारः	प्र.
(First future)	कर्तासि	कर्तास्थः	कर्तास्थ	म.
	कर्तास्मि	कर्तास्वः	कर्तास्मः	उ.
लृट्	करिष्यति	करिष्यतः	करिष्यन्ति	प्र.
(Simple future)	करिष्यसि	करिष्यथः	करिष्यथ	म.
	करिष्यामि	करिष्यावः	करिष्यामः	उ.
लोट्	करोतु,कुरुतात्	कुरुताम्	कुर्वन्तु	प्र.
(Imperative mood)	कुरु, कुरुतात्	कुरुतम्	कुरुत	म.
	करवाणि	करवाव	करवाम	उ.
लङ्	अकरोत्	अकुरुताम्	अकुर्वन्	प्र.
(Past imperfect)	अकरोः	अकुरुतम्	अकुरुत	म.
	अकरवम्	अकुर्व	अकुर्म	उ.

Regarding Sanskrit vocabulary, there is a dictionary of the root words and prefixes and suffixes called dhatu path at the end of Ashtadhyayi. It has an abundance of words and furthermore, Sanskrit grammar is capable enough for creating any number of new words for a new situation or concept or thing.

Hence the Sanskrit language fulfills the first requirement of an Inter-lingua language that it should have word, explicit in representation.

Second requirement that is an inter-lingua should be universal, can be proved by the Sanskrit language has been in its perfect form since thousands of years earlier even before the infancy of the earliest prime languages of the world like Hebrew, Greek, and Latin.

These languages have adopted many words from Sanskrit and have undergone many changes as they passed from one stage to another whereas there has never been any kind, class or nature of change in the science of Sanskrit grammar.

The sound of each of the 36 consonants and the 16 vowels of the Sanskrit language are precise and fixed since its inception. The words of Sanskrit language were never changed, improved, altered or modified in any way. All the words of Sanskrit language used to be pronounced in the same way as they are pronounced today. Also the Sanskrit vowel system has also been immune to any kind of alteration. The reason to this immunity is that Sanskrit was the first language of the world and that it attained its absolute perfection by its nature and formation

When a language changes its form and shape to some extent when it is spoken by unqualified people and people of other origin, is known as 'Apbhransh'. For example Sanskrit word 'matri', with a long 'a' and 'soft' 't', became 'mater' in Greek, and 'mother' in English. It represents that English and Greek languages are 'apbhransh' form of Sanskrit. Such 'apbhranshas' of Sanskrit words are found in all the languages of the world, which proves that Sanskrit was the mother language of the world.

As such Sanskrit language has the capacity to represent all forms of representation of a word, from morphological to semantic, from pragmatic to discourse representation.

4 Conclusion

Considering all the above points as explained above, it is quite evident that Sanskrit is the source of all other languages of the world and not a derivation of any language. As such, it can represent any other language thus qualifying as the inter-lingua language that has the capacity to represent the content of any source language. Hence it qualifies as an inter-lingua language which can be used in the mapping of multiple source languages to multiple target languages. As opposed to the KANT system [7], which produces a source F-structure as the inter-lingua language, the proposed system may be easier to implement as each transfer from source language to Sanskrit language will be governed by the well known grammatical rules of Sanskrit which can be further transferred to any other language.

References

1. Sinha RMK, Jain A (2003) AnglaHindi: an English to Hindi machine aided translation system. http://anglahindi.iitk.ac.in, MTS-2003
2. Dave S, Parikh J, Bhattacharyya P (2001) Inter-lingua-based English-Hindi machine translation and language divergence. Mach Transl 16:251–304
3. Nyberg EH (1996) Controlled Language and Knowledge Based Machine Translation: Principles and Practice. In: First International workshop on controlled language applications, Katholieke University, Leuven, 26–27 March 1996
4. Adusumilli KK Natural languages translation using an intermediate language. IAENG Int J Comput Sci 33:1, IJCS_33_1_20
5. Lampert A (2004) Inter-lingua in machine translation. Technical Report, 2004
6. Al Ansary S Inter-lingua-based machine translation Systems: UNL versus other inter-linguas
7. Mitamura T, Nyberg EH, Carbonell JG (1991) An efficient inter-lingua translation system for multi-lingual document production. In: Proceedings of machine translation summit III, Washington D.C., 2–4 July 1991

Scalability in Virtualization

Chandrika Prasad, H.M. Varun, M.T. Vijay kumar, K. Yashaswini and G. Suhas

Abstract Virtualization is the process of creating a virtual version of an operating system, a server, a storage device or network resources. The main objective of this paper is developing virtualization infrastructure using open source solution for enabling virtual machine for a selected or developed application. This system makes use of bare-metal virtualization in which Virtual Machine Monitor runs directly on physical hardware. The system allocates resources dynamically via virtualization based on the application demand. As the demand increases the hypervisor dynamically creates virtual guest operating system and shutdown the guest operating system as demand decreases, thus achieving scalability. Hypervisor distribute the resources equally among the guest operating system.

1 Introduction

Virtualization has emerged as the wave of the future. Virtualization reduces the number of devices which in turn result in reduced physical machine costs [1]. Devices emit heat as the by-product of their operations. High temperature may result in inoperability of the devices. Cooling systems are used to maintain the temperature. Less number of devices results in lesser cooling systems. Reduced physical machine costs, cooling costs, energy bills and complexity in administration are the driving forces in unexpected rise of virtualization in IT landscape. Virtualization is rapidly improving data center efficiency.

This paper proposes a system which allocates resources dynamically via virtualization based on the application demand, thus reducing number of resources and energy costs of idle resources [2]. As the demand increases the hypervisor dynamically creates virtual guest operating system and shutdown the guest operating system as demand decreases, thus achieving scalability (Tables 1 and 2).

C. Prasad (✉) · H.M. Varun · M.T. Vijay kumar · K. Yashaswini · G. Suhas
Department of Computer Science and Engineering, M.S Ramaih Institute of Technology, MSR Nagar, Bangalore, Karnataka, India
e-mail: chandrika@msrit.edu

K.R. Attele et al. (eds.), *Emerging Trends in Electrical, Communications and Information Technologies*, Lecture Notes in Electrical Engineering 394,
DOI 10.1007/978-981-10-1540-3_4

Table 1 Load distribution among virtual machines (Increasing application load)

No of instances	VM1 status	VM2 status
00	Running	Idle
10	10 instances running	Idle
20	20 instances running	Idle
30	30 instances running	Idle
40	40 instances running	Idle
50	50 instances running	Idle
60	50 instances running	10 instances running

Table 2 Load distribution among virtual machines (Decreasing application load)

No of instances	VM1 status	VM2 status
60	50 instances running	10 instances running
50	50 instances running	Idle
40	40 instances running	Idle
30	30 instances running	Idle
20	20 instances running	Idle
10	10 instances running	Idle
00	Running	Idle

2 Background

The following are the components used to create virtualization environment:

A. *Virtual Machine (VM)*:
   A virtual machine is an environment in isolation which appears to be a whole computer but has access to a portion of resources. A virtual machine shares hardware resources but isolates application or the operating system from other users to avoid interruptions and hence improving the experience of the end users.
B. *Virtual Machine Monitor (VMM)*:
   A Virtual Machine Monitor is the software layer that supports one or more Virtual Machines. It is also called as hypervisor.
C. *Guest Operating System*:
   A guest operating system is an operating system which runs on a Virtual Machine rather than directly on hardware. Guest operating system runs in user mode.
D. *Kernel Virtual Machine*:
   Kernel-based Virtual Machine (KVM) is a Virtual Machine Monitor [3] which is developed as a full virtualization solution for Linux on ×86 hardware containing virtualizations extensions such as Intel VT or AMD-V. KVM is bare metal hypervisor which runs directly on physical hardware.

E. *Quick Emulator (QEMU)*:
QEMU is an open source machine emulator and virtualizer. QEMU can execute guest code directly on the host CPU. QEMU supports virtualization when it is executed using the KVM kernel module.

F. *Libvirt*:
Libvirt is an open source virtualization API which supports KVM-QEMU Linux hypervisor.

G. *Virt-manager*:
Virt-manager is an open source application which acts as a virtual machine manager. It is a desktop user interface. It manages KVM Virtual Machines through Libvirt.

3 Design

The architecture i.e. is deployed is as shown in the Fig. 1. Host operating system have VMM (Virtual Machine Monitor) on its top which manages and monitors all the virtual operating systems. All the guest operating systems are on top of VMM each one forming a virtual machine. Application is deployed on each guest operating system. GUI is deployed on host operating system for load balancing.

To check the scalability, the GUI is designed known as myappln i.e. shown in the below snapshot, which allows user to start the analyses of the load handled by the virtual machines. The start button is used to start the analysis and the refresh button is used to reload the contents when there is some problem or delay. The stop button is used to stop the analysis of the load on the virtual machines started by the virtual machine monitor (Fig. 2).

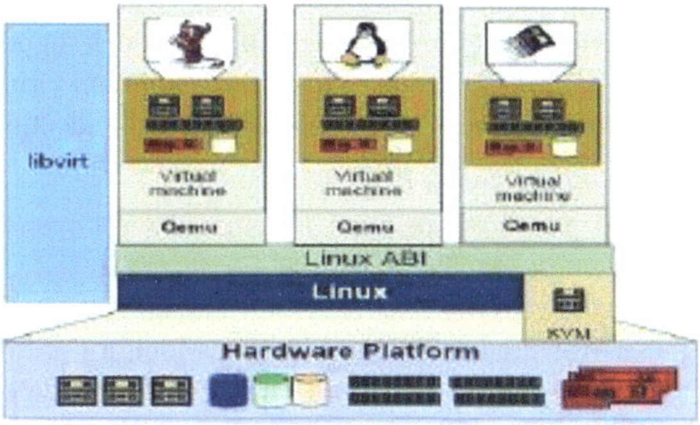

Fig. 1 Architecture design [4]

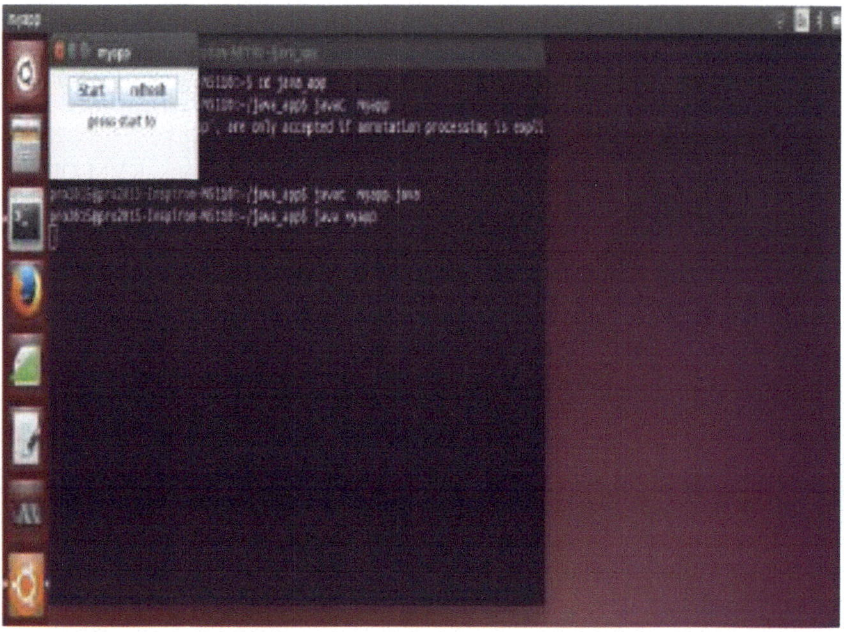

Fig. 2 Snapshot of GUI myappln

The text view box is used to display the contents, information about the virtual machine, the operations or the process carried out when there is an increase in load or decrease in load.

4 Implementation

In order to implement the concept of virtualization [3, 5–8], the system should support virtualization. Virtualization extensions support such as IntelVT or AMD-V need to be enabled. In order to deploy minimum virtualization setup in data centers six modules are necessary. Following are the identified modules.

A. *Installing Ubuntu*:
   To deploy host operating system and guest operating system, Ubuntu operating system is adopted.
B. *Installing Kernel-based Virtual Machine (KVM) and its packages [3]*:
   KVM is used as Virtual Machine Monitor (VMM) for our system. QEMU can make use of KVM when running a target architecture that is same as the host architecture. Libvirt is used as an API and management tool for managing platform virtualization. It manages KVM. Virt-manager is used as a desktop user interface for managing virtual machines through libvirt.
C. *Creating guest operating system*:
   Five Ubuntu 14.04 guest operating systems are installed.

D. *Network interface between guest and host*:
   A folder has to be created in host and share with the guest operating system by mounting a folder in guest operating system using system commands.
E. *Deploying application*:
   Once the virtualization is setup, elective management application is deployed on guest operating system. This application can be accessed only by corresponding guest operating system.
F. *Graphical User Interface (GUI)*:
   GUI continuously runs on host operating system and makes cautious observation on number of guest operating system running at a specific instance.

5 Algorithms Used

The below algorithms explains about the scale up and scale down features of virtual machine [9]. Initially file is opened in read mode and this file is constantly read by load variable so that when load in virtual machine exceeds 50, new virtual machine is started and its usage information is recorded. In the same manner when the load decreases to less than 1, current virtual machine is shut down and its status is recorded.

Scale up algorithm	Scale down algorithm
A thread is created to monitor the scale up process of virtual machine, The run function is called when the thread is started and it keeps running until the thread is destroyed. When the load of the particular virtual machine is more than the threshold value, the new virtual machine is started by calling detectvm function	A thread is create a thread to monitor the scale down process of virtual machine, The run function is called when the thread is started and it keeps running until the thread is destroyed. When there is no load on the particular virtual machine, the virtual machine is shutdown

class scaleupmonitor implements runnable// creates thread for scale up
mutex=1
scaleupmonitor(String name)
memorypath = name;
t = new Thread(this,memorypath);
t.start();
run()
while(mutex==1)
if(textData==null)
continue;
if(textData>10)
Virtualscale object = new Virtualscale();//starts new thread
object.detectvm(); //to check for free vm and start virtual machine
if(Virtualscale.busywait==0)
mutex=0; //release lock for busy wait
else
Thread.sleep(5000);
	class scaledownmonitor implements Runnable //create thread for scaledown
	mutex = 1;
	scaledownmonitor(String name)
	
	memorypath = name;
	t = new Thread(this,memorypath);
	t.start();
	run()
	while(mutex==1)
	if(textData==null)
	Thread.sleep(3000);
	continue;
	if(textData==0)
	mutex=0;
	virtualmachine-->stop virtual machine //stops virtual machine

6 Test Results and Snap Shots

The following are the snap shots which have been taken to show the activities of scale up and scale down functions in virtual functions.

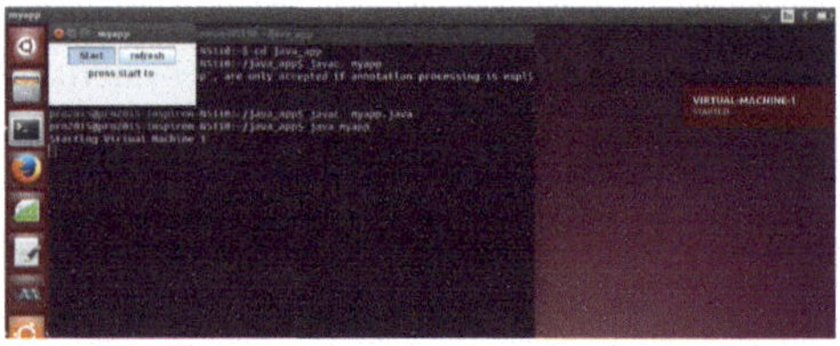

When start button is pressed, by default virtual machine 1 is started

VMM showing the status of all virtual machines

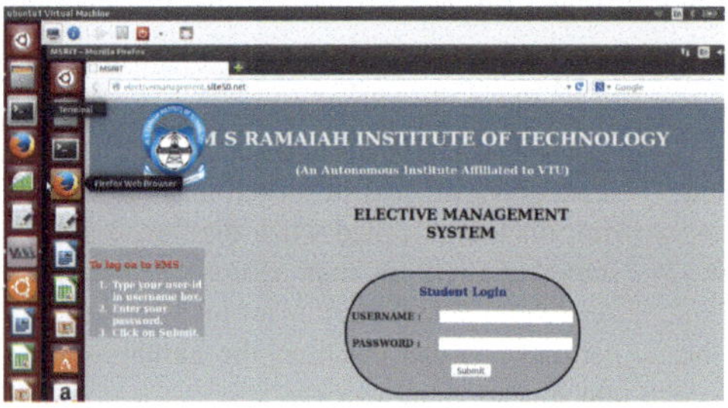

Elective management application which is used to check the load of virtual machine

When the load in virtual machine 1 exceeds threshold then virtual machine 2 will be started

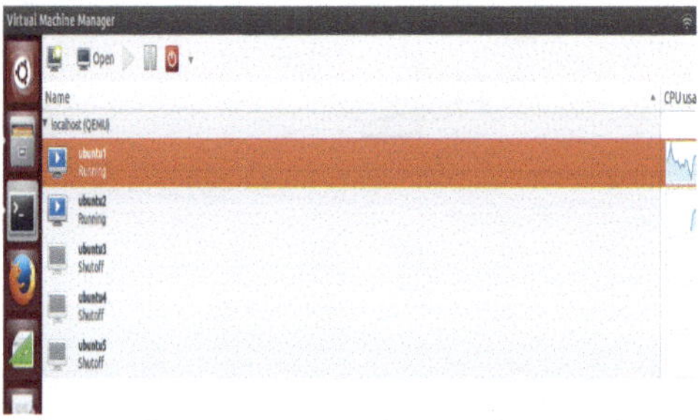

VMM showing that virtual machine 2 is running

virtual machine 2 shutting down when there is no load

Display showing all virtual machine are running

The following are the tables and graphs which indicates the load distributions among virtual machines.

Case 1: Increasing application load
From the graph (Fig. 3, Table 1) it can be noted that a Virtual Machine can handle 50 instances of web browser, so a new Virtual Machine is started when there are more than 50 requests. New Virtual Machine will now handle those further requests.

Case 2: Decreasing application load
From the graph (Fig. 4, Table 2) it can be noted that a Virtual Machine will shut down when the number of instances reduces to zero. It should be noted that Virtual Machine 1 will always be running to handle requests.

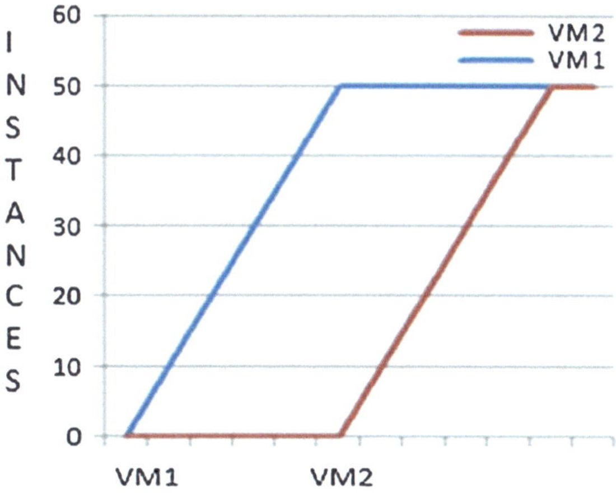

Fig. 3 Load distribution metrics

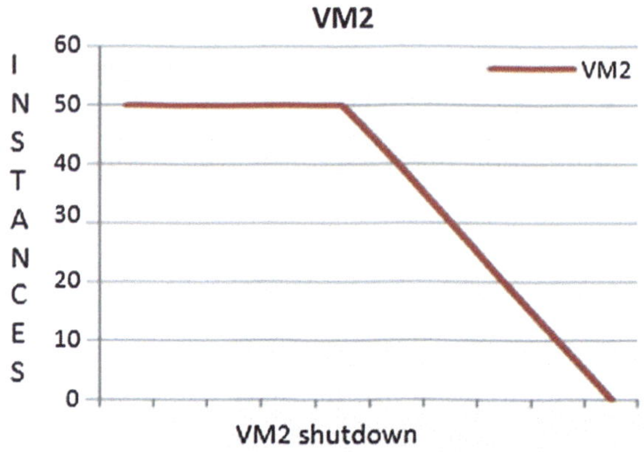

Fig. 4 Load distribution metrics

7 Conclusion and Future Scope

This paper focuses on scaling of virtual machines when any application is deployed in it. Based on the load of the application in virtual machine it can be switched from one to another efficiently. When the load of a particular application increases in a virtual machine then a new virtual machine will be started to take further request of that application. When there is no request for the application on a virtual machine then that virtual machine will be shut off. So now a particular server will not be overloaded due to more requests from an application or won't be ideal without handling any requests.

Establishment of large organization will not cost too much due to deployment of virtual machines. Data security issues will be fixed easily by storing the data on the virtual machines. Productivity of the organization will be increased. This paper proposes an idea that can be extended to data storage in cloud technology and can be used to handle request of any websites like IRCTC website, any exam result websites or even E-commerce website.

References

1. Scroggins R (2013) Virtualization technology literature review. Capella University
2. Qutier B, Neri V, Cappello F (2007) Scalability comparison of four host virtualization tools. J Grid Comput 5(1):83–98
3. KVM virtualization. http://www.linuxnix.com/2013/02/kvm-virtualization-install-kvm-hypervisor-in-ubuntudebian-linux.html
4. http://encrypted-tbn2.gstatic.com/images?q=tbn:ANd9GcSd-BQZjQFQZtZ8t9P5taTFwfu5xhfgbUwW23Ff_xesLMjCa3ZK
5. The Top 5 Enterprise Type 1 Hypervisors You Must Know. http://www.virtualizationsoftware.com/top-5-enterprise-type-1-hypervisors/kvmon-ubuntu-server.html

6. KVM virtualization in Linux. http://www.linuxnix.com/2013/02/what-is-kvm-virtualization-in-linux.html
7. King ST, Dunlap GW, Chen PM (2013) Operating system support for virtual machines. University of Michigan
8. Crash course: virtualization with KVM on Ubuntu Server. http://www.itworldwith-kvmon-ubuntu-server.html
9. Supreeth S, Biradar S .com/article/2827037/virtualization/ crash-course–virtualization-scheduling virtual machines for load balancing in cloud computing platform. Int. J. Sci. Res. India Online ISSN: 2319-7064. http://www.ijsr.net/archive/v2i6/IJSROFF2013291.pdf

Blind Spectrum Sensing Techniques in Cognitive Radio-Survey

S. Vasundara, D. Raghavaraju and D. Venkatesh

Abstract With the always expanding interest for remote correspondences, the spectrum has turned into a rare asset. Spectrum lack is turning into a major obstacle for the advancement of new remote advances and presentation of new applications and administrations. To keep away from this lack issue, the thought of cognitive radio (CR) was proposed. It permits the utilization of spectrum in a proficient way. The thought of cognitive radio is in view of compelling spectrum use. The spectrum is apportioned to essential client or authorized client. At the point when essential client is not using the assigned band, optional client can guarantee for that empty band of essential client. For that spectrum sensing is needed. At the point when the essential client comes then auxiliary client all will be dropped. Spectrum sensing is a standout amongst the most difficult issue. This paper concentrates on diverse Blind spectrum sensing procedures.

Keywords Cognitive radio • Blind spectrum sensing • Frequency • Cyclostationary

1 Introduction

The requirement for higher information rates is expanding as a consequence of the move from voice-just interchanges to Multimedia sort applications. Given the constraints of the Natural recurrence spectrum, it gets to be evident that the present

S. Vasundara (✉)
CSE Department, JNTUACEA, Anantapuramu, India
e-mail: vasundaras@rediffmail.com

D. Raghavaraju
SVIT, Anantapuramu, India
e-mail: raghava.digala@gmail.com

D. Venkatesh
CSE Department, GIT, Gooty, India
e-mail: deancseit@gmail.com

K.R. Attele et al. (eds.), *Emerging Trends in Electrical, Communications and Information Technologies*, Lecture Notes in Electrical Engineering 394,
DOI 10.1007/978-981-10-1540-3_5

Static recurrence allotment plans can't suit the necessities of an expanding number of higher information rate Devices. Therefore, imaginative methods that can offer better approaches for misusing the accessible spectrum are required. Cognitive radio emerges to be an introducing so as to entice answer for the otherworldly Congestion issue pioneering use frequency groups, which are not strongly involved by authorized clients [1, 2]. When no concurrence on the formal Definition of cognitive radio starting now, the idea has developed as of late to incorporate different implications in a few connections [3]. By using definition, cognitive radio is a framework, which identifies its equipped electromagnetic atmosphere and progressively and self-sufficiently alter Its radio working parameters to adjust framework operation, Such as amplify throughput, relieve impedance, encourage Interoperability, access auxiliary markets [2]. Henceforth, one Main part of cognitive radio is identified with independently Exploiting by regional standards unused spectrum to give new ways to Spectrum access. In cognitive radio phrasing, Primary clients can be characterized as the clients who have higher Priority or legacy rights on the use of a particular piece of the Spectrum. Then again, auxiliary clients, which have Lower need, misuse this spectrum so as to not bring about impedance to essential clients. In this way, optional clients require to be having CR abilities, for example, sensing range to make sure whether it is used by a client and to modify the radio parameters to abuse the Unused piece of the spectrum. Classification of blind spectrum sensing techniques is shown in Fig. 1.

The broad categorization of spectrum sensing algorithms based on previous in sequence is as follows:

1. Transmitted noise and signal
2. Environmental noise
3. Signal and noise without information

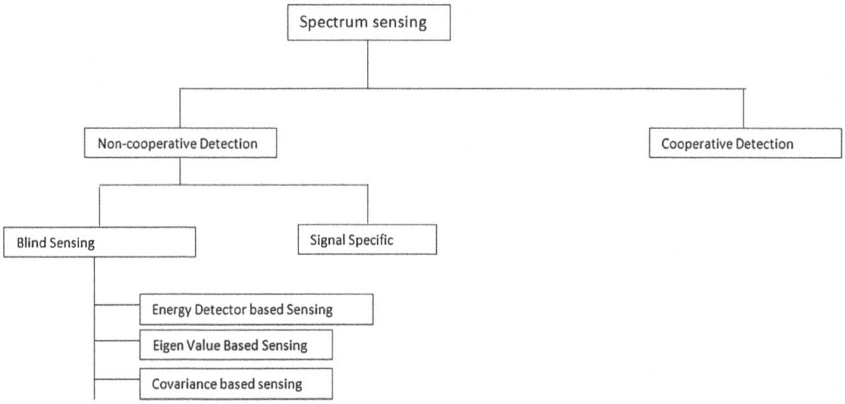

Fig. 1 Spectrum sensing techniques

These systems are vitality discovery, coordinated channel, cyclostationary, wavelet coordinating, Eigen quality based and other various methodologies. Out of which vitality location tech is adjusted in a significant number of exploration environment because of its effortlessness and simplicity in configuration [4].

2 Issues

Authors [5] expressed that, few elements make spectrum sensing basically difficult. The obliged SNR for location may be low. Case in point, regardless of the information is an essential transmitter is close to an optional client (the identification hub), broadcasted sign is essential client would be profound blurred so the essential signs signal noise ratio at the auxiliary collector beneath −20 dB. Notwithstanding, the auxiliary client yet requires to recognize the essential client and abstain from utilizing channel in light of the fact that it might firmly meddle with the essential recipient on the off chance that sends. A viable situation is a remote mouthpiece working in TV groups, which just sends along power under 50 mW and a data transfer capacity under 200 kHz. On the off chance that an optional client is a few 100 m as of the mouthpiece gadget, got signal to noise ratio beneath −20 dB. Also, different ways of blurring along time scattering of the remote guides confuse the sensing issue. Multipath blurring may bring about the sign energy to vary as much as 30 dB. Then again, obscure time scattering in remote channels will revolve the understandable detection inconsistent. The commotion/obstruction stage might alter with moment and area, it gives way the clamor control instability issue for location.

As in [6] the creators gave challenges in SS(spectrum sensing), where a few open exploration issues, which should be researched in the advancement of SS systems:

- interfering warmth estimation: outstanding to the absence of collaborations between essential systems and CR systems, by and large a CR client can't be mindful of the exact areas of the essential beneficiaries. Hence, new methods are obliged to gauge or assessment the obstruction temperature at adjacent essential recipients.
- SS in user arranges: The client atmosphere, comprising of different CR clients and essential clients, makes it more hard to sense spectrum gaps and evaluation impedance. Subsequently, spectrum sensing capacities ought to be produced considering the multi-client environment.
- Spectrum-productive sensing: Sensing can't be performed while transmitting bundles. Thus, CR clients ought to quit transmitting while sensing, which diminishes spectrum productivity. Thus, adjusting spectrum proficiency and sensing precision is an essential issue. Additionally, in light of the fact that sensing time specifically influences transmission execution.

3 Blind Spectrum Sensing Techniques

Spectrum sensing strategies can be characterized into two classifications, i.e. blind spectrum sensing and non-daze spectrum sensing. The upside of visually impaired spectrum sensing is that it doesn't require any data of the source air interface. Samples of visually impaired spectrum sensing procedures would be vitality location, wavelet based identification, and eigenvalue based discovery and for the non-dazzle recognition coordinated separating and cyclostationary based discovery can be brought up. In class of non-visually impaired sensing coordinated separating is known not the ideal technique for identification.

The greater part of the traditional helpful spectrum sensing strategies are in light of vitality discovery. It obliges and clamor learning to perform spectrum sensing. Be that as it may, exact clamor estimation at low commotion levels is difficult to get. So sensing strategies that does not oblige commotion information are hotly debated issue of examination. Sensing calculations that don't oblige commotion information to perform spectrum sensing are alluded as visually impaired spectrum sensing algorithms [7]. The data to the calculation is cyclostationary in nature. The visually impaired sensing calculation utilized as a part of this paper takes a shot at the premise of direct expectation and QR decomposition [7] of the got signal. In this strategy two sign statics are registered from the got signal and the proportion of these statics demonstrates vicinity or unlucky deficiency of the essential client. This calculation does not oblige clamor power or channel information to perform sensing. The guideline of this technique is in view of the way that the sign specimens indicates exceptionally solid connection properties yet not the noise. This property empowers sensing even at low SNR.

As in [8] creators proposed new visually impaired spectrum calculation is in light of second request factual examination which separates commotion and transmitted sign. A choice on whether a sub-band is used by another client or not is made in light of PSD estimation which is joined in the proposed strategy. There is a technique that can beat the impact of clamor instability. In particular it doesn't require any data of the source air-interface while it can accomplish a decent exchange off in the middle of inertness and dependability. As in creators [9] proposed another SS structure, which utilizes kurtosis estimation inside the visually impaired basis partition calculations. This structure enhances the SS identification execution uncommonly when the cr (cognitive radio) and the essential system are in process at the same time. We likewise proposed to perform spectrum combining so as to sense visually impaired source partition with routine irregular framework hypothesis based spectrum sensing. Creators [10] displayed a methodology that has the capacity remake the visually impaired groups signals for CRNs without knowing the groups area data, which can enhance the spectrum sensing effectiveness and diminish the sensing time. As in [11] methods taking into account the eigenvalues of the example covariance network of the got signals are discussed. Most recent arbitrary network speculations are utilized to set the edges and get the likelihood of identification. The systems can be utilized for different sign location uses excluding learning of sign, channel and commotion power.

Creators [12] examined the relationship between the conduct of the incline of sign measurement bend and the move from an involved band to a neighboring free band (and the other way around). Creators [13] initially proposed the SITC sensing calculation. This calculation altogether lessens the computational many-sided quality without losing any location execution contrasted and the current ITC-based sensing calculation. Also, it empowers a more trackable logical study on the location execution. From that point, applying the late advances in irregular framework hypothesis, we have inferred mathematical equations, likelihood of false alert and the likelihood of identification, which can firmly rough the real results in reenactment. Creators [14] proposed a spectrum-sensing strategy for OFDM-based cognitive radio frameworks has been produced in view of the GLRT system. The key element in our advancement is to expressly consider the structure (imperative) of the covariance lattice of the hidden OFDM flag so that the ML estimations of obscure parameters are enhanced, which prompts hearty and proficient

Table 1 Existing techniques review

Title	Technique	Advantage
Blind spectrum sensing techniques for cognitive radio system	Cyclostationary feature based detection technique	It improves spectrum sensing performance without significant increasing in overall complexity
An improved blind spectrum sensing technique for cognitive radio systems	Kurtosis BSS spectrum sensing	The SS detection mechanism is improved specially when the cr and the main network are in operation at a time
Performance analysis of blind spectrum sensing in cooperative environment	Blind spectrum sensing based on QR decomposition and linear prediction	Improves the reliability of spectrum sensing
Blind spectrum sensing for OFDM-based cognitive radio systems	Covariance matrix of the underlying OFDM signal	Robust and efficient spectrum-sensing
Blind spectrum sensing by information theoretic criteria for cognitive radios	Simplified information theoretic criteria sensing algorithms	Reduces the computational complexity without losing any detection performance
Blind spectrum sensing for cognitive radio based on signal space dimension estimation	Spectrum sensing method based on Sliding window technique	Good performance in spectrum holes detection
A novel blind spectrum sensing approach for cognitive radios	Blind spectrum sensing based on second order statistical analysis which differentiate noise and transmitted signal	It does not require any information of the source air-interface while it can achieve a good trade-off between latency and reliability

spectrum-sensing tests. Creators proposed [4] a spectrum sensing strategy in light of visually impaired source division procedure by utilizing Kurtosis measurements that can sense the spectrum while auxiliary transmitters are in operation.

Table 1 shows different blind spectrum sensing techniques. It also shows the techniques previously used and also advantages corresponding to those techniques. Kurtosis based technique is another SS method, which performs better than energy based SS techniques [15].

4 Conclusion

Spectrum sensing is a standout amongst the most vital parts in the execution of each cognitive framework. Cr works like answer of wasteful use of spectrum groups. SS is a standout amongst critical issues in each cognitive radio framework. Traditionally utilized SS methods oblige that the cognitive transmitter wouldn't in process as distinguishing the vicinity/unlucky deficiency of the essential sign amid the sensing time. This article portrays a few visually impaired spectrum sensing strategies, which are now existed. This overview is extremely useful to the inquires about those are enthusiastic about working Cognitive radio range particularly daze spectrum sensing.

References

1. Mitola J, Maguire GQ (1999) Cognitive radio: making software radios more personal. IEEE Personal Commun Mag 6(4):13–18
2. Federal Communications Commission (2005) Notice of proposed rule making and order: facilitating opportunities for flexible, efficient, and reliable spectrum use employing cognitive radio technologies, ET Docket No. 03–108, Feb. 2005
3. Neel JO (2006) Analysis and design of cognitive radio networks and distributed radio resource management algorithms, Ph.D. dissertation, Virginia Polytechnic Institute and State University, Sept. 2006
4. Rana MD, Patel HR (2014) Review on spectrum sensing and sharing techniques of cognitive radio system. Int J Futuristic Trends Eng Technol 4(01)
5. Y¨ucek T, Arslan H (2009) A survey of spectrum sensing algorithms for cognitive radio applications. IEEE Commun Surv Tutorials 11(1), first quarter
6. Akyildiz IF, Lee W-Y, Vuran MC, Mohanty S Georgia Institute of Technology. A survey on spectrum management in cognitive radio networks
7. De P, Liang Y-C (2008) Blind sensing algorithms for cognitive radio networks. IEEE Trans Veh Tech 57(5)
8. Cheraghi P, Ma Y, Tafazolli R A novel blind spectrum sensing approach for cognitive radios
9. Khajavi NT, Sadeghi S, Sadough SM-S An improved blind spectrum sensing technique for cognitive radio systems
10. Li S, Wang C, Zhou X, Wang J Efficient blind spectrum sensing for cognitive radio networks based on compressed sensing

11. Zeng RY, Senior Member, IEEE, Liang Y-C, Senior Member (2009) Eigenvalue-based spectrum sensing algorithms for cognitive. IEEE Trans Commun 57(6)
12. Zayen B, Hayar A, Kansanen K Blind spectrum sensing for cognitive radio based on signal space dimension estimation
13. Wang R, Tao M Member, IEEE, Blind spectrum sensing by information theoretic criteria for cognitive radios
14. Bokharaiee S, Student Member, IEEE, Nguyen HH, Senior Member, IEEE, Shwedyk E (2011) Blind spectrum sensing for OFDM-based cognitive radio systems. IEEE Trans Veh Technol 60(3)
15. Suresh S, Prakriya S, Bhatnagar MR (2012) Kurtosis based spectrum sensing in cognitive radio. Phys Comm 5:230–239

Offline Data Synchronization with Occasionally Connected Databases Using Smart-IPMS

R.M. Jagadish, L. Swarna Jyothi and Rohini Patil

Abstract We live in a world with an increasing number of connected computing resources. However, in many cases we cannot expect one hundred percent connectivity throughout. Applications may not be able to access network resources all the time without good network connectivity. A requested service could be busy, down, or just temporarily unavailable. The worldwide network is increasing rapidly. The devices connected to network are vastly different from desktop computers, because they are meant for different purposes. Their main purpose is to connect people to information. Social media, their work information and their emails are information sources. Offline data synchronization plays a vital role in ensuring efficiency in communication between the client devices and the web server in an environment with limited internet connection. This paper presents an algorithm for data synchronization in Insurance Policy Management System.

Keywords IPMS · Offline mode · Client · Server · Mobile database

1 Introduction

The client-server system relays on the availability of some form of network connection for proper functioning. A client-server depends on the internet, largely because all the resources it needs are external and they are accessed remotely. This

R.M. Jagadish (✉)
Faculty of Engineering College, Department of Computer Science,
BITM, Bellary, Karnataka, India
e-mail: rm.jagadish@gmail.com

L. Swarna Jyothi
RRCE, Bengaluru, Karnataka, India
e-mail: swarna_jy@yahoo.com

R. Patil
BITM, Bellary, Karnataka, India
e-mail: patilsrohini09@gmail.com

K.R. Attele et al. (eds.), *Emerging Trends in Electrical, Communications and Information Technologies*, Lecture Notes in Electrical Engineering 394,
DOI 10.1007/978-981-10-1540-3_6

means that without a network connection, the transaction progress may fail, or it may not go according to the way it is planned and most likely without intimation. In terms of the threshold of the kind of system acceptability and reliability, the client-server system does not guarantee the user the kind of standards that would be normally be expected in computing [1].

A conceptual model for this argument would be indicated as shown below (Fig. 1):

The conceptual model displays occasionally connected system in the upper part. They provide the user access to the system through the user interface and the GUI. The interface provides access to the backend.

The user interface facilitates adding a new record of a new client by field agent and processing is done by the user interface logic. A service agent is what checks for the connectivity with the database and makes the necessary calls like add, update or delete the record in the backend. When there is no active connection, the processing will not take place [2].

IPMS will be used as base example for the whole argument. In today's scenario the number of notebook computers, mobile phones that run on complex but reliable platforms has been growing and hence there is a compelling need to use them other than depend on the traditional bulky computers like the desktops [3]. The bulky computers have now been out-numbered in all the fields, and the smaller portable

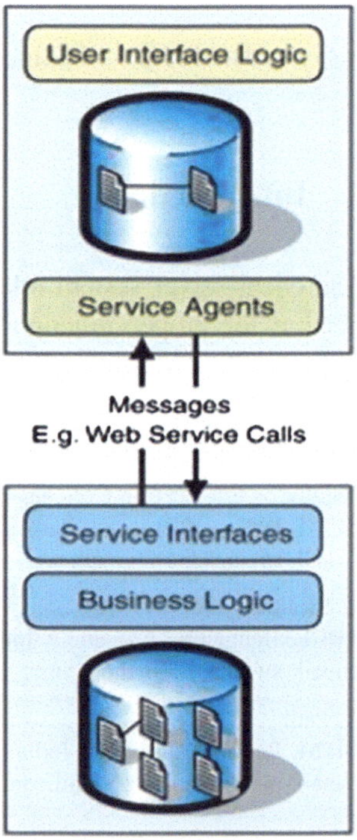

Fig. 1 A Conceptual model for occasionally connected systems

gadgets have become the order of the day in today's business and computing sectors and platforms.

The trend is very much useful in designing an efficient system for insurance sector, and similar other sectors. The insurance sector is largely dependent on the area of data processing, and the profitability of the sector is tied to the earnings from the sales realized from the payments of policies. So the management of policies needs to be done on the fly. The security and the efficiency of how that is handled are of paramount importance. The system, which handles the policies, is what can be referred to as the IPMS [4, 5]. This kind of a system allows the insurance details to be created, updated and to be expunged from the records. It is also possible to view the details of all the resellers, the agents, the policy holders of the company and all the information about the premium payments made by the clients. The only fundamental thing to be considered, is the user levels and to enforce authentication constraints to avoid the abrasion of data security per se.

2 Methodology

Proposed Smart-Insurance Policy Management System (Smart-IPMS) design forms part of the solution. This work will include the system's ability to manage the policy information just from anywhere instead of being tied to a single office location. The system will be aimed to reduce all the possible labors by a customer and the agent too. There will be a commendable reduction in the burden of a development officer/insurance policy agent in terms of reduced paperwork and in maintaining all policy information directly and into some form of electronic form. The field agents will get a connection to the database, and then synchronization will be initiated, and the handheld device will enable the field agent or whoever is using the device to upload the database automatically. The process will overcome data consistency problems. The final aspect that will be implemented in the design in addition to the 24/7 support is the provision to auto switch from offline mode to online mode [3].

Applications that are meant to use a data-centric approach are designed and coupled with a relational database management system (RDBMS). The RDBMS will be installed on a local client, and the system depends on the built-in capabilities of that database system which propagates the local data changes on the back-end to the server [6]. The database has to match the order so that the synchronization process updates data uniformly. It takes care of detection and resolution of any data conflicts in this form of architecture. The service-oriented architecture approach focuses more on the storage of information in the form of messages arranged in queues when the device is offline and it gets updated when the device has internet connection [7]. The approach is operational by the Google Gmail app on Android. The app updates all the emails in the inbox in database form, so that the client may read all mail updates even with no network connection. Soon after the network connection is reinstated, the queued messages are sent to the server for reprocessing and the newer emails are added to the database.

The approach is data-centric which is more accommodative for an IPMS solution [8]. The major requirement for an occasionally connected system, in addition to the specified, supporting deliverables will be to design a Windows application that will sit on the client side. An application is installed to manage the user requests and those in the local database that stores the insurance policy data. The application synchronization will be enforced with the complex objects with write permission. The complex objects will come in handy to change and update data on both the sides [9].

Conclusively, it is vital to note that the trend in computing will have to be accommodative to the concept of occasionally connected systems architectures by all means. The Modules that work together to make it possible to realize much cheaper and more efficient IPMS. The proposed solution will be dependent on the concept of occasionally connected systems architecture. The initial objective has been attained through this efficacy and investigative report. Next stage will be implemented through design of a web-server, database, and a Windows application. The system is tested with appropriate data to prove efficacy.

3 Offline Data Synchronization Algorithm

Data synchronization is updating two or more databases with each other's changes. Client-server systems rely heavily on reliable network connections for it to function effectively. The need for connected network cannot be over emphasized in the client-server set up. Internet is an important component for any update to take place. Advancement in the information technology industry has revolutionized the way data synchronization is done.

The success of data exchange in an offline setup strongly depends on the following key issues of both the client and the server side system.

(1) Similarity of the database structure. If the client database structure corresponds to the server machine, then data synchronization will be possible whenever connection is established.
(2) How often is the offline data need to synchronized
(3) The cost and effort the one is willing to offer in establishing data synchronization
(4) The effort need for each synchronization session.

Offline data synchronization needs a system where in the backend and the front end has the same database structure and schema for data update and exchange to take place at both the ends of the system. Synchronization criteria of the offline data should use either time or internet connection factor. In time factor synchronization, the offline system should have an algorithm that will ensure the configuration of user machine to synchronize data to the online database of the system. For instance an algorithm that will enable client machine to set a specific time in which synchronization of the daily business transaction with the online database; mostly synchronization is set to take place in the midnight when all the business transactions of the day are over.

This periodic synchronization of data is very useful especially when dealing with aggregate data. Aggregate data should be synchronized once so that discrepancies are avoided and data integrity is ensured. This will generate a reliable and accurate report in the daily business transactions.

Data synchronization based on internet availability can also be used to synchronize offline data. This can be achieved by designing an algorithm that will check on the network interface card of the client machine to detect any internet signal that is being received. Once the connection is detected data synchronization is initialized. Once synchronization has been initialized the data is organized into small jobs. The synchronized part will be aborted when internet connection is terminated abruptly while the job is still in process. Synchronized data should only be committed if and only if the job is completed successfully. The concept of jobs also require a well structured algorithm that can organize the process into smaller jobs that takes less time to commit data.

The algorithm shown below will perform data synchronization in occasionally connected systems.

//Algorithm for data synchronization in occasionally **connected systems**
DATA SYNCHRONIZATION PROCEDURE (CHECK, DETECT, START, SYNCHRONIZE); DETECT CONNECTIVITY;
IF CONNECTION DETECTED, THEN ESTABLISH SYSTEM CONNECTION;
IF ONLINE, THEN CHECK CHANGES;
IF CHANGES = TRUE, THEN
PROMPT: DO YOU WANT TO OVERRIDE IT WITH YOUR LAST CHANGES? IF OVERRIDE = YES, THEN
ORGANIZE DATA INTO SMALL JOBS;
IF JOB = DONE, THEN
START JOB SYNCHRONIZATION
IF JOB SYNCHRONIZATION COMPLETED = TRUE, THEN UPDATE CHANGES ELSE
TERMINATE THE PROCESS TO WAIT FOR CONNECTION,
RETURN;

IPMS enable both firms and insurance clients communicate effectively. IPMS provide quick response to any issues arising within the business line of the firm. A client will be able to confirm the value of his or her insurance cover. An insurance client can request his or her claims remotely using their mobile devices; the main challenge in doing this is the absence of a reliable internet connection in most remote places.

With IPMS, insurance clients will be able to request for their claims. In the normal scenario the requests to client's claims mostly use a computer machine that is connected via the internet to the central server of the insurance firm. Alternatively the customer uses a mobile device to access the services remotely. When mobile devices are used to access the services, offline data synchronization is very important. Since there are so many such devices that might be trying to access the server simultaneously this will cause delay due to data traffic caused by the many client requests. Data traffic is not the only reason for offline data synchronization, many places lack in internet connection and this can overcome by offline data synchronization for no matter how many requests are made. The data will be stored in the local database of the system that will synchronize with the online database when connection is detected.

4 Experimental Design and Implementation

The proposed system has three modules

- Admin module
- Agent module
- Customer module

The admin must login to the system. Admin is given privilege to add new schemes of insurances, edit and update the existing schemes. Admin is given privilege to add agents and also to edit or modify and update the agent details. Once the registration process is completed the Admin logs out of the system shown in Fig. 2. Agent must login to the system. Agent is given privilege to add customer or edit customer details regardless of network failure. Agent gets details of schemes introduced by admin and each time schemes are updated in local database of agent. Once the registration process completes the Agent logs out of the system shown in

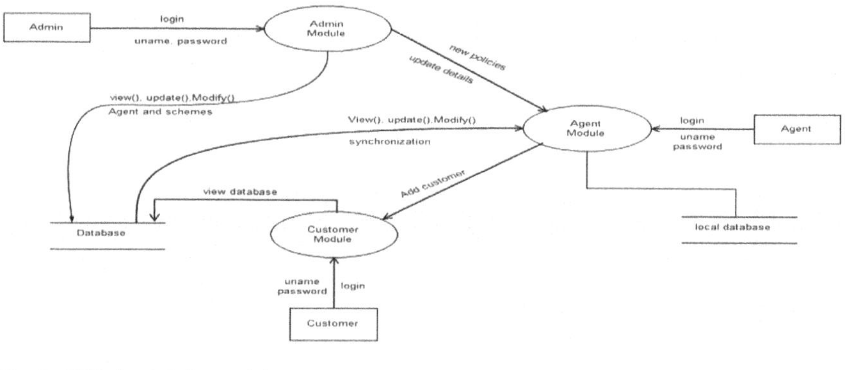

Fig. 2 DFD for Admin module

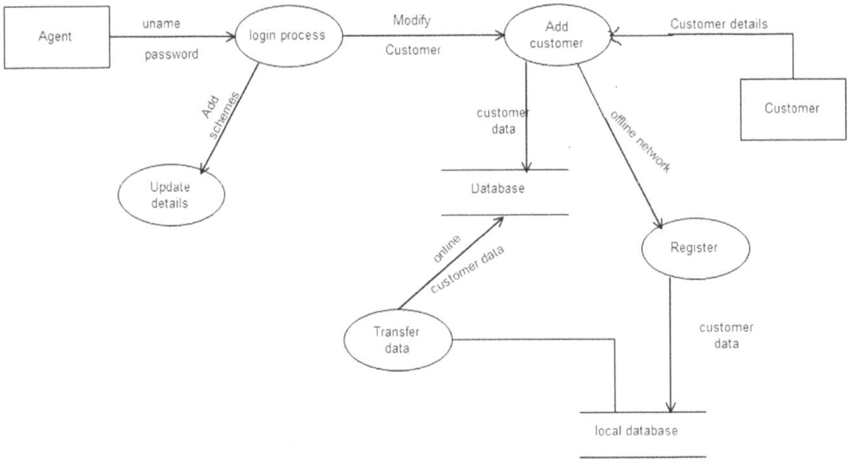

Fig. 3 DFD for Agent module

Fig. 3. Customer must login to the system. Customer after login can inquire about his account details such as schemes details by providing customer-id. Customer logs out of the system after request is processed.

Figure 4 Snapshot of customer registration during offline mode shows agent registering customer though there is no availability of internet.

Figure 5 Snapshot of database before synchronization shows that when registration is done offline the data i.e., customer details is not updated to database. But it is stored in local database.

Figure 6 Snapshot of database after synchronization shows that as and when the network is available the data i.e., customer details stored in local database is synchronized to database.

4.1 *Availability Evaluation*

The existing IPMS will depend completely on internet connection. In the proposed system the clients/agents continue to get service when internet/network goes off or server goes down. The system will switch to offline mode and the data given by the agent is stored in local database or as text files in the system which is updated as and when the network connectivity is available. The proposed system is 98 % available and efficiently utilizes network and time.

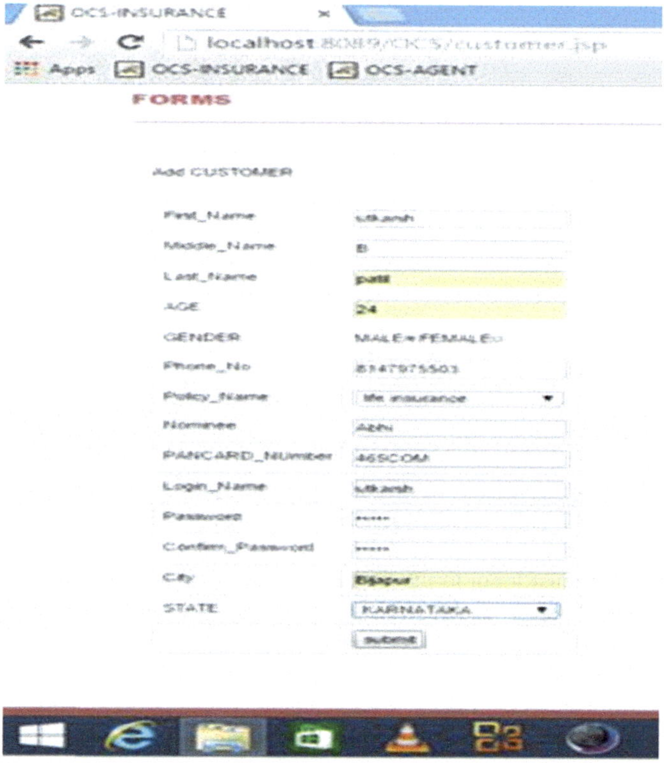

Fig. 4 Snapshot of customer registration during offline mode

c_id	fname	mname	lname	pwd	age	gen...	phone	policy...	nominee	pancard	city	state	lgname	agent...
183	vishwa	s	patil	vish	24	male	986378...	car	yuvaraja	567gw...	Bijapur	KARNA...	vish	2
184	shwetha	s	patil	shwet...	26	fem...	726329...	life	rohini	73792...	Bijapur	KARNA...	shwetha	2
187	anand	s	patil	anand	26	male	27889384	life	rohini	4834jndf	Bijapur	KARNA...	anand	2
188	rohini	s	patil	patil	24	fem...	741137...	jeevan	rohini	4256bhj	Bijapur	KARNA...	rohini	5
189	akshatha	k	k	aksha...	24	fem...	742289...	jeevan	rohini	62382...	Bijapur	KARNA...	aksha...	5
190	vishwa	s	Patil	vishwa	24	male	965350...	car	yuvaraj	234jbyb	Bang...	KARNA...	vish	5
191	ganaga	shank...	Patil	anand	45	fem...	983792...	life	shanka...	2382jjd	Bijapur	KARNA...	ganga	5
192	Shankar...	D	Patil	patil	50	male	782394...	home	rohini	788BHI	Bijapur	KARNA...	shank...	5
193	savitri	s	Patil	patil	26	fem...	962034...	jeevan	anand	23ahu	Bagal...	KARNA...	savitri	5

Fig. 5 Snapshot of database before synchronization

c_id	fname	mname	lname	pwd	age	gender	phone	policy_n...	nomin...	pancard	city	state	lgname	agent_id
183	vishwa	s	patil	vish	24	male	98637...	car	yuvar...	567gw5e	Bijapur	KARNAT...	vish	2
184	shwetha	s	patil	shwe...	26	female	72632...	life	rohini	73792...	Bijapur	KARNAT...	shwetha	2
187	anand	s	patil	anand	26	male	27889...	life	rohini	4834jndf	Bijapur	KARNAT...	anand	2
188	rohini	s	patil	patil	24	female	74113...	jeevan	rohini	4256bhj	Bijapur	KARNAT...	rohini	5
189	akshatha	k	k	aksh...	24	female	74228...	jeevan	rohini	62382...	Bijapur	KARNAT...	akshat...	5
190	vishwa	s	Patil	vishwa	24	male	96535...	car	yuvaraj	234jbyb	Bang...	KARNAT...	vish	5
191	ganaga	shan...	Patil	anand	45	female	98379...	life	shank...	2382jjd	Bijapur	KARNAT...	ganga	5
192	Shank...	D	Patil	patil	50	male	78239...	home	rohini	788BHI	Bijapur	KARNAT...	shank...	5
193	savitri	s	Patil	patil	26	female	96203...	jeevan	anand	23ahu	Baga...	KARNAT...	savitri	5
194	utkarsh	B	patil	patil	24	male	81479...	life	Abhi	465COM	Bijapur	KARNAT...	utkarsh	5

Fig. 6 Snapshot of database after synchronization

Table 1 Comparison of existing IPMS and proposed Smart-IPMS

	Existing system	Proposed system
Availability of systems	Not 100 %	98 %
Reliability and efficiency of system	Not 100 %	High
Can system switch from online mode to offline mode and vice versa	No	Yes
Offline application execution	No	Yes
Offline content available	No	Yes

4.2 System Comparison

See Table 1

5 Conclusion

Usefulness of offline data synchronization cannot be over emphasized. The existence of distributed systems has been made possible due to the capability of devices being able to send data from remote places to synchronize with data in an online server. This has been a mile stone in the information technology industry since multiple devices can synchronize offline data to one online server. This has greatly reduced inconvenience to the customer in data communication services since we do not need a dedicated network for data exchange and synchronization.

References

1. Blonda E (2003) Consolidate storage without losing control. Comput Technol Rev. http://www.thefreelibrary.com/Consolidate+storage+without+losing+control.-a0109082340
2. Bosworth A (2010) Occasionally-connected application server
3. Chia-Chi T, Brown K, Caro C, Nielsen W, Wells J (2012) A service oriented livestock management system using occasionally connected mobile-cloud architecture. IEEE
4. Dahan U (2007) Entity framework: disconnected problems and solutions. http://www.udidahan.com/2007/03/30/entity-framework-disconnected-problems-solutions/
5. Dahan U (2007) Occasionally connected systems architecture. http://www.udidahan.com/2007/04/04/occasionally-connected-systems-architecture/
6. Shaban B, Dika A (2012) Solving problems in software applications through data synchronization in case of absence of the network. Int J Comput Sci Issues (IJCSI) 9(Issue 1, No 3) (2012)
7. Rao X, Yang L (2012) Occasionally connected research and application under the architecture of smart client. In: The 7th International conference on computer science & education (ICCSE 2012), Melbourne, Australia

8. Preston B (2003) Connecting stranded servers. West World Productions, Inc. http://www.thefreelibrary.com/Connecting+stranded+servers.-a0109082342
9. George C, Jean D, Tim K, Gordon B (2011) Distributed system, concepts and design, 5th edn. Addison Wesley
10. Bosworth A (2006) Data model for occasionally-connected application server. http://www.google.com/patents/US20060026168

An Empirical Analysis of Unsupervised Learning Approach on Medical Databases

Ritu Chauhan, Harleen Kaur and Roma Puri

Abstract The early prediction of disease from diverse clinical features is among the critical job for health care practitioners. The clinical database are generally integrated from various sources such as electronic health records, administrative health records, and monitoring facilities, including CT scan or ultra sonic images. Thus it employs numerous efforts by clinical and data mining specialists to discover knowledge from large and complex clinical databases for future medical diagnosis. The discovery of patterns should be an automated process as data is voluminous and complex in nature. To discover hidden and novel information from such databases a proficient methodological technique must be involved. In this article we have laid emphasis on diabetes mellitus II dataset to discover clusters of variant shape and size. In current approach we have initially preprocessed datasets to reduce missing, noise and inconsistent values from database. Further preprocessed data is clustered using Density Based Spatial Clustering of Applications with Noise (DBSCAN) algorithm; however the key parameters are controlled in DBSCAN clustering algorithm to discuss the comparative results for efficient discovery of clusters with variant shapes and size. The study relatively determines the clusters with variants shapes and size from diabetes mellitus II datasets for future medical diagnosis of disease.

Keywords Data mining · Density based spatial clustering of applications with noise (DBSCAN) · Preprocessing · Diabetes mellitus II · Clustering

R. Chauhan (✉)
Amity University, Noida, India
e-mail: rituchauha@gmail.com

H. Kaur (✉)
Hamdard University, Delhi, India
e-mail: harleen_k1@rediffmail.com

R. Puri (✉)
College of Educators, Ontario, Canada
e-mail: roma.chauhan@gmail.com

K.R. Attele et al. (eds.), *Emerging Trends in Electrical, Communications and Information Technologies*, Lecture Notes in Electrical Engineering 394,
DOI 10.1007/978-981-10-1540-3_7

1 Introduction

In India the nationwide prevalence of diabetic is as high as 9 % and in some parts it's relatively as 20 % according to the International Diabetes Federation (IDF) and Madras Diabetes Research Foundation [1]. This disease is one of the major causes of several other diseases such as heart attack, kidney failure, eye diseases and other complications. To overcome the generalized scenario Indian government has declared significant amount of funding for diabetes disease. However number of emphasis has been laid down by India's National Program for Prevention and Control of Diabetes, and Cardiovascular Diseases and Stroke (NPCDS) which was launched in 2008, to implement prevention techniques and plans to educate people about risk factors for diabetes. Moreover, diabetes is lifelong disease and data generated for individual cases might be too voluminous and complex to understand by human capabilities alone. To overcome flaws of complexity among medical databases several data mining techniques are involved to discover hidden and novel information for knowledge discovery process [2, 3, 4].

The data mining techniques are applied in varied application areas of medical domains to discover knowledge from complex and high voluminous databases [5, 6, 2]. Usually real world medical databases comprise of missing, noise and inconsistent data records if processed can discover knowledge with no meaningful benefits to health care practitioners [5, 7, 8]. To deal with negative flaws of real word datasets several preprocessing and transformation techniques can be applied to discover novel information from large and complex databases [7, 9]. Consequently, it will predict the effective and early detection of disease which further improvise the medical decision making. For this reason varied data mining techniques are utilized such as classification, clustering, association rules and outlier detection to generate patterns which can benefit the end users for prediction of disease [10]. However, application area of data mining is gaining momentum in several other research areas, including marketing, customer relationship management, remote sensing, future forecasting, and business oriented goals to retrieve hidden and novel information for future prediction.

In this article we have utilized the diabetes Mellitus Type II data sets from Department of Medicine, University of Virginia School of Medicine [11]. The data was studied for individual patient records where several attributes tends to significant for cause for increase in diabetes mellitus. We have utilized data mining techniques to discover hidden and novel patterns from diabetes mellitus databases which can detect patterns for future medical diagnosis of diabetes mellitus II.

The rest of the paper is organized as follows. Section 2 briefly discusses literature survey for mining of diabetes dataset for detecting patterns. In Sect. 3, the experimentation and results are discussed for diabetes mellitus datasets II to retrieve clusters of variant shapes and size and conclusion is finally referred in last Section.

2 Extraction of Patterns from Diabetes Mellitus II Dataset

The extraction of hidden and novel information from diabetes mellitus II dataset is performed using appropriate data mining tools. Initially the data warehouse is preprocessed to reduce missing and inconsistent values. Further DBSCAN clustering algorithm is applied with training dataset to discover clusters of variant shapes and size [12]. The recovered knowledge can benefit health care practioners in development of future medical diagnosis of Diabetes Mellitus II.

The data consists of irrelevant, redundant and noisy features which can relatively produce inappropriate results for knowledge discovery process. So, cleaning and filtering task should be performed an automated process with respected to data mining technique. The preprocessing technique comprises of certain basic tasks such as removal of redundant values, transformation of data, and removal of missing and inconsistent data values. In our current approach we have utilized preprocessing technique to remove redundant, missing and inconsistent values from Diabetes Mellitus II datasets. Further, DBSCAN clustering technique is utilized which has a density based approach to develop clusters of arbitrary shapes and size. The Density Based Spatial Clustering of Applications with Noise (DBSCAN) is a clustering technique approached for density connected points. It starts with an arbitrary point p and retrieves all points density-reachable from p wrt. *Eps* and *MinPts* [12]. If p is a core point, this procedure yields a cluster wrt. *Eps* and *MinPts*. If p is a border point, no points are density-reachable from p and DBSCAN visits the next point of the database, Ester, et al., 1996. In density-based clustering each object of a cluster in the neighborhood of a given radius should have at least a minimum number *MinPts* of objects, i.e. the cardinality of the neighborhood has to exceed a given threshold [12]. With so many advantages DBSCAN still suffers from a few drawbacks. Namely they are sensitivity to input parameters (the clustering result very much depends on the 'epsilon' parameter of the algorithm) and in some cases algorithm is not able to correctly identify clusters that are close to each other [12].

This algorithm requires minimal knowledge of domain to determine the Input parameters, because appropriate values are often not known in advance when dealing with large databases. They are capable of finding the clusters of arbitrary shapes.

3 Experimentation

The proposed experiment explores diabetes Mellitus Type II data sets from Department of Medicine, University of Virginia School of Medicine. The data was studied for individual patient records where several attributes tends to significant for cause for increase in diabetes mellitus II [11]. The dataset was multivariate in nature which contains 234 female and 169 male patients' records. We have implemented our proposed approach in Java to discover hidden and significant patterns for future medical diagnosis [13]. In this study we have first preprocessed our dataset by removing duplicate records and supplying missing values. The approach was to discover significant patterns for future diagnosis for diabetes mellitus II. The preprocessed data set is further utilized for unsupervised clustering approach that is DBSCAN clustering algorithm to retrieve clusters of variant shapes and size.

The features retrieved from preprocessing techniques tend to be significant to retrieve clusters of variant shapes and size. Once the data was processed we then used a comparative study to disseminate the constant features and relatively determine the accurate Epsilon and min points for retrieval of effective and efficient patterns. The study was conducted in two phases, where the first phases relatively determine the epsilon and min point value for both training data and the target class data. The second phase retrieves the patterns from the determined epsilon and min points for effective and efficient retrieval of novel patterns.

We first evaluated the epsilon and Min point values for training datasets where the numbers of clusters were instantiated with respect to the time taken for unclustered instances. The value of epsilon was increased and decreased with respect min point and the time taken by DBSCAN clustering algorithm was measured in relation to unclustered instances. We significantly found that at 0.8 epsilon and min pts 4 the training datasets attains the maximum accuracy with respect to unclustered instances, clusters formed and the time taken. We then calculated the same epsilon and min pts values with respect to target class for retrieval of effective and efficient patterns. The context of increment and decrement of values for epsilon and min pts were followed with respect to generated clusters, incorrectly clustered instances and the time taken for each iteration. We extensively found that for target class DBSCAN works efficiently at 0.4 epsilon and min pts 8.

The results of the DBSCAN clustering algorithm with variant epsilon and min pts for training data have been considered for analysis on the Diabetes Mellitus Type II data have been depicted in Table 1. The portrait results display the accuracy of the clusters formed in relative to variant aspects with the computation time. For, Target class data DBSCAN clustering algorithm with variant epsilon and min pts for training data have been considered for analysis on the Diabetes Mellitus Type II data have been depicted in Table 2. The portrait results display the accuracy of the clusters formed in relative to variant aspects with the computation time.

The Fig. 1a, b overlays the graphical representation of training data and the target class data with variant epsilon and min pts of DBSCAN clustering algorithm.

Figure 2 represents the clusters retrieved by DBSCAN clustering algorithm. There are generalized 12 clusters formed where each cluster similarity is measured in respect to training data.

Table 1 DBSCAN values for training data/target class

DBSCAN values for training data				
Epsilon	Min points	Generated clusters	Time	Unclustered instances
0.2	2	1	1.05	401
	4	1	1.09	403
	6	1	1.09	403
	8	1	1.17	403
0.4	2	44	1.11	250
	4	11	1.05	335
	6	3	1.16	381
	8	0	1.19	403
0.6	2	20	1.08	68
	4	13	1.17	88
	6	12	1.12	107
	8	10	1.25	134
0.8	2	13	1.19	17
	4	12	1.02	21
	6	12	1.08	22
	8	12	1.12	23
1	2	12	1.08	4
	4	12	1.08	4
	6	12	1.17	4
	8	12	1.06	4

Table 2 DBSCAN values target class

DBSCAN values for target class				
Epsilon	Min points	Generated clusters	Time	Incorrectly clustered instances (%)
0.2	2	1	1.02	0
	4	1	1.09	0
	6	1	1.09	0
	8	1	1.17	0
0.4	2	27	1	39.206
	4	8	1.02	21.34
	6	6	1.01	15.3846
	8	5	1.03	9.93
0.6	2	15	0.95	61.54
	4	4	1.01	58.31
	6	5	1.06	56.82
	8	5	1.05	55.09

(continued)

Table 2 (continued)

DBSCAN values for target class				
Epsilon	Min points	Generated clusters	Time	Incorrectly clustered instances (%)
0.8	2	6	1	66.25
	4	5	1.05	66.25
	6	4	1.05	64.76
	8	4	1	64.27
1	2	6	1.06	67.25
	4	4	1.02	67.25
	6	4	1	67.25
	8	4	1.05	67.25

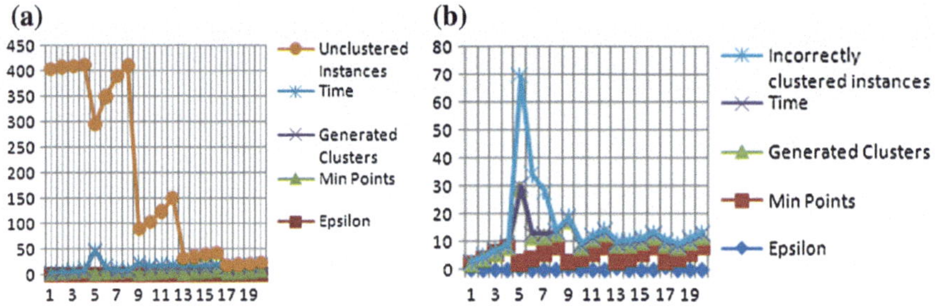

Fig. 1 **a.** Training data, **b**. Target class data

Figure 3 represents 420 data subjects taken into consideration while target data class results where we have utilized the patient frame as one of nominal attribute. The frame is usually calculated in accordance to the obesity level of each patient where different attributes were taken in concerns such as age, height, blood glucose level and weight. We have found that largest number of clusters (around 30 % of cases studied) was formed under medium frame whereas the maximum chances of cases (around 21 %) among the large frame have higher chances of suffering from Diabetes Mellitus II. The Fig. 3 below represents the number of clusters formed with DBSCAN clustering algorithm for target class.

The study reveals that maximum number of cases discussed is in between age group 35 and 45 years among the highest those who were suffering from diabetes Mellitus II and their frame defined as medium where the total cholesterol level can be high as 400. In our approach we have identified clusters of variant shapes and size in accordance to the target class whereas there can be other relative patterns can be discussed in our future studies.

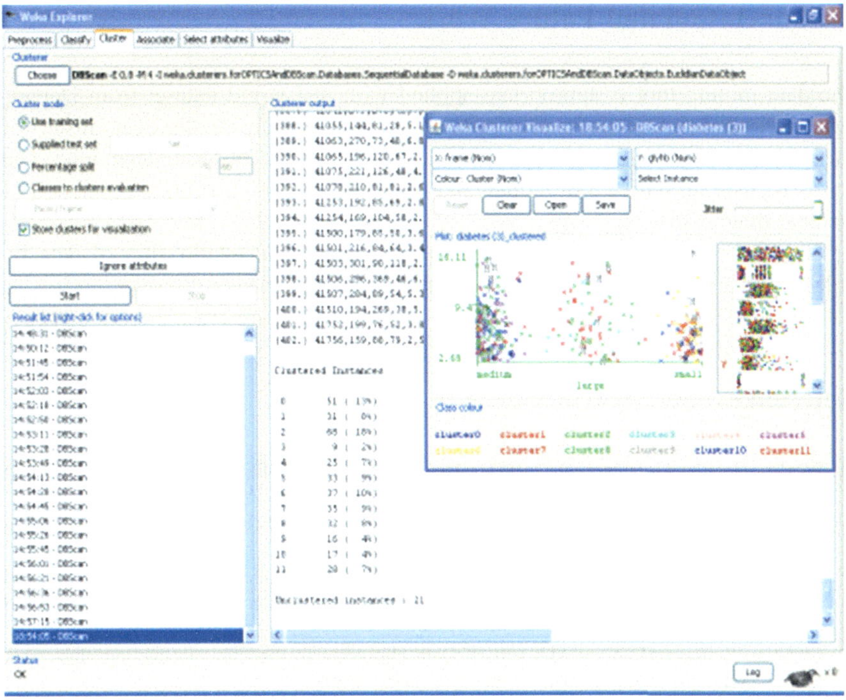

Fig. 2 Represents the clusters retrieved by DBSCAN clustering

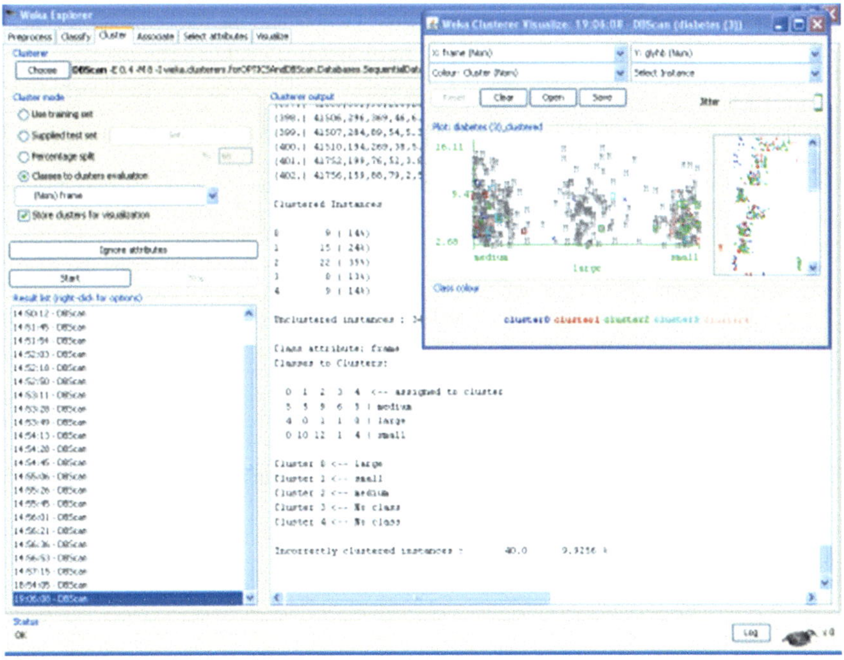

Fig. 3 Represents the clusters retrieved by DBSCAN clustering

4 Conclusion and Future Work

The data mining tool is appropriate applied with real world dataset from Department of Medicine, University of Virginia School of Medicine to discover clusters of variant shape and size using preprocessing and DBSCAN clustering technique. The article focuses on retrieval of hidden and novel information for future prediction of medical diagnosis.

The future study will be extended to real space time relationship of data by utilizing data mining tools to discover patterns for knowledge discovery.

References

1. World Health Organization (1985) Diabetes Mellitus: Report of a WHO Study Group. Geneva: WHO, Technical Report Series 727
2. Guyon I, Weston J, Barnhill S, Vapnik V (2002) Gene selection for cancer classification using support vector machines. Mach Learn 46:389–422
3. Dy JG, Brodley CE (2000) Feature subset selection and order identification for unsupervised learning. In: Proceedings of the 17th international conference on machine learning. Stanford University, CA, pp 247–254
4. Chauhan R, Kaur H, Alam MA (2010) Data clustering method for discovering clusters in spatial cancer databases. Int J Comput Appl 0975−8887
5. Agrawal R, Gehrke J, Gunopulos D, Raghavan P (1998) Automatic subspace clustering of high dimensional data for data mining applications. In: Proceedings of the 1998 ACM SIGMOD International Conference on Management of Data, Seattle, Washington, pp 94−105, June 1998
6. Kaur H, Chauhan R, Alam MA (2012) SPAGRID: a spatial grid framework for medical high dimensional databases. Int Conf Hybrid Artif Intell Syst HAIS 2012(1):690–704
7. Blum AL, Langley P (1997) Selection of relevant features and examples in machine learning. Artif Intell 97:245–271
8. Bellazzi R, Zupan B (2008) Predictive data mining in clinical medicine: current issues and guidelines. Int J Med Inform. 77(2):81–97
9. Kaur H, Chauhan R, Alam MA (2010) An optimal categorization of feature selection methods for knowledge discovery. In: Zhang, Segall, Cao (eds.) Visual analytics and interactive technologies: data, text and web mining applications. IGI Publishers Inc.
10. Smith JW, Everhart JE, Dickson WC, Knowler WC, Johannes RS (1988) Using the ADAP learning algorithm to forecast the onset of diabetes mellitus. In: Proceedings of 12th symposium on computer applications and medical care. Los Angeles, California, USA, pp 261–265
11. http://Vanderbilt.edu
12. Borah B, Bhattacharyya DK (2004) An improved sampling-based DBSCAN for large spatial databases. In: Proceedings of the international conference on intelligent sensing and information, p 92
13. Weka-3–4-10jre: data mining with open source machine learning software © 2002-2005 David Scuse and University of Waikato

Applying Agile Programming and Design Patterns in IT Domain

V. Dattatreya, K.V. Chalapati Rao and V.M. Rayudu

Abstract Agile software development methods have recently become popular in software engineering. Agile methods emphasize simplicity and the elimination of work that is not absolutely necessary. These techniques give preference for working software which satisfies all stake holders, and leads to a successful completion in shorter time frames, giving much importance to real-time communication. Design Patterns have been widely used by many software developers. Recent statistics reveals that patterns can have a beneficial impact on software quality. In brief, what is needed for today's software development scenario is a combination of Agile Methodologies for faster development and usage of Design Patterns for maximizing design reusability. The combined approach applied to the software development leads to robust, secure and almost risks free software which has adaptability to accommodate rapid requirements changes. Section 1 of this paper explains the Agile Development Methodologies. Section 2 outlines the Design Patterns, while 3rd and 4th sections elaborate case studies for the combined approach.

1 Introduction

Agile Development Methodologies play a vital role in building quality software with predictability [1]. The main aspect of Agile programming is building the adaptable programs in a simple and localized way so that they are free from rebuilding, refactoring, and retesting [1]. One of the advantages of Agile

V. Dattatreya (✉) · K.V. Chalapati Rao (✉)
Department of CSE, CVR College of Engineering, Mangalpally,
RR District, Telangana, India
e-mail: dattatreya.valiveti@gmail.com

K.V. Chalapati Rao
e-mail: chalapatiraokv@gmail.com

V.M. Rayudu (✉)
DTA, Government of A.P, Hyderabad, India
e-mail: madhava.rayudu@gmail.com

K.R. Attele et al. (eds.), *Emerging Trends in Electrical, Communications and Information Technologies*, Lecture Notes in Electrical Engineering 394,
DOI 10.1007/978-981-10-1540-3_8

programming is that there is no necessity to work out each and everything in advance from beginning to end. It is a methodology leading to flexibility as it builds software in short iterations. It is also one of the flexible methodologies in meeting organizational challenges. Agile technologies adopt recurrent delivery of smaller and worthy increments instead of waiting till the end. This approach helps in high quality of software. In the formative years, there was prejudice against the decision to take up the agile approach and typically a project teams decided to adopt or criticize the transition 'on its own terms'. Progressively, suppliers, consultants, partners and customers and even public sector bodies are insisting on the use of agile methodology, either through a requirement to do so or to ensure inter-organizational process alignment [2].

1.1 Agile Practices [3]

Coding Standards are easy to execute as it is a clear-cut approach to teach and impose the standards through pair—programming.
Refactoring is one of the hardest techniques to teach. Agile modeling techniques are useful to discover applications that could go through more extensive refactoring.
Pair programming is very valuable to teach developers testing and refactoring techniques. It helps them in learning the language avoiding syntax mistakes and compilations problems.

Agile methodologies (XP [4], Scrum [5]), are used to improve organization or project team ability to manage projects. One of the solutions for efficient product development is functional increment. The purpose of Agile project management is to develop products simply and effectively based on customer demands. It needs assessment of planning in the areas of improvement of the organization's current software development mapping with the most suitable agile practices.

1.2 Extreme Programming (XP) [4]

XP is a methodology which is based on basic principles like rapid feedback, incremental change and quality work. Extreme programming projects make use of a minimum of forthright design. The approach focuses on building good team spirit through team working on personal communication and morale rather than on documentation. It is essential to maintain release-quality code at all times while the success of small release is a worthy discipline. **Respect and trust** are the secrets of XP and they play vital role in retaining the pair programming activities and also to retain actual association with the communication, simplicity, feedback and courage. Although XP is seen to be beneficial to small and medium sized companies during use, expensive Refactoring must be considered when problems in design are discovered. In other words XP's de-emphasis analyses and design in the very beginning.

2 Design Patterns

2.1 Introduction

Software development has become increasingly large, diverse and complex in recent years. Because of this, extending the life-span of software has become a challenging problem [6]. Object-Oriented technology offers one solution. Improving the modularity and readability of software helps us to understand the structure of the software and to maintain, modify and extend the software developed. However, achieving well defined object-oriented designs is a skilled task that requires both time and experience. Design patterns help to overcome this problem. They are abstract descriptions of object oriented designs which appear repeatedly and which have succeeded as solutions to past design problems. They are useful in achieving flexible and extensible design making it easy to effect future changes and modifications. A systematic methodology is required to apply design patterns effectively to construct flexible and extensible software. Design patterns, elements of reusable object-oriented software, are a familiar tool for designers to use to converse, document, and investigate design alternatives [7]. They allow us to converse about them at a privileged level of abstraction than design notation or programming language and make a system look less complex. Familiarity with design patterns makes easier to understand existing object-oriented systems. Further, large object-oriented systems exploit these design patterns. Hence people learning object-oriented programming must be acquainted with design patterns as they have to work with use of inheritance in complicated ways. Otherwise they have to face so many difficulties to follow the flow of control. An apprentice can also act more like an expert by gaining knowledge of these patterns as they provide solutions to common problems. Design patterns play key role in turning an analysis model into an implementation model. For a smooth transition, Analysis models must be frequently redesigned to make them flexible and reusable by adding new objects. The programming language and class libraries also impinge on the design [8]. All these themes are filed by many of the design patterns, justifying the term design patterns. Similarly, there can also be analysis patterns, user interface design patterns, or performance-tuning patterns for a full-blown method.

2.2 Basic Elements

1. Design patterns are descriptions of communicating objects and classes that are adapted to solve a general design problem in a particular context.
   The essential elements of a pattern are:
   - The "pattern name" as denoting the concept.
   - The "problem" as description of the class of situations when the concept is applicable.

- The “solution” which gives an abstract description of a generic system of elements and their relationships that solves the problem.
- The “consequences” that result from the application of the solution, for example trade-offs, costs, and benefits.

2. **Design options**: Applying Design patterns may be an approach of proven practitioner. During software development, the selection of a pattern is simply one option among many alternate patterns. The approval of these options is based on the comprehensive and exact quality goals of the software project. Therefore the choice must be based on the quality goals of the project keeping in mind the favor or against the use of a pattern. One key apprehension of patterns is flexibility. In [8] Gamma et al. note: The changes may involve the redefinition of a class, reimplementation, client modification, and retesting. Redesigning causes unexpected changes which are consistently expensive when it affects many parts of the software system. Design patterns ensure that a system can change in an exact way. The important point here is to ensure that the pattern supports the desired aspect of flexibility or not.
3. **Cost**: Gamma et al. note that the flexibility initiated by patterns becomes worthy in view of the mechanism of delegation that is used by many patterns: The main disadvantage of Delegation is that it makes software more flexible by sharing with other techniques. There are also run-time inefficiencies, but the human inefficiencies are more important in the long run.
4. **Removal**: It is essential to balance three contradictory issues viz., possible benefit of the pattern, possible extra cost of the patterns and the cost of its removal in case of removal of a pattern from code [9].

3 Ruby on Rails

3.1 Ruby

In 1993 Ruby was introduced by Yukihiro Matsumoto. It is widely known as Matz. Ruby is a dynamic interpreted language and has many strong features. Ruby is an Object-Oriented programming language which has single inheritance like Java or C++. It has a special feature called mixins [10]. Using mixins we can easily from multiple classes. Ruby is also scripting language. Ruby helps to maintain the reliability of programming and security of the code due to the concept of Object Oriented concept. Ruby is open source language. Because of this feature, it is widely used by the Web developers. Ruby follows the concepts of Object Oriented programming languages and it has objects, methods and classes. Ruby has classes and Instances. In Ruby all algorithms are written on objects are placed safely in methods. All code which is written in Ruby is nothing but the method of one class or another. Ruby provides the programmer can keep his code in a distributed form over the network using portability. A security concept in Ruby is the Lock Down concept that enables the programmer to decide which part of the code or data may be a threat to the remaining part of the code.

3.2 *Rails*

Rails is a framework written in Ruby language. It is easier to develop, deploy, and maintain Web applications. It is a best choice for implementation of Web 2.0 applications. All Rails applications are implemented by using MVC architecture. Any Web developer can express his ideas naturally and cleanly in Ruby. Rails is based on three philosophies:

1. Code is short and readable.
2. DRY means every piece of knowledge in a system should be expressed in a one place.
3. Convention Over Configuration means that Rails what we want to do and how we are going to do it.

Rails uses Active Record which supports Object Relational Mapper (ORM). Using Active Record to manage table relational ships and is in the process cover Create, Read, Update, Delete operations commonly referred as CRUD methods.

3.3 *Agile Approach in Rails [11]*

Agile Manifesto has a set of preferences as given below:

(a) Individuals and interactions over processes and tools: Rails focuses more on individuals and interactions rather than processes and tools. Rails contains simple set of tools without any complex configurations and without emphasizing on processes. The code written by groups of Ruby developers is reflected immediately to the customers, providing full fledged transparency. It is basically an interactive process.

(b) Working software over comprehensive documentation: Rails is not just confined to documentation and specifications but it can deliver working software early in the development cycle itself.

(c) Customer collaboration over contract negotiation: The project implemented in Rails project can quickly respond to changes required by the customers. This makes the customers convinced that the team can deliver what is required, not just what has been requested.

3.4 *Rails Support Object—Relational Mapping*

Object-Relational Mapping (ORM) [10] is a programmable tool which converts data between incompatible types in object-oriented programming languages. ORM libraries consist of tools that can map database entities to object oriented paradigm classes. For an illustration, let there be a class defined with name *Order* in Rails framework, then the ORM generated database consists of a table called *orders*. Each row of *orders* table is mapped to an object of the class—a specific order is represented as an object of class *Order*. The columns of the table are the object attributes. ORM generated tables converts the class methods as table-level operations where each method operates at row level. Figure 1 illustrates the functionality of ORM, according to Patterns of Enterprise Application Architecture (P of EAA) by Martin Fowler.

Ruby on Rails (RoR) had some success in developing websites and applications [12]. RoR Websites inspire us to take the implementations from social networks of local tutors to tools allowing business partners to work together on projects. To show the advanced RoR implementations the following are the best web sites and tools in today's environment.

(a) **Business**:

- Track RSS feeds, Web traffic levels and exchange information internally.
- Useful to estimate the frames, resources and employment issues prior to beginning a project.
- Provide a tool for organizing the employee's schedules.

(b) **Travel**:

- Using these sites we are saving money, time and stress.
- Travelers share a detailed route with each other in order to plan more adventurous trips.
- Provide Google Maps solutions for creating maps to describe locations like restaurants, traffic hazards.

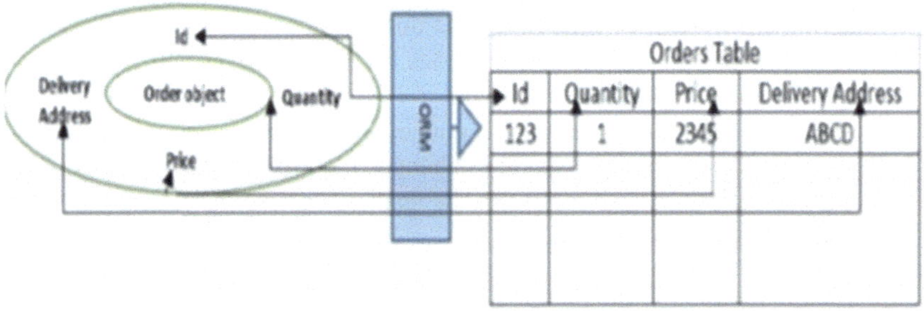

Fig. 1 Ruby on rails in IT domain

(c) **Web Developers**:

1. These sites monitor the performance, downtime, network problems.
2. Includes so many tools to test the cookies and forms.
3. Best resources for developing the Websites for the beginners.
4. We can build template based Websites which includes blogs, forms, and content management system.

(d) **Multimedia**:

- These are excellent resources for Internet media.
- Contains a collection of videos, where users save and share their favorite videos.
- Offers for finding internet radio stations, MP3's and audio's.

(e) **Social Networks**:

- In these Websites where the users tell the world what they are doing.
- They pose questions and receive feedback from others on any subject.

(f) **Miscellaneous**:

- These websites offer different solutions for different things.
- Websites simulate public markets for various topics ranging from who will win the next presidency to which corporations will merge next.
- Maintains statistics for the television programming which includes most popular actors, titles and more.
- Provides a useful tool for developing the detailed charts by offering online spread sheets and sharing them over a social network.

4 Conclusion

Design Patterns are abstract descriptions of object oriented designs which appear over and over and which have succeeded solutions to the previous design problems. Design patterns improve extensibility and flexibility of software, the most important parts of building the evolutionary software. Design Patterns improves the program quality. Agile Programming is a software methodology which is better suited for shorter time frames. By combining these two approaches one can get the advantages of shorter time frames and reusable designs with working software. Ruby On Rails is a framework which supports Agile Programming techniques and Design Patterns. Developing a web application on Rails is faster than that of a Java based web application framework. Ruby On Rails based framework that can be used for easy database integration using Active Record. Ruby On Rails had huge success due to its elegant, built in testing, powerful and conciseness. Rails starts project in three different environments namely Developing, Testing and Production environments. All three environments behave differently from each other and each complement the software development life cycle.

References

1. Thomas D (2005) Agile programming: design to accommodate change. IEEE 22(3)
2. http://www.cio.com/article/491489/ Agile_Development_s_Benefits_and_Challenges
3. Jeffries R, Hendrickson C, Anderson A (2001) Extreme Programming Installed. Addison-Wesley
4. Beck K (2005) Extreme programming explained: embrace change, 2/e. Addison-Wesley, Boston, MA
5. Schwaber K, Beedle M (2001) Agile software development with SCRUM. Prentice-Hall, Upper Saddle River, NJ
6. Masuda G, Sakamoto N, Ushijima K (2000) Redesigning of an existing software using design patterns. In: International symposium on principles of software evolution, p 165
7. Zhang C, Budgen D (2012) What do we know about the effectiveness of software design patterns? IEEE Trans Softw Eng 38(5):1213–1231
8. Gamma E, Helm R, Johnson R, Vlissides J (1995) Design patterns: elements of reusable object-oriented software. Addison Wesley
9. Wendorff P (2001) Assessment of design patterns during software reengineering: lessons learned from a large commercial project. In: Proceedings of the fifth european conference on software maintenance and reengineering, pp 77–84
10. http://www.rubyist.net/~slagell/ruby/
11. Dattatreya V (2012) Agile programming and design patterns in web development—a case study. Int J Softw Eng Appl 3(1), 37–45
12. http://www.softwaredeveloper.com/features/best-ruby-on-rails-061307/

A Novel Approach to Improve the System Performance by Proper Scheduling in Memory Management

Jisha P. Abraham and Sheena Mathew

Abstract Virtual memory techniques are available in modern operating system will permits the execution of a program even if it is partially available in memory thus providing an illusion of very large memory to the user and freeing the user from the concern of large program size. The primary storage size has increased by multiple order of magnitude. With several gigabytes of primary memory, algorithms that require a periodic check of each and every memory frame are becoming less and least practical. Some basic assumptions used by the traditional page replacement algorithms were invalidated, resulting in a revival of research area. Currently the usage of the object oriented programming is much higher than the structured programming. Hence the locality of reference of user software has weakened. Due to the above problem for the page replacement methods also we have to think about new methods. In this paper mainly focused on the improvement of the processor performance. In order to achieve this, the number of page faults are tried to reduce, by using the page replacement method in the correct way. In modern operating systems mainly make use of the Least Recently Used (LRU) page replacement method, in which only the arrival time of the pages to the page frame are make used. Here we made an attempt to consider both the arrival time and the number of reference done on the page is considered. By consider these two parameter we developed a new algorithm Least Recently Used and Least Frequently Used (LRU-LFU) which will give a better performance than LRU page replacement method.

J.P. Abraham (✉)
M A College of Engineering, Kothamangalam, India
e-mail: jishapa@mace.ac.in

S. Mathew
Cochin University of Science and Technology, Kochi, India
e-mail: sheenamathew@cusat.ac.in

K.R. Attele et al. (eds.), *Emerging Trends in Electrical, Communications and Information Technologies*, Lecture Notes in Electrical Engineering 394,
DOI 10.1007/978-981-10-1540-3_9

1 Introduction

The concept of virtual memory in operating system provide an illusion to a user having a very large amount of main memory available and allowing him/her to execute a program to be partially there in main memory [1]. The memory management module of the virtual memory of the operating systems are implemented with the concept of demand paging. There are several advantages of using this concept like less I/O, efficient resource utilization etc. The overall performance of the system is affected by the choice of page replacement technique used in virtual memory. There are several page replacement techniques suggested by different researchers. O'Neil1 et al. uses LRU-K page replacement algorithm or database disk buffering [2]. Khajoueinejad et al. used fuzzy cache replacement policy [3] and so on. Among the various page replacement which is proven to be the best, optimal technique, will face the problem that it cannot be implemented because it requires future knowledge of the reference string. Here comes the LRU policy and that is the reason different variations of LRU have been implemented. It is well known however that there are many situations where LRU behaves far from optimal [4]. Under LRU an allocated memory page of a program will become a replacement candidate if the page has not been accessed for a certain period of time under two conditions: (a) the program does not need to access the page; and (b) the program is conducting page faults (a sleeping process) so that it is not able to access the page although it might have done so without the page faults. However, LRU page replacement implementations do not discriminate between two types of LRU pages and treat them equally [5]. So it means that LRU can be improved. Also other page replacement policies are First in First out (FIFO), First in Last out (LIFO), Most Recently Used (MRU), Optimal method and LRU k method are given below.

It has been observed that LRU approximates optimal so LRU and its variants are quite common in use in operating system as a reasonable choice of page replacement algorithm. In this paper we presents a simple modified LRU algorithm with some new method were the time stamp is mixed with the concept of frequency parameter in terms of distance. The result shows that the new algorithm gives better results than LRU.

2 Page Miss Handling in Memory Management

Memory management module of the operating system tries to handle the page fault by making the required page accessible at a location in physical memory or terminates the program in the case of an illegal access. The processor is the one which detect the page fault in the memory management. The software used for exception handling in case of page fault is generally part of the operating system. Page faults are not always errors, that are common and necessary to increase the amount of memory available to programs in any operating system that utilizes virtual memory, including OpenVMS, Microsoft Windows, Unix-like systems (including Mac OS X, Linux, *BSD, Solaris, AIX, and HP-UX), and z/OS [6–8]. Microsoft uses the term hard fault in more recent versions of the Resource Monitor (e.g., Windows Vista) to mean 'page fault'.

The mechanism used by an operating system to increase the amount of program memory available on demand is done with the help of major page fault. The loading of the program from the disk to the memory is delayed by the operating system until a page fault is generated. If the page is not loaded in memory at the time of the fault, then it is called a major or hard page fault. In order to handle this problem the page fault handler in the OS needs to find a free location: either a page in memory, or another non-free page in memory. This latter might be used by another process, in which case the OS needs to write out the data in that page (if it hasn't been written out since it was last modified) and mark that page as not being loaded in memory in its process page table. Once the space has been made available, the OS can read the data for the new page into memory, add an entry to its location in the memory management unit, and indicate that the page is loaded. Thus major faults are more expensive than minor faults and add disk latency to the interrupted program's execution.

The execution time of a program is related to the number of major and minor fault inside the system. Execution time of programs are measured in terms of user time, system time and elapsed time. The user time is the time spent by the operating system in user mode. The system time is the time spent by the kernel executing in system mode on behalf of the process e.g.: executing system calls. And elapsed time is the time taken by the system in order to execute the program (system preemption). By analysing these various time parameters we are able to detect the system performance using the system command time./a.out in Linux.

3 Experimental Setup of Page Replacement

The page replacement module is implemented for Ubuntu 12.04 and checked for suitability and efficiency. Ubuntu is composed of many software packages, the majority of which are free software. Free software gives users the freedom to study, adapt/modify and distribute it. Ubuntu can also run proprietary software. When the end user related to the particular module is satisfied with the performance, the next step of implementation is preceded. The various modules present in Ubuntu 12.04 can be done in parallel. During the program execution, the records are stored in the workspace in order to handle unexpected errors. This helps to recover the signal status of the records from any accidental updating or unintentional deletion of records. The proposed system linked with the following modules: Kernel Compilation, Page Replacement Policies, Algorithm design, Comparison.

3.1 Page Replacement Policies

Advanced version of Least Recently Used (LRU) page replacement policy is made used in Ubuntu 12.04. That is the pages frequently used over a short period in the recent past are likely to be used in the future. The LRU algorithm is based on the converse assumption that pages not used recently will not be needed frequently in the immediate future. Such pages are therefore likely candidates for swap-out when memory is scarce. The fundamental principle used in LRU may be simple, but it is difficult to implement it appropriately. In the paper we implemented the different page replacement policies.

3.1.1 MRU

In the existing system for the implementation of Least Recently Used algorithm we make a reference to each of the pages with the help of a pointer. Here the way in which the page can be replaced is modified in such a way that:

if ((_page) → lru.prev != _base)

which is pointed from the first page and in MRU it is pointed from the last page and its next page is being replaced

if ((_page) → lru.next != _base)

3.1.2 LFU

For the implementation of LFU a parameter freq is added as unsigned long data type to the struct page. This is used as the counter or a frequency parameter. Whenever a new page is being referenced the value of freq is being incremented. Whenever the function [isolate_lru_pages], to identify the vitim page, is being called the pages are being sorted with the help of an inbuilt merge sorting mechanism depending on the frequency parameter freq. Front page is then being replaced in the LFU page replacement algorithm.

3.1.3 MFU

To implement this we added a parameter freq as unsigned long data type to the struct page. This is used as the counter or a frequency parameter. Whenever a new page is being referenced the value of freq is being incremented. The function [isolate_lru_pages] which is used to identify the victim page, is being called the pages are being sorted with the help of an inbuilt merge sorting mechanism depending on the frequency parameter freq. Last page is then being replaced in the MFU page replacement algorithm.

3.1.4 LFU-LRU

The combination of LFU-LRU, we need to consider the existing parameters available in LRU along with new parameter is used for the frequency i.e. freq. Here we have introduced another data to the struct page and is named as "compound". Compound (which is a concatenated value of 'freq' the new parameter and the time which is already used in the LRU) is used to take care of the above combination in a well-defined manner. Majority of the operations are carried out in the [mark_-page_accessed], which is used to add a page to the list.

The boot time of the system is obtained using the function [get_monotonic_-boottime (*ptr)] which is used in the above function i.e. [mark_page_accessed]. Initially the freq parameter inside the page structure is being incremented as is assigned to compound and then it is being left shifted by six bits in order to increase the frequency parameter. Finally the compound parameter will be having the boot time along with the above value summed up. The "compound" parameter is used for the page replacement of pages which is least frequently and recently being accessed inside the system.

3.1.5 LRU-MFU

In the case of a combination of LRU-MFU, the above mentioned compound parameters such as time stamp and frequency are reused. The operation that were performed in the [mark_page_accessed] function is being changed to meet the new needs. The boot time of the system is obtained using the function [get_mono-tonic_boottime] (*ptr) which is used in the function [mark_page_accessed]. The "compound" parameter is initialized with the "freq" parameter that is being incremented is reduced from the maximum range of unsigned long, which is then summed to the boot time of the system. This parameter is sorted and is used for this page replacement approach.

3.1.6 LFU-MRU

In LFU-MRU page replacement algorithms it is required to consider the existing parameters along with the new parameter used for the frequency i.e. freq. Here we have introduced another data to the struct page and is named as "compound". Compound is used to take care of the above combination in a well-defined manner.

The operation that were performed in the mark_page_accessed function is being changed according to meet the new needs. The boot time of the system is obtained using the function get_monotonic_boottime (*ptr) which is used in the above function i.e. mark_page_accessed. The "compound" parameter is initialized with the "freq" parameter that is being incremented is reduced from the maximum range of unsigned long, which is then summed to the boot time of the system. This parameter is sorted and is used for the above page replacement approach.

3.1.7 MFU-MRU

The combination of the MFU-MRU page replacement algorithms required existing parameters along with the new parameter, freq, which is used for the frequency. Here we have introduced another data to the struct page and is named as "compound". Compound is used to take care of the above combination in a well-defined manner. The pages are sorted inside the new function named [freq_sort] which sorts the compound values associate with each of the pages.

if((page_a → compound) < (page_b → compound))return 1;
else if((page_a → compound) > (page_b → compound)) return −1; else return 0;

Then the entire list is being sorted according to this value. The boot time of the system is obtained using the function [get_monotonic_boottime] (*ptr) which is used in the above function i.e. mark_page_accessed. Initially the freq parameter inside the page structure is being incremented as is assigned to compound and then it is being left shifted by six bits. Finally the compound parameter will be having the boot time along with the above value summed up. This parameter is used for the sorting.

4 Result Analysis

The comparison between each of these algorithms can be done with the help of finding the page faults for different programs of varying sizes. Here we have used inbuilt programs as well as user created programs. The page fault can be in terms of major fault or minor fault. The time of execution of the program is measured in terms of user time and system time. The different programs with same file size are considered in order to analyze the effect of backward and forward loop reference and also how do they reflect in the fault rate and in the execution time. For final conclusion one programme from each category (small, medium and large) is considered and shown in Tables 1, 2, 3, 4 and 5 as file selected.

The work is about implementing different page replacement policies in Ubuntu 12.04. Ubuntu version 12.04 has been used for the work due to its stability and support. LRU page replacement algorithm has been modified with MRU, LFU, MFU and their respective combinations. From Table 1 and 2, it is clear that there is difference in the value of the number of page faults both in the major and minor fault, which is generated by the system as well as the user program code. This small

Table 1 Details of major page fault

Average-major-fault									
File size (kb)	LRU	MRU	LFU	MFU	LFU-LRU	LRU-MFU	LFU-MRU	MFU-MRU	File selected for analyses
2.7	10.6	10.5	10.4	10.6	10.7	10.5	10.7	10.7	Selected
7.1	0	0	0	0	0	0	0	0	
7.2	0	0	0	0	0	0	0	0	
7.2	0	0	0	0	0	0	0	0	
7.3	0	0	0	0	0	0	0	0	
7.3	0	0	0	0	0	0	0	0	
7.3	0	0	0	0	0	0	0	0	
26.8	0.4	0.3	0.4	0.5	0.6	0.5	0.3	0.5	
36.4	0.4	0.5	0.5	0.3	0.2	0.4	0.5	0.3	
59	8.7	23.4	9.9	3.4	3.6	2.9	2.6	3.1	Selected
59	4.1	0.3	4.4	0.5	4.4	4.6	4	0.4	
59	0.5	4.2	0.4	4.3	0.5	0.4	0.8	4.5	
59	5.7	1.6	1.9	8.5	8.1	8.8	9.1	8.5	
686.7	4.6	4.6	5.5	4.6	5.5	5.5	5.5	5.5	Selected

Table 2 Details of minor page fault

Average-minimum-fault									
File size (kb)	LRU	MRU	LFU	MFU	LFU-LRU	LRU-MFU	LFU-MRU	MFU-MRU	File selected for analyses
2.7	24600.3	24583.4	22533.2	24611	23486.7	24165	24599.6	23516.9	Selected
7.1	147.3	148.1	148	147.5	147.9	147.6	148.8	147.7	
7.2	146.4	147.8	147.8	147.8	147.2	147.5	147.8	147.6	
7.2	146.1	146.8	147.1	147.1	147.1	146.8	147.4	147.9	
7.3	146.9	147.2	147.1	147.1	147.6	147.1	147.4	147.7	
7.3	147	147.7	147.8	147.8	148.3	148.3	148.6	148.6	
7.3	145.7	146.6	146.5	146.8	146.8	146.5	146.5	146.4	
26.8	687.8	689.9	691.7	692.4	690	689.7	689.6	691.9	
36.4	937.2	937	936.6	936.9	937.3	936.9	937	936.5	
59	39216.6	39377.5	40813.5	36729.1	38217.3	37961	35205.5	39705.2	Selected
59	32902.2	36491.6	31687.8	33033.2	33096.3	31724.1	37528.6	34899.8	
59	37033.5	34356.5	38700.9	37462.2	37336	38838.1	32686.6	35858.3	
59	38578.4	38847.3	35997	40167.1	38724.9	38573.2	41507.7	37269.6	
686.7	5717.7	5786.6	5837.4	6077	5817	5844.9	5694.7	5774.2	Selected

Table 3 Details of user time

Average_User_Time								
		MRU	LFU	MFU	LFU-LRU	LRU-MFU	LFU-MRU	MFU-MRU
2.7	2.211	2.017	2.001	1.906	2.147	2.785	2.512	2.357
7.1	0	0	0	0	0	0	0	0
7.2	0	0	0	0	0	0	0	0
7.2	0	0	0	0	0	0	0	0
7.3	0	0	0	0	0	0	0	0
7.3	0	0	0	0	0	0	0	0
7.3	0	0	0	0	0	0	0	0
26.8	0.018	0.014	0.013	0.015	0.027	0.031	0.03	0.039
36.4	0.032	0.032	0.045	0.03	0.033	0.031	0.039	0.039
59	1.326	1.269	2.13	1.313	1.605	1.589	1.36	1.986
59	1.36	1.779	1.639	1.439	1.759	1.539	2.378	1.746
59	1.456	1.109	1.928	1.627	1.782	1.947	1.124	1.577
59	1.338	1.347	1.214	1.595	1.543	1.708	1.951	1.575
686.7	1.693	2.0	0.648	5.714	2.776	3.082	2.004	2.989

Table 4 Details of system time

Average_System_Time								
File size (kb)	LRU	MRU	LFU	MFU	LFU-LRU	LRU-MFU	LFU-MRU	MFU-MRU
2.7	2.718	2.388	2.112	2.433	2.059	2.617	2.313	2.26
7.1	0.004	0.197	0.182	0.171	0.162	0.152	0.175	0.169
7.2	0	0.236	0.192	0.132	0.162	0.155	0.19	0.181
7.2	0.002	2.317	2.341	1.605	1.962	1.998	2.027	2.167
7.3	0	1.981	1.747	1.688	1.584	1.444	2.083	1.83
7.3	0.004	1.398	2.139	1.687	1.892	1.83	1.502	1.736
7.3	0	1.964	1.93	1.873	2.07	2.107	2.432	1.919
26.8	0.138	1.331	1.152	2.239	1.042	0.991	1.045	1.134
36.4	0.201	0.006	0.004	0.007	0.004	0.005	0	0
59	1.966	0	0	0	0.001	0	0	0
59	1.473	0.001	0.003	0.002	0.002	0.001	0	0
59	1.628	0.001	0.002	0	0	0.001	0	0
59	1.764	0.003	0	0.006	0.003	0	0	0
686.7	0.968	0.001	0	0.002	0.003	0	0	0

difference can be taken as a delay. Experimental results of the above page replacement policies and their respective graphs are shown in Figs. 1 and 2.

In Table 1 it can be found that for large programme size especially by using system program the page fault number will be almost same in case of MRU, MFU and LRU, but we can observe that for the user program (medium size) the major page

Table 5 Details of average elapsed time

Average_Elapsed_Time								
File size (kb)	LRU	MRU	LFU	MFU	LFU-LRU	LRU-MFU	LFU-MRU	MFU-MRU
2.7	13.919	13.539	12.824	13.463	12.091	17.348	13.744	13.682
7.1	1.6097	1.4309	1.1618	1.3554	1.2223	1.1512	0.8219	1.4358
7.2	0.6971	0.6868	0.7167	0.7163	0.7823	0.7657	0.6917	0.7019
7.2	1.3743	1.1691	1.191	1.3314	1.4252	1.1378	0.7858	0.7236
7.3	1.1988	0.9586	1.084	1.3594	0.8947	1.3219	0.8906	0.8788
7.3	1.5008	1.3494	1.032	1.417	1.289	1.0524	0.9341	0.7316
7.3	0.5302	0.5472	0.5542	0.5559	0.5452	0.5525	0.5219	0.5249
26.8	17.925	17.467	17.383	16.759	15.079	20.598	17.282	16.538
36.4	19.427	20.263	19.173	19.101	16.87	22.328	19.602	18.541
59	15.303	14.011	22.18	15.165	15.845	16.379	14.524	19.461
59	16.829	20.945	17.978	16.134	15.02	13.838	24.189	16.425
59	14.346	10.868	19.275	17.088	16.048	17.397	8.668	14.229
59	14.709	13.805	12.837	16.281	14.734	14.667	19.441	13.821
686.7	16.069	16.122	14.726	15.222	13.417	19.036	15.712	15.00667

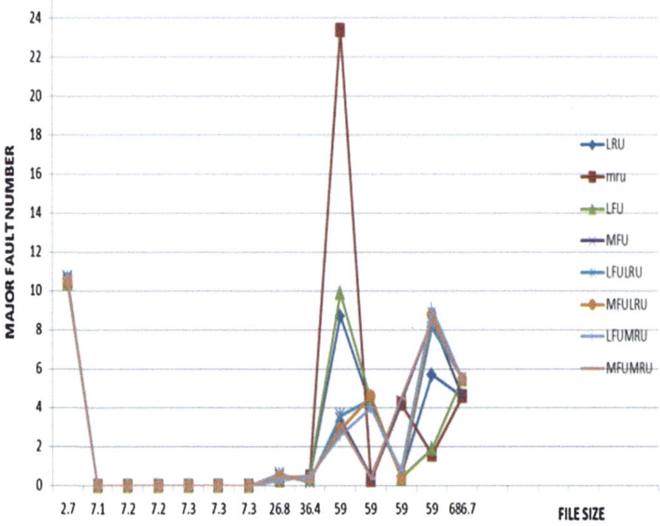

Fig. 1 Page faults (minor fault) versus file size

fault rate is shooting up in the case of MRU. In case of LFU there is a slight increase in major fault and due to this the performance is degraded when compared with the LRU. In case of the combinations of the various methods the result of LRU-MFU and LFU-MRU give almost the same result. LFU-LRU performance slightly lies

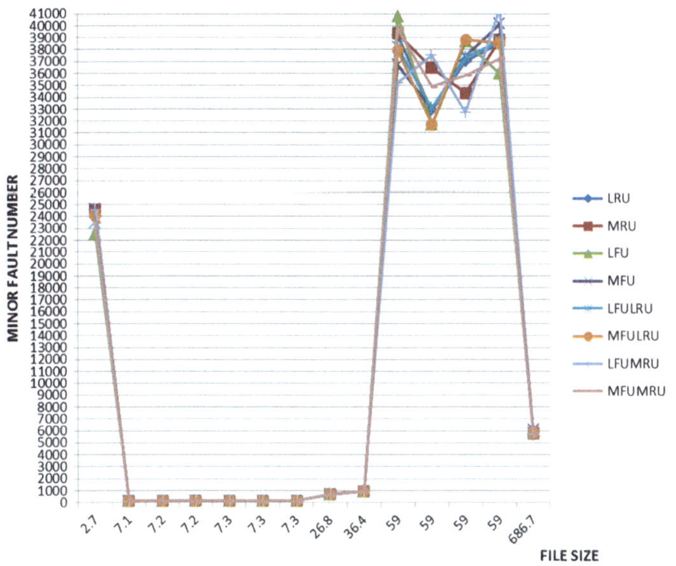

Fig. 2 Page faults (minor fault) versus file size

down the above two methods in user program. In case of MFU-MRU we can see that the number of major page fault is going to be getting reduced in the case of medium size program when compared with the LFU-LRU and is getting down much compared with LRU algorithm. And it can also observed that for medium size programs the combination will always give a better performance. Even though we are not able to make any far-reaching change in the number of major fault, the number of processor cycles required for each major fault is in the range of 2000000 cycles [9]. Hence a small degradation in the major fault will improve the overall system performance. The graphical representation of Table 1 is given in Fig. 1.

In Table 2 the analysis of the minor fault describes the fact that for small user programs, even LRU, MRU and LFU-MRU shows the same result, but the number of minor fault is getting reduced in the case of LFU, LFU-LRU and MFU-MRU. When the size of the program get increased the existing LRU will give better performance. In the case of LFU-LRU and LRU-MFU the performance of the system is going to get degraded for large size programs. LFU-MRU is able to reduce the minor fault for the small, medium and large size programs when compared to the existing LRU method.

Tables 3, 4 and 5 give the data regarding the various time parameters. Which are extracted from the system using the system built command over a set of programs those are used for the measurement of page faults. The analysis of these result will give a clear idea about the relationship of the execution time and the page fault number.

The graphical representation of Table 2 is given in Fig. 2.

In Table 3 the analysis of the user time is given. It shows that the existing LRU is giving the better user time considering small, medium and large size programme. MFU is the one which consumes the maximum user time when the size of the program getting increased. The time taken by the LFU-LRU and MFU-MRU are almost the same and is closer to the user time of LRU.

The system time analyse is shown in the Table 4. All methods except LRU consume less amount of system time for programs with average and large size. It is noted that when the system commands are going to get executed inside the system, system time is almost zero when replacement is done with any other method than LRU. LFU-LRU gives the better performance while considering the small, medium and large program compared to other methods.

The graphical representation of the above Table 4 is given in Fig. 3.

In elapsed time calculation the system pre-emption condition is also considered because the pre-emption condition may occur due to the unavailability of the data. Hence we have considered the user time, system time and elapsed time measurements for the evaluation.

In order to conclude the result analysis we will make use of the parameters major fault, minor fault, user time, system time and elapsed time. Table 6 will give a comparative study of these various parameters with the existing LRU method along with LFU-LRU, LRU-MFU, LFU-MRU and MFU-MRU. The average number of major fault, minor fault, user time, system time and elapsed time were calculated for the Tables 1, 2, 3, 4 and 5 respectively and shown in Table 6. We can see that the elapsed time is getting reduced in the case of LFU-LRU and MFU-MRU compared to LRU method. Also observed that the no of major fault and minor fault is getting reduced in these two methods. The graphical representation of above Table 5 is given in Fig. 4.

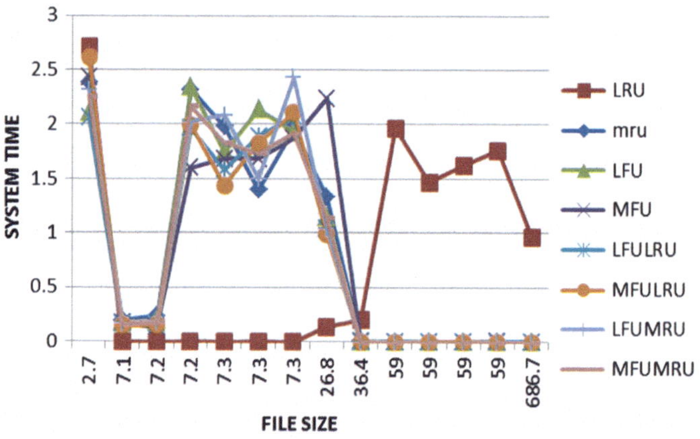

Fig. 3 System time versus file size

Table 6 Comparison details

Parameters	LRU	LFU-LRU	LRU-MFU	LFU-MRU	MFU-MRU
Major fault	0.033915	0.032558	0.032558	0.032461	0.032461
Minor fault	174.9546	173.6341	174.0472	174.1626	173.9712
User time	0.009141	0.01131	0.012318	0.011045	0.011926
System time	0.010529	0.010607	0.010951	0.011402	0.011043
Elapsed time	0.131238	0.121379	0.142997	0.133535	0.128586

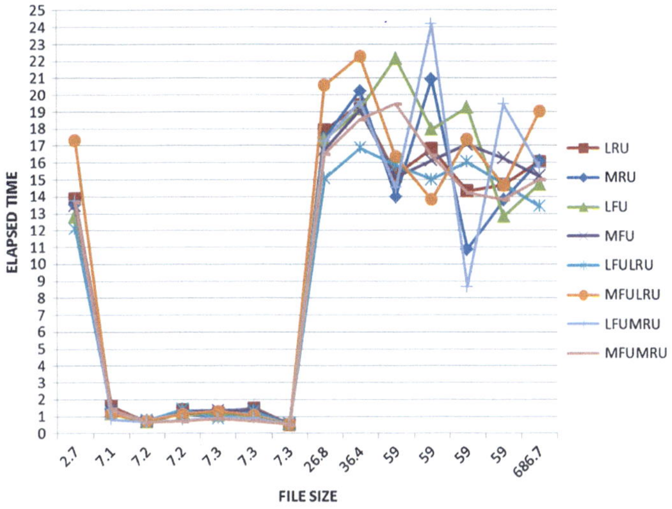

Fig. 4 Elapsed time versus file size

5 Conclusion

It is observed that during the replacement of the data from the various memory level, using the page replacement algorithm, we have to take care of the frequency parameters along with time stamp which is currently being used. The execution time of a file depends on the number of page faults which the system under goes. Hence if we are able to reduce the page fault number inside the system the execution time can reduce and the system performance can increase. Hence we have considered existing LRU, MRU, LFU, MFU and combination of all these methods. It is observed that, in the case of LFU-LRU the major fault and minimum fault was less compared to existing LRU. Due to the reduction in major fault and minor fault the elapsed time also reduced for LFU-LRU.

References

1. Silberschatz A (1999) Peter bear. In: 1999 operating systems concepts, 5th edn. Wiley, New York
2. O'Neill EJ, O'Neill PE, Weikum G (1993) The LRU-K page replacement algorithm for database disk buffering SIGMOD. ACM, Washington, DC, USA
3. Khajoueinejad S, Sabeghi M, Sadeghzadeh A (2006) A fuzzy cache replacement policy and its experimental performance assessment. IEEE
4. Juurlink B (2004) Approximating the optimal replacement algorithm. CF04 April 14–16 2004. ACM 158113/149/04/0004
5. Jianga S, Zhangb X (2004) Token-ordered LRU: an effective page replacement policy and its implementation in Linux systems. Elsevier
6. Sobell MG A practical guide to solaris. Addison Wesley
7. Tanenbaum AS, Woodhull AS Operating system design and implementation. PHI
8. Das S UNIX concept and applications, 4th edn. Tata McGraw-Hill
9. Bryant RE, O'Hallaron D Computer systems a programmer's perspective. Pearson Education

Query Performance Prediction Using Joint Inverse Document Frequency of Multiple Terms

P. Viswanath, J. Rohini and Y. C. A. Padmanabha Reddy

Abstract In an information retrieval system, predicting query performance, for keyword based queries is important in giving early feedback to the user which can result in an improved query which in turn results in a better query result. There exists clarity score based and ranking robustness score based techniques to solve this problem. Both these, eventhough shows good performance, suffers from high computational time needs and are post-retrieval methods. In contrast to this, there do exist several pre-retrieval parameters which can judge the query without executing it. Pre-retrieval parameters based on distribution of information in query terms, which basically depends on inverse document frequency (*idf*) of query terms, are shown to be good predictors. Among these, the standard-deviation of *idf* values of query terms is known to be better. This paper generalizes this and proposes to use joint *idf* for a set of terms together, than using each term's *idf* individually. Empirical studies are done using some standard data sets. The parameters based on the proposed method are shown to be better than the previous method which is nothing but a special case of the proposed method.

1 Introduction

In any information retrieval (IR) system, predicting the query performance is shown to be important, not only in giving feedback to the user who might refine the query, but in improving the effectiveness of the IR system by choosing appropriate retrieval strategy [4] and in doing query expansion [2].

P. Viswanath
Department of Computer Science and Engineering, Indian Institute of Information Technology - Chittoor, Sri City, Chittoor District, Andhra Pradesh, India
e-mail: viswanath.p@iiits.in

J. Rohini (✉) · Y.C.A.P. Reddy
Department of Computer Science and Engineering, Madanapalle Institute of Technology and Science, Madanapalle, Chittoor District, Andhra Pradesh, India
e-mail: rohini454@gmail.com

Y.C.A.P. Reddy
e-mail: ananthayc.reddy@gmail.com

K.R. Attele et al. (eds.), *Emerging Trends in Electrical, Communications and Information Technologies*, Lecture Notes in Electrical Engineering 394,
DOI 10.1007/978-981-10-1540-3_10

Mostly, query performance prediction is done in two ways, *viz.*, (i) based on clarity score, and (ii) based on ranking robustness score. The clarity score of a query is the Kullback-Leibler distance (relative entropy) between the query language model and the document collection language model [4]. Ranking robustness score measures the tolerance of the query result to the presence of noise [12]. Empirically it is shown that ranking robustness score works better than the clarity score for some data sets [12].

But the above mentioned methods are post-retrieval, meaning, the IR system after doing the retrieval finds out the query performance. It uses the query, corpus (database) and the result given by the IR system and does not use User's feedback or any other relevance information. So, they are time consuming. In contrast to these, there exists pre-retrieval methods which uses some precomputed statistics, called predictors of query performance, to measure the query performance [7]. Widely used predictors are, (i) query length, (ii) the distribution of information amount in query terms, (iii) query clarity, and (iv) query scope. The distribution of information amount in query terms are captured by two predictors, *viz.*, standard-deviation of inverse document frequency (*idf*) values of individual query terms, and ratio of maximum *idf* value to minimum *idf* value. The former one, i.e., the standard-deviation is found to be better than the latter one. The paper proposes a generalization of these predictors where instead of *idf* of individual terms, joint *idf* of multiple terms are used.

The rest of the paper is organized as follows. Section 2 reviews existing pre-retrieval predictors. Section 3 describes the proposed predictors which are based on joint *idf* values. Section 4 describes about the experimental studies carried out which shows the superiority of the proposed predictors. Finally, Sect. 5 concludes the paper.

2 Existing Pre-retrieval Query Performance Predictors

Existing pre-retrieval predictors [7] are as follows.

1. Query Length: The query length is the number of non-stop words in the query [7]. This is found to have strong influence on the smoothing methods used in the language modelling approaches [11], and in general, found to influence the length normalization methods in the probabilistic models [6].
2. Distribution of Information Content in Query Terms: Two parameters used in [7], based on inverse document frequency (*idf*) of individual terms are (i) standard-deviation of *idf* of query terms (γ_1), and (ii) max-min ratio of *idf* of query terms (γ_2). That is,

$$\gamma_1 = \sigma_{idf} \tag{1}$$

where σ_{idf} is the standard-deviation of the *idf* of the query terms.

$$\gamma_2 = \frac{idf_{max}}{idf_{min}} \tag{2}$$

where idf_{max}, idf_{min} are respectively, maximum and minimum *idf* of query terms.

Inverse document frequency of a term t, $idf(t)$ is found from the following equation [1].

$$idf(t) = \frac{\log_2(N + 0.5)/N_t}{\log_2(N + 1)} \tag{3}$$

where N is the total number of documents in the corpus, and N_t is the number of documents in which the term t is present.

These parameters, *viz.*, γ_1 and γ_2 are found to influence the effectiveness of the retrieval performance [8]. Among these two, γ_1 is shown to be better [7]. This paper basically generalizes this predictor and is explained in Sect. 3.

3. Query Clarity: Clarity score [4], which measures the relative entropy between query terms and the entire corpus is a post-retrieval parameter. A simplified version of the clarity score called *Simplified Clarity Score (SCS)*, as given in [7], is a pre-retrieval parameter.

$$SCS = \sum_{Q} P_{ml}(t|Q)\log_2\frac{P_{ml}(t|Q)}{P_{coll}(t)} \tag{4}$$

where Q is the query, i.e., the set of query terms, $P_{ml}(t|Q)$ is the maximum likelihood of getting term t in the query Q (defined below), $P_{coll}(t)$ is the maximum likelihood of the term t in the entire corpus.

$$P_{ml}(t|Q) = \frac{q^tf}{ql} \tag{5}$$

where q^tf is the frequency of the term t in the query and ql is the query length.

$$P_{coll}(t) = \frac{tf_{coll}}{token_{coll}} \tag{6}$$

where tf_{coll} is the frequency of term t in the entire corpus and $token_{coll}$ is the number of distinct terms (when stopwords are filtered off) in the corpus.

4. Query Scope: This parameter, as defined in [9] is

$$\omega = -\log(n_Q/N) \tag{7}$$

where n_Q is the number of documents that contains at-least one query term, and N is the total number of documents in the corpus.

3 Proposed Method: Predictors Using Joint *Idf*

The paper proposes an improvement over the predictor related to 'Distribution of Information Content in Query Terms', i.e., it proposes to use $\gamma_1^{(i)}$, where $i \in \{1, 2, \ldots, |Q|\}$, which is a generalization of γ_1 (See Eq. 1). The proposal mainly differs in the way in which *idf* is calculated. Instead of taking *idf* for each term, *idf* for a set of terms is taken jointly. For example, for the set of two terms $\{t_1, t_2\}$, its joint *idf* (called "joint-*idf*") is found from

$$joint\text{-}idf(\{t_1, t_2\}) = \frac{\log_2(N + 0.5)/N_{\{t_1,t_2\}}}{\log_2(N + 1)} \tag{8}$$

where N is the total number of documents in the corpus, and $N_{\{t_1,t_2\}}$ is the number of documents in which the terms t_1 and t_2 are present. In general, for a set of k terms $\{t_1, t_2, \ldots, t_k\}$, its joint idf is found by replacing $N_{\{t_1,t_2\}}$ by $N_{\{t_1,t_2,\ldots,t_k\}}$ where the later means the number of documents which contain the terms $\{t_1, t_2, \ldots, t_k\}$ jointly.

Frequent itemset mining methods available in association rule mining [5] can be used to find $N_{\{t_1,t_2,\ldots,t_k\}}$. In experimental studies the a priori method available in [5] is used.

The proposed predictor $\gamma_1^{(i)}$ is defined as follows.

$$\gamma_1^{(i)} = \sigma_{joint-idf} \tag{9}$$

where $\sigma_{joint-idf}$ is the standard-deviation of the set of $joint - idf$ values of the query term sets of cardinality 1 to i. Note, $\gamma_1^{(1)} = \gamma_1$. In the experiments the value of i used is from $1, \ldots, |Q|$. But, it is found that $i = 2$ or $i = 3$ is better than other higher values of i.

4 Experimental Study

We did a similar experimental study as done in [7]. TREC disks 4 and 5 are used. Test queries are topics 351–450 from TREC. Stop-word filtering and stemming are done in a standard way. As done in [7], three types of queries, *viz.,* short queries (only titles are used), normal queries (only the descriptions, i.e., abstracts are used) and long queries (all the three fields, i.e., title, description, and narrative are used) are used in the experimental studies.

Average precision (AP) value for each query is found based on methods PL2 [3] and BM25 [10]. For each query the proposed or existing predictor's value is found. The r and the corresponding p-value of the linear dependence between these two value lists, i.e., AP values and the predictor's values, measures the effectiveness of the predictor.

Table 1 Parameter settings

Parameter	Short query	Normal query	Long query
Query length	2.45 ± 0.31	7.61 ± 3.22	22.34 ± 8.61
c for PL2	5.90	1.61	1.73
b for BM25	0.09	0.25	0.64

Table 2 Results: Showing r and p-values

Parameter	$\gamma_1^{(1)}$	$\gamma_1^{(2)}$	$\gamma_1^{(3)}$	$\gamma_1^{(4)}$	$\gamma_1^{(1)}$	$\gamma_1^{(2)}$	$\gamma_1^{(3)}$	$\gamma_1^{(4)}$
	PL2, short query				BM25, short query			
r	0.245	**0.382**	—	—	0.173	**0.275**	—	—
p-value	0.166	**0.072**	—	—	0.064	**0.021**	—	—
	PL2, normal query				BM25, normal query			
r	0.302	**0.452**	0.345	0.311	0.294	0.387	**0.391**	0.212
p-value	0.003	**0.001**	0.013	0.016	0.004	0.002	**0.002**	0.011
	PL2, long query				BM25, long query			
r	0.323	0.484	**0.556**	0.351	0.283	0.372	**0.491**	0.331
p-value	0.003	0.002	**0.001**	0.011	0.004	0.007	**0.001**	0.007

Query length and other parameters used for PL2 and BM25 method are given in Table 1

For various values of i, Table 2 gives the correlation value r and the p-value against respective AP values. Many short queries have less than or equal to two query terms. Hence, for $i = 3$ and higher values, for short queries the results are unreliable, hence is omitted.

It is observed that for short queries, $i = 2$ gives better results than the conventional predictor (which is nothing but $\gamma_1^{(1)}$). For Normal length queries, it is observed that $\gamma_1^{(2)}$ for PL2, and $\gamma_1^{(3)}$ for BM25 are performing well. For long queries, for both PL2 and BM25, $\gamma_1^{(3)}$ has good correlation with AP.

5 Conclusion

Among pre-retrieval query predictors, the one that is based on the distribution of information content in query terms is shown to be better. This predictor uses individual query term's *idf* values and finds its standard-deviation. The paper proposed to use multiple query terms jointly instead of individual query terms and proposed generalized predictors accordingly. It is experimentally found that the proposed predictor which uses joint *idf* has better correlation with average precision of the retrieval. One future extension is to use similar generalizations with other relevant pre-retrieval predictors.

Acknowledgments This work is supported by a UGC-SERO Minor Project with Reference No. MRP-4609/14.

References

1. Allan J, Ballesteros L, Callan JP, Croft WB, Lu Z (1995) Recent experiments with inquery. In: Proceedings of the 4th Text Retrieval Conference, pp 49–64
2. Amati G, Carpineto C, Romano G (2004) Fondazione Ugo Bordoni. Query difficulty, robustness, and selective application of query expansion. In: ECIR, vol 4. Springer, pp 127–137
3. Amati G, Van Rijsbergen CJ (2002) Probabilistic models of information retrieval based on measuring the divergence from randomness. ACM Trans Inf Syst (TOIS) 20(4):357–389
4. Cronen-Townsend S, Zhou Y, Croft WB (2002) Predicting query performance. In: Proceedings of the 25th annual international ACM SIGIR conference on Research and development in information retrieval. ACM, pp 299–306
5. Jiawei Han, MK, Pei J (2011) Data mining: concepts and techniques: concepts and techniques. Elsevier
6. He B, Ounis I (2003) A study of parameter tuning for term frequency normalization. In: Proceedings of the twelfth international conference on Information and knowledge management. ACM, pp 10–16
7. He B, Ounis I (2004) Inferring query performance using pre-retrieval predictors. In: String processing and information retrieval. Springer, pp 43–54
8. Pirkola Ari, Järvelin Kalervo (2001) Employing the resolution power of search keys. J Am Soc Inf Sci Technol 52(7):575–583
9. Plachouras V, Ounis I, van Rijsbergen CJ, Cacheda F (2003) University of glasgow at the web track: dynamic application of hyperlink analysis using the query scope. In: TREC, vol 3, pp 636–642
10. Robertson SE, Walker S, Jones S, Hancock-Beaulieu MM, Gatford M, et al (1995) Okapi at trec-3. NIST SPECIAL PUBLICATION SP, pp 109–109
11. Zhai C, Lafferty J (2001) A study of smoothing methods for language models applied to ad hoc information retrieval. In: Proceedings of the 24th annual international ACM SIGIR conference on Research and development in information retrieval. ACM, pp 334–342
12. Zhou Y, Croft WB (2006) Ranking robustness: a novel framework to predict query performance. In: Proceedings of the 15th ACM international conference on Information and knowledge management. ACM, pp 567–574

Removal of High Density Salt and Pepper Noise from the Image Using CMA

S. Vijaya Kumar and C. Nagaraju

Abstract The quality of the image plays a vital role in numerous image processing applications such as medical image analysis, pattern recognition, satellite image processing, etc. One of the most important noises that affect the quality of the image is impulse noise. This noise alters the value of the pixels to either extreme. An efficient noise reduction algorithm is required to improve the quality of the image by detecting the noisy pixels and then replacing it with the appropriate value. The algorithm should have high noise reduction efficiency and computational efficiency especially when dealing with high noise density images. This paper proposes an improved algorithm which detects the noisy pixels using Cloud Model method and replaces the value of the corrupted pixel by Cloud Model Average (CMA) method, which improves computational efficiency by 2.94 % without compromising the noise reduction efficiency compared to the existing methods in the literature.

Keywords Cloud model · Image quality · Noise reduction · PSNR

1 Introduction

In Digital image processing, removing the noise from gray image and retaining the key details is still a challenging problem for researchers over the past several decades. When salt and pepper impulsive noise is introduced in 8-bit image, a pixel value is modified either to 0 or to 255 randomly. The noise can be quantified in terms of percentage of pixels corrupted [1]. There are several filtering techniques available in the literature. The classical Median Filter is widely used filter for removing the impulsive noise [2–4]. The Median Filter is simple to implement, it

S.V. Kumar (✉)
Research scholar,Department of CSE, JNTU, Hyderabad, India
e-mail: svksr105@gmail.com

C. Nagaraju
Associate Professor,Department of CSE, Yogivemana University
college of Engineering, Proddituru, Andhra Pradesh, India

K.R. Attele et al. (eds.), *Emerging Trends in Electrical, Communications and Information Technologies*, Lecture Notes in Electrical Engineering 394,
DOI 10.1007/978-981-10-1540-3_11

replaces the central pixel value with the median of the chosen neighborhood. However, one of the main disadvantages of the median filter is that the image loses the key details like thin lines and not suitable for high noise density images. Adaptive Median Filter (AM) and Rank-Order Based Adaptive Median Filter addresses the disadvantage of the classical median filter [5] by checking the correctness of the median by varying window sizes adaptively. These filters work well compared to the classical median filter for noise levels below 50 %, but suffer the same disadvantage of losing key image details and blurring for noise levels more than 50 %. Both the classical median and adaptive median filters apply the filtering to all the pixels. A number of researchers proposed Switching Median Filters to selectively apply filters to corrupted pixels [5–12]. They became popular because of their best performance and simple implementation. The switching median filters work in two stages. First, corrupted pixels are identified by comparing the value of a pixel to the median value in its neighborhood. If the difference is greater than a given threshold, the pixel is identified as corrupted. In the second stage, the classical median filter is applied to replace the value of the corrupted pixel with the median value of the uncorrupted pixels in its neighborhood. In 2012, a method for removing impulsive noise from the images was proposed using the cloud model [13]. The Cloud model (CM) filter uses the cloud model for identifying the corrupted pixels and replaces the corrupted pixel with center weighted mean value (CWM) [14]. It gives better results compared to all the other existing methods, even images corrupted by 95 % of impulsive noise.

This paper discusses the characteristics of the cloud model in Sect. 1, in Sect. 2 we discuss the identifying the informative pixels using the cloud model, in Sect. 3 discuss the filtering CMA (cloud model average method) and CMRWM (cloud model robust weighted mean method) filters, next Sect. 4 we focus on results and discussion and finally conclusions found in the Sect. 5.

1.1 Cloud Model and Its Characteristics

Boca Raton proposed a cloud model in "Artificial intelligence with uncertainty" [13]. Cloud model is a novel uncertainty theory developed from probability, statistic and fuzzy set theory. This model provides the transition of uncertainty between qualitative and quantitative data. Let U be a Universal set that describe by particular numbers and C be the qualitative concept correlated to U. If number v Є U which randomly realize the concept C and certainty degree of v for C is μ(v) Є [0, 1] is a random value with stabilization tendency i.e.

$$\mu\colon \mathrm{U} \to [0,1]\ \text{for all}\ \mathrm{v} \in \mathrm{U}\ \mathrm{v} \to \mu(\mathrm{v}) \tag{1}$$

then the distribution of v on U is defined as cloud and every v define as cloud drop. Normal cloud can characterize by the three digital characters [13], excepted value Ex, entropy En and hyper entropy He. Ex value is positioned at U representing the center of the gravity of the cloud. Entropy is a measure of the uncertainty of quality concept, which is characterized by fuzzy, and randomness of concept. Hyper entropy is the entropy of entropy of En which is a measure of distribution of cloud drop. The following steps are used to generate the cloud

Calculate the Ex, En and Hn of the random number v

Calculate the membership value

$$\mu(v) = e^{-(v - ex(v))2/2(En(v))2} \quad (2)$$

v with a certainty degree of μ(v) is the cloud drop in the concept

Repeat the steps (i) to (iii) until n cloud drop are generated.

2 Identifying the Informative Pixels Using the Cloud Model

The main aim of this paper is to suppress the non-informative details and preserving the informative details of an image. This paper uses the cloud model for identifying the pixel informative or non-informative, because of the advantage of the fuzziness and randomness of the cloud model.

Let $W_{(i,j)}$ represents the N x N window centered about $x_{(i,j)}$, i.e.

$W_{(i,j)} = \{ x_{i-1,j-1}, x_{i-2,j-2}, \ldots x_{(i,j)} \ldots\ldots, x_{i-+1,j+1}, x_{i+2,j+2}\}$

N = 3, 5, 7….up to maximum window size.

i, j = 1, 2, 3, 4……up to size of the image

in order to find the $x_{(i,j)}$ is a informative pixel or not the algorithm as follows

i. Initialize the count = 0, smax = 9, threshold T, where count represents the no of pixels uncorrupted in given window and smax represent the maximum window size up to which the window increment.
ii. Calculate the mean value of all the pixels in the window $W_{i,j}^{2N}$

$$Ex = \frac{1}{n} \sum_{x_{i+s,j+t} \in W_{i,j}^{2N+1}} x_{i+s,j+t} \quad (3)$$

iii. Compute the entropy En i.e.,

$$En = \sqrt{\frac{\pi}{2}} \times \frac{1}{n} \sum_{x_{i+s,j+t} \epsilon W_{i,j}^{2N+1}} |x_{i+s,j+t} - Ex| \quad (4)$$

iv. Compute the boundaries b1 and b2 using the cloud 3n rule i.e.,

b1 = min (W_{max}, Ex + 3En)
b2 = max (W_{min}, Ex − 3En)

where W_{max}, W_{min} are the maximum and minimum values of the window (sub image) respectively.

v. if b1 < $x_{i,j}$ < b2 then $x_{i,j}$ is an informative pixel else go to step vi.
vi. To find the other pixels in the window which are informative or not. For the pixels $x_{i+s,j+t}$, if b1 < $x_{i+s,j+t}$ < b2 then D(i, j) = $x_{i+s,j+t}$ and count = count + 1.
vii. if b1 >= $x_{i,j}$ or $x_{i,j}$ <= b2 with count < T, then increase the window size and go to step ii.
else $x_{i,j}$ is a corrupted pixel.

the above method uses the mathematical 3n rule which covers the normal distribution of the sample space so that it detect the extreme high and low extreme values consider as non-informative pixels.

3 Methods for Replacing the Non-informative Pixels

A. **CMA filter:** After identifying the non-informative pixel its value is replaced with the average of the informative pixels within the window by using Eq. 5.

$$Y(i,j) = \sum_{i=1}^{count} D(i) \times \frac{1}{count} \tag{5}$$

where count gives the no of pixels which are informative in the given window.

B. **CMRWM Filter**

a. Compute the mean of all the informative pixels in the local window D

$$MEAN1 = \sum_{i=1}^{count} D(i) \times \frac{1}{count} \tag{6}$$

where the count = no of pixels which are informative.

b. Compute the standard deviation of the informative pixels in the given local window

$$\sigma = \sqrt{\frac{\pi}{2} \times \frac{1}{n}} \sum_{D_{i+s,j+t} \epsilon W_{i,j}^{2N+1}} \left| D_{i+s,j+t} - MEAN1 \right| \tag{7}$$

c. Compute the weight

$$W(\mathrm{r},\sigma)_k = \frac{2\sigma}{(\sigma + r_k^2)^2} \tag{8}$$

where k = 1, 2, … count, r_k = *MEAN1-D(k)* and σ is the standard deviation of the informative pixels in the window(sub image)

a. Compute the weighted average

$$C(i,j) = \frac{\sum_{k=1}^{count} W(\mathrm{r},\sigma)_k}{count} \tag{9}$$

b. Replace the non-informative pixel with the *C*(i, j).

4 Results and Discussion

The performance of the CMA and CMRWM methods is quantified by evaluating Peak Signal-to-Noise Ratio (PSNR). The PSNR value in decibels can be expressed mathematically using the following equation.

$$PNSR = 10\log_{10}\left(\frac{255^2}{MSE}\right)dB \tag{10}$$

where,

$$MSE = \frac{1}{MN}\sum_{i=1}^{M}\sum_{j=1}^{N}(B_{(i,j)} - I_{(i,j)}) \tag{11}$$

M = Number of rows of the image
N = Number of columns of the image
$B_{(i,j)}$ = The filtered image
$I_{(i,j)}$ = The original image

The PSNR values of the CMA, CMRWM, CM, and AM filters for four different images (“mandrill” and “mammogram”) across 10–90 % noise levels are shown in Tables 1 and 2, and are also plotted in Fig. 3. As can be seen from Tables 1, 2 and Fig. 3, the proposed CMA and CMRWM methods significantly outperform the AM filter and are much closer to the CM filter in PNSR measurement. The results

Table 1 Comparison of restored Mandrill image in PSNR (in decibels)

Filter/% of noise	10 %	20 %	30 %	40 %	50 %	60 %	70 %	80 %	90 %
CMA	35.22	34.85	34.30	33.65	32.94	32.27	31.54	30.80	29.96
CMRWM	33.18	32.84	32.52	32.10	31.73	31.32	30.77	30.10	29.24
CM Filter	33.23	32.88	32.54	32.08	31.75	31.29	30.71	30.08	29.26
AM	35.53	34.87	34.11	33.26	32.52	31.73	30.94	30.24	29.54

Table 2 Comparison of restored Mammogram image in PSNR (in decibels)

Filter/% of noise	10 %	20 %	30 %	40 %	50 %	60 %	70 %	80 %	90 %
CMA	46.16	45.41	44.74	43.69	42.69	41.6	40.29	38.89	37.21
CMRWM	44.03	43.73	43.54	43.31	42.83	41.99	41.2	40.06	37.34
CM Filter	43.89	43.66	43.16	42.93	42.52	41.99	41.26	39.67	37.2
AM	38.47	33.37	31.41	30.25	30.32	30.26	29.63	29.19	28.63

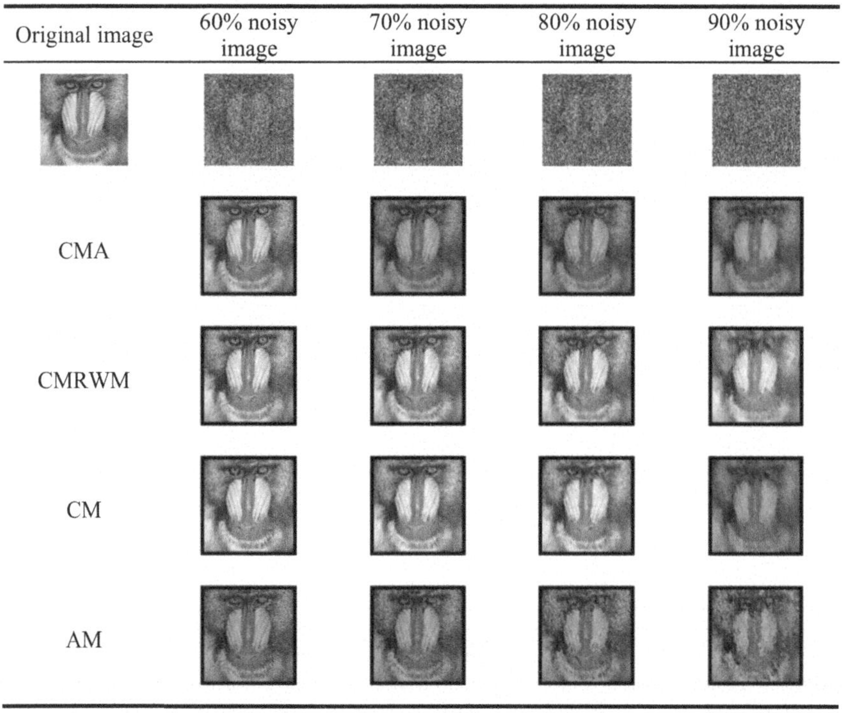

Fig. 1 Restoration details of different filters. *Row1* Original and corrupted Mandril Image with 60, 70, 80 and 90 % Noise. *Row2*, *Row3*, *Row4* and *Row4* are restoration images of CMA, CMRWM, CM, and AM filters respectively

indicate that the CMA and CMRWF methods are robust across a wide range of the salt and pepper noise densities. The subjective visual comparison of the noise removal and perception of the image details for the two test images; “mandrill” and “mammogram” are presented in Figs. 1 and 2, respectively. The CMA and CMRWM filters are clearly superior in terms of retaining critical details of the image compared the CM and AM filters (Fig. 3).

Runtime analysis The runtimes (seconds) for the different filters using INTEL (R), CORE(TM)2DUO, 2.93 GHz processor, 2 GB RAM with MATLAB 7.9.0 (R2009b) are documented in the Tables 3 and 4 for different images across 10–90 % noise range, and are also plotted in Fig. 4. The results show that the runtime for the CMRWF method is close to the CM filter. The runtimes for the CMA method are lowest compared to the CM, AM, and CMRWM filters. For example, the runtime for the CMA filter for the “mammogram” image with 10 % noise image is 2.94 times less than the CMRWM filter, 2.96 times less than the CM filter, and 1.46 times less than the AM filter. A similar trend continues at other noise levels too.

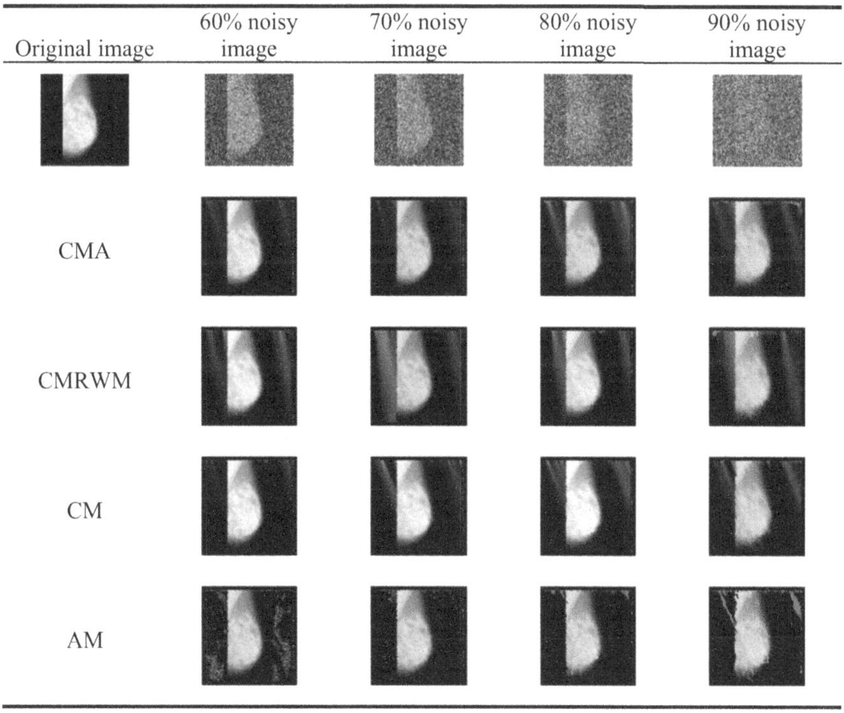

Fig. 2 Restoration details of different filters. *Row1* Original and corrupted Mammogram Image with 60, 70, 80 and 90 % Noise. *Row2*, *Row3*, *Row4* and *Row4* are restoration images of CMA, CMRWM, CM, and AM filters respectively

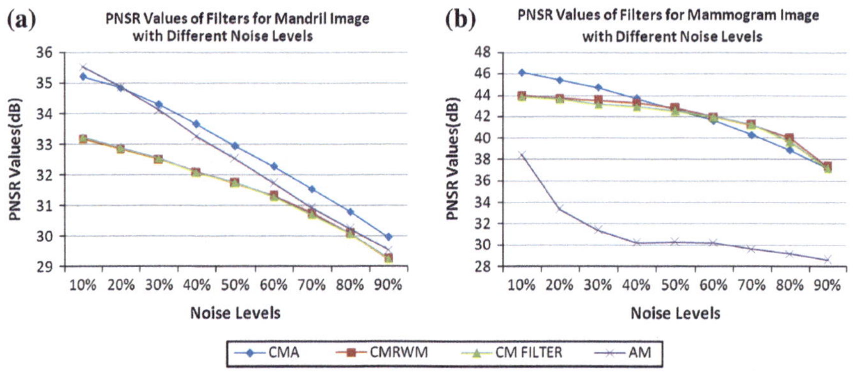

Fig. 3 PSNR Values for different filters operating on the images at various noise levels: **a** Mandrill. **b** Mammogram

Table 3 Elapsed time values for different filters operating on Mandril image at various noise levels

Filter/% of noise	10 %	20 %	30 %	40 %	50 %	60 %	70 %	80 %	90 %
CMA	2.41	2.46	2.49	2.55	2.61	2.70	2.78	2.88	2.99
CMRWM	6.25	6.48	6.85	7.17	7.52	7.86	8.27	8.64	9.08
CM Filter	6.28	6.56	6.89	7.18	7.71	7.92	8.42	8.78	9.14
AM	3.32	3.30	3.27	3.32	3.33	3.35	3.40	3.46	3.60

Table 4 Elapsed time values for different filters operating on Mammogram image at various noise levels

Filter/% of noise	10 %	20 %	30 %	40 %	50 %	60 %	70 %	80 %	90 %
CMA	3.99	3.64	3.38	3.26	3.14	3.11	3.06	3.04	3.48
CMRWM	11.75	11.18	10.69	10.13	9.73	9.24	9.11	9.24	9.43
CM Filter	11.84	11.29	10.70	10.19	9.81	9.59	9.60	9.30	9.44
AM	5.84	5.79	5.69	5.57	5.30	4.26	3.64	3.52	3.51

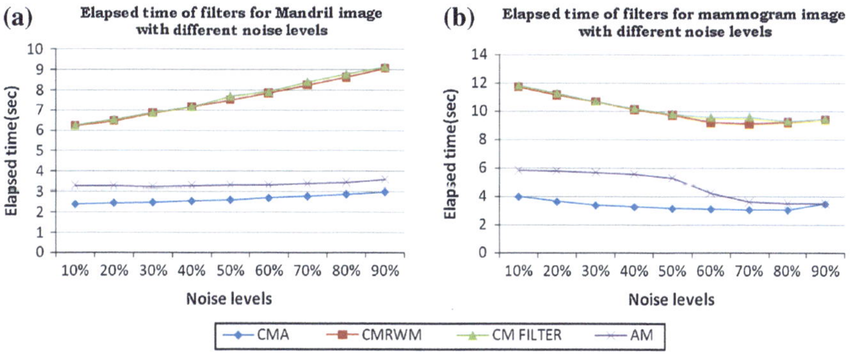

Fig. 4 Elapsed time Values for different filters operating on the images at various noise levels: **a** Mandrill, **b** Mammogram

5 Conclusions

We are judging the filter performance based on how filter improves the subjective visual quality, preserving the key details of the image and based on the computational time. This paper concludes with all the three factors respectively the CMA filter gives the best results compared to the remaining tested filters but, failing in keeping the edge details of the images in some areas, which would be addressed, in next paper.

References

1. Gonzalez RC, Woods RE (2002) Digital image processing. Prentice Hall
2. Pratt WK (1975) Median filtering Tech. Rep., Image Proc. Inst., Univ. Southern California, Los Angeles, Sep 1975
3. Pitas I, Venetsanopou A (1990) Nonlinear digital filters: principles and application. Kluwer, Norwell, MA
4. Astola J, Kuosmanen P (1997) Fundamentals of nonlinear digital filtering. CRC, Boca Raton, FL
5. Sun T, Neuvo Y (1994) Detail-preserving median based filters in image processing. Pattern Recogn Lett 15:341–347
6. Florencio, Schafer R (1994) Decision-based median filter using local signal statistics. In: Proceedings of the SPIE international symposium on visual communications image processing. Chicago, Sept 1994
7. Hwang H, Haddad RA (1995) Adaptive median filters: new algorithms and results. IEEE Trans Image Process 4(4)
8. Wang Z, Zhang D (1999) Progressive switching median filter for the removal of impulse noise from highly corrupted images. IEEE Trans Circuits Syst II, Analog Digit Signal Process 46 (1):78–80
9. Zhang S, Karim MA (2002) A new impulse detector for switching median filters. IEEE Signal Process Lett 9(11):360–363

10. Eng H-L, Ma K-K (2002) Noise adaptive soft-switching median filter. IEEE Trans Image Process 10(2)
11. Ng PE, Ma KK A switching median filter with boundary discriminative noise detection for extremely corrupted images. IEEE Trans Image Process 15(6):1506–1516
12. Srinivasan KS, Ebenezer D (2007) A new fast and efficient decision-based algorithm for removal of high-density impulse noises. IEEE Signal Process Lett 14(3):189–192
13. Li Y, Du Y (2007) Artificial intelligent with uncertainty. CRC Press
14. Zhou Zhe (2012) Cognition and removal of impulse noise with uncertainty. IEEE Trans Image Process 21(7):3157–3167

Neural Network—Based Diesel Engine Emissions Prediction for Variable Injection Timing, Injection Pressure, Compression Ratio and Load Conditions

M. Shailaja and A.V. Sita Rama Raju

Abstract The present study investigates the use of artificial neural network modelling for prediction of emission parameters of a four stroke single cylinder variable compression ratio diesel engine. ANN model was developed to predict emissions namely CO, NO_X and HC. Emission data was collected by conducting experiments by varying compression ratio, Injection time, and injection pressure in four steps and load in five steps. Two training algorithms traingd and trainlm with hidden nodes varying from 3 to 20 in step of one were developed and trained. Best network from 36 networks was selected based on MSE, regression coefficients for training, validation, testing and correlation coefficient for prediction of unseen data. The best model was found to be Levenberg–Marquardt algorithm with 17 neurons and regression coefficients for training, validation and testing are 0.99628, 0.99561, 0.99472 and 0.99577 respectively. The correlation coefficient R for training data is 0.99643 and for unseen data is 0.99322. The regression coefficients for prediction of training sets of CO, NO_X and HC are 0.99643, 0.99486 and 0.99601 respectively. The average % error for prediction of CO, NO_X and HC are -0.16178, -0.38814 and 0.7459 respectively which are less than 1. It is found that artificial neural networks serve as an excellent tool for prediction of emissions from diesel engine under variable operating and design parameters.

M. Shailaja (✉)
Department of Mechanical Engineering, JNTUH College of Engineering, Jagtial, India
e-mail: shailaja324@rediffmail.com

A.V. Sita Rama Raju
Jawaharlal Nehru Technological University, Hyderabad, India

K.R. Attele et al. (eds.), *Emerging Trends in Electrical, Communications and Information Technologies*, Lecture Notes in Electrical Engineering 394,
DOI 10.1007/978-981-10-1540-3_12

1 Introduction

The demand for diesel engines on road and in other fields like industry, agriculture etc. is continuously increasing which results in increase of emissions from diesel engines. Ever increasingly stringent emissions requirement, from diesel engines, motivated researchers to focus their investigations on techniques to reduce emissions. Many researchers proposed and worked with various bio-diesels and reported satisfactory results [1–4].

However statistics shows that demand and supply of bio- diesel are not matching and even biodiesel is not produced up to the available potential. They can only substitute for a fraction of petro diesel in use. In this context authors are concerned to study the behavior of emissions with variable engine design and operating parameters so that emissions can be reduced. Interestingly emissions have shown significant variation with variation of design and operating parameters. The future technology may advance in such a way that electronic control units may control, many design and operating parameters. In such case, experiments need to be conducted at numerous sets of conditions of operation and results are to be analyzed for better designs. But it is very costly and time consuming to conduct such a huge number of experiments. As the relation among engine parameters and emission parameters is highly non-linear, it is highly difficult to model the relation with conventional numerical techniques. In the recent years ANNs have emerged as a powerful tool for prediction of parameters with highly complex relations. Existing literatures show the efficient role of ANNs in prediction of parameters in the field of IC engines.

Taghavifar et al. [5] evaluated the potential of ANN to predict CO_2, soot and NO_x with crank angle, equivalence ratio, temperature, pressure, liquid mass evaporated and O_2 as inputs. A back propagation neural network with Levenberg Marquardt algorithm is developed and trained by varying number of neurons in hidden layer and reported that network with 18 neurons in the hidden layer shows highest performance with MSE 0.0001086 and R^2 0.9976, 0.9995, and 0.9951 were obtained for CO, soot and NO_x emissions, respectively. Roy et al. [6] developed neural network model to predict brake specific fuel consumption, brake thermal efficiency, CO_2, NO_x, and particulate matter, the inputs being load, fuel injection pressure, EGR and fuel injected. Levenberg–Marquardt algorithm with log-sigmoid transfer function and 4-10-10-5 architecture is able to predict the parameters with good accuracy with correlation coefficient (R) values ranging from 0.987 to 0.999. Roy et al. [7] attempted to predict brake specific fuel consumption, brake thermal efficiency, NO_x, HC and particulate matter, the inputs being load, fuel injection pressure and CNG energy share. ANN was developed with Levenberg–Marquardt algorithm with log-sigmoid transfer function, 2 hidden layers. Number of neurons in hidden layer was varied from 2 to 25 and found network with 3-8-8-5 was able to predict with high accuracy (correlation coefficients' within the range of 0.99833–0.99999). Rezaei et al. [8] attempted to develop ANN model to predict performance, combustion and emission parameters. Feed forward and radial basis function neural networks are developed and trained with hidden layer neurons ranging from 2 to 20. 12 training algorithms were applied to feed forward neural network and best performance was reported by Levenberg–Marquardt approach with log-sigmoid transfer function and 15 neurons in hidden layer. Mohammadhassania et al. [9] tried to develop a neural network model to predict soot and NO_x of a diesel engine. Two different networks were developed to predict soot and NO_x separately. Best results for NO_X were obtained with conjugate gradient back propagation (*cgb*) algorithm with 36 neurons in hidden layer and MSE for testing and training are 0.0019 and 0.014 respectively. Soot was predicted best by a network with training algorithm conjugate gradient propagation (*cgp*) algorithm with 38 neurons in hidden layer and MSE for testing and training are 0.0010 and 0.0025, respectively.

Vinay Kumar et al. [10] employed ANNs for prediction of brake specific fuel consumption, brake thermal efficiency, CO, HC and NOx emissions for engine load, fuel type and type of coating as inputs. Numerous trials resulted in best network with Levenberg–Marquardt algorithm and 6 neurons in hidden layer. When ANN predicted values were compared with experimental values average relative error was found to be 6.8 % which shows good performance of ANN. The work of Javed et al. [11] investigated the use of ANNs for prediction of brake thermal efficiency, brake specific fuel consumption, CO, O_2, CO_2, NO_X, HC and exhaust gas temperature for load, blends of biodiesel and hydrogen flow rates as inputs. Networks were trained for seven training algorithms and for each algorithm five transfer functions with two hidden layers were considered. Best performance was achieved by Levenberg-Marquardt back propagation training algorithm with log-sigmoid & tan-sigmoid transfer function with 16 neurons in each hidden layer. The regression value and MSE are 0.99360 and 0.0011 respectively for the best network.

Most of the researchers used ANNs for prediction of performance, combustion and emission parameters for different types of inputs. However there is no published research for prediction of emission parameters with compression ratio, injection pressure, injection timing and load as inputs. Hence the objective of the present work is to develop an artificial neural network to predict diesel engine emissions at different design and operating conditions, to avoid costly, time consuming and strenuous experimentation.

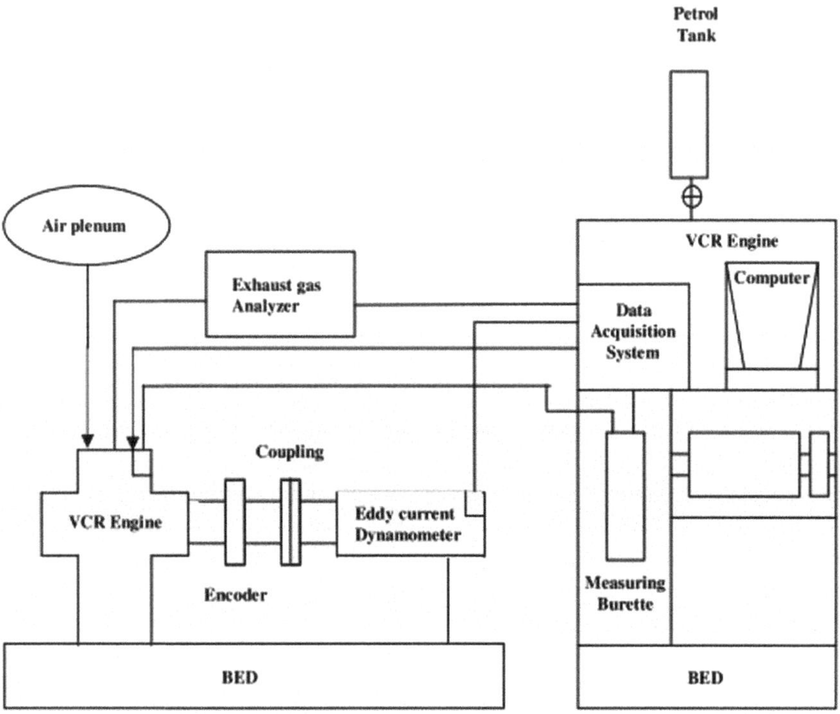

Fig. 1 Block diagram of experimental setup

Table 1 Specifications of Engine

Engine type	Kirloskar
Number of cylinders	Single (01)
Combustion	Direct injection
Bore	80 mm
Stroke	110 mm
Compression Ratio	Variable (15–20)
Rated Speed	1500 rpm
Power	5 hp
Type of cooling	Water cooling
Fuel injector opening pressure	200 bar
Fuel injection timing	24° before TDC
Type of loading	Electrical loading

Table 2 Specifications of 5-gas analyser Indus make

Exhaust gas	Measurement range	Resolution	Accuracy
CO	0–15.0 % vol	0.01 % vol	$\pm$ 0.06 % vol
CO_2	0–20.0 % vol	0.01 % vol	$\pm$ 0.5 % vol
HC	0–30000 PPM (Propane) 0–15000 PPM (Hexane)	1 ppm vol	$\pm$ 12 PPM
O_2	0–25.0 % vol	0.01 % vol	$\pm$ 0.1 % vol
NO_X	0–5000 PPM	1 ppm vol	$\pm$ 50 % vol

2 Materials and Methods

In the present work a single cylinder, four stroke and variable compression ratio diesel engine is employed to carry out the experiments. The compression ratio is varied by a provision at the top of the engine cylinder head. Injection pressure is varied with the help of an injection pressure test rig. Fuel injection timing is changed by inserting/removing the metal shims placed under fuel injection pump. Load is applied on the engine with the help of an eddy current dynamometer. Lay out of the experimental set up is shown in Fig. 1 and specifications in Table 1. An Indus make 5-gas analyzer is used to measure emissions. Specifications of 5-gas analyzer are presented in Table 2.

2.1 Experimental Procedure

Initially base line experiments are conducted at injection timing 24° BTDC and injection pressure 200 bar. At this setting compression ratio is set to 15 and experiments are conducted at 0, 20, 40, 60 and 80 % of full load. The experiments are repeated for compression ratios 16.5, 18 and 19. Similarly experiments are conducted at four injection timings, 22°, 24°, 26° and 28° BTDC. At each injection timing, four injection pressures 180, 200, 220 and 240 bar are considered. At each injection pressure, four compression ratios 15, 16.5, 18 and 19 are considered. Finally at each compression ratio four loads 0, 20, 40, 60 and 80 % of full load are considered. In this procedure total 320 experiments are conducted. At each experiment values of CO, NO_X and HC emissions are noted.

3 ANN Approach

ANNs are widely used for solving complex problems which are difficult to solve by conventional modelling methods because ANNs doesn't require any explicit mathematical equations for modelling of physical phenomena in a complex system. Human brain may be considered as a highly complex, non-linear and parallel computer which inspired to the development of ANN concept.

ANNs can be trained with some data by providing inputs and outputs and later new/unseen inputs are given to get outputs. The basic processing element in ANN is a neuron. A typical architecture of ANN consists of an input layer, hidden layer(s) and output layer. Neurons of one layer are fully connected to neurons of subsequent layers. The connections between the neurons are called weights. The weights may be assigned some random values or desire values before training and are adjusted during learning process which is called learning algorithm. The learning procedure adjusts weights and biases, minimise an error function between given outputs and network outputs. Each input is multiplied to its connection weight and sum of products of all neurons connected to one neuron is transformed through a transfer function to generate output of the neuron. The learning process may be supervised where both input and output are presented to the network and weights are adjusted to produce desired output whereas in unsupervised learning outputs are not supplied. A popular algorithm for prediction problems is back propagation algorithm. This back propagation algorithm has many variants like gradient descent, gradient descent with momentum are quite slow. Moreover successes of these depend on parameters such as learning rate and momentum constant. Some of the faster methods are there of which Levenberg-Marquardt algorithm is fast and shows better convergence. The mathematical background, the procedures for training as shown in Fig. 2 and testing the ANN and account of its history can be found in the text by Haykin [12].

In the present work a feed forward network with back propagation, Levenberg-Marquardt algorithm and gradient descent algorithm with tan-sigmoid transfer function for hidden layer and purelin transfer function for output layer are adopted. The four input parameters are compression ratio, injection pressure, injection timing and load. The outputs are CO, NOX, and HC. Number of neurons in hidden layer is varied from 3 to 20. The performance of network is

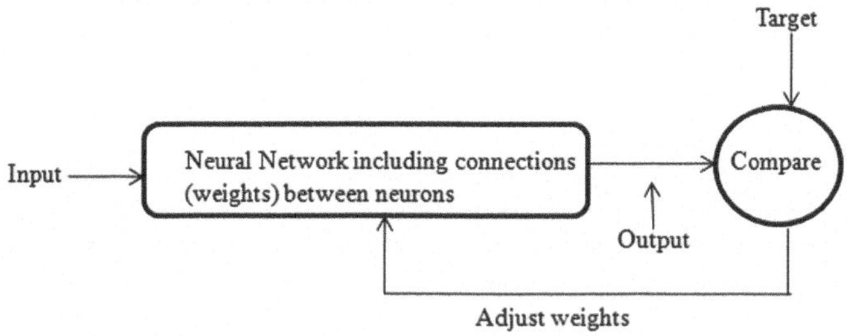

Fig. 2 Procedure in ANN

measured/compared by regression coefficients for training, validation and testing, MSE, correlation coefficient. Out of 320 data sets obtained from experiments are divided randomly for training (85 % of data, 272 sets) and testing (15 % of data, 48 sets). Before submitting data to the network, it is normalized in the range (−1 to +1) using following formula.

$$y = \left(\frac{y_{max} - y_{min}}{x_{max} - x_{min}}\right) * (x - x_{min}) + y_{min} \quad (1)$$

Mean squared error is calculated using following formula

$$MSE = \frac{1}{N} \sum_{i=1}^{N} (y_i - y_k)^2 \quad (2)$$

where, yi is the predicted value of i th pattern, yk is the target value of i th pattern and N is the number of pattern.

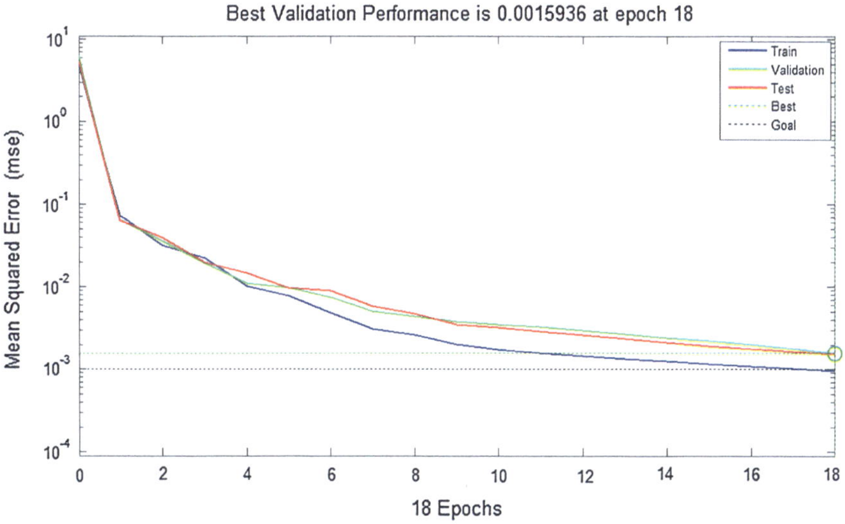

Fig. 3 Variation of MSE with number of epochs

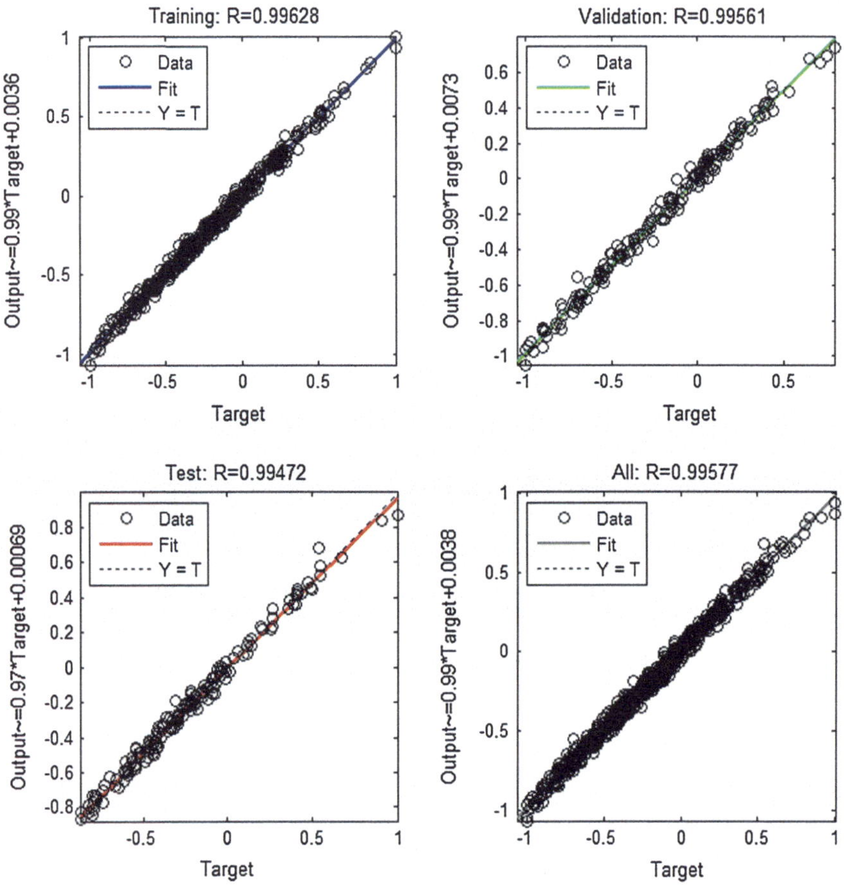

Fig. 4 Regression plots for training, validation, testing and all

4 Results and Discussions

The objective of this work is to develop an ANN model to predict CO, NO_X and HC for compression ratio, injection timing, injection pressure and load as inputs. Training and testing are done by using Matlab 7.7 version software. 320 experiments are conducted and data sets are collected. 272 data sets (85 % of data) are used for training the network and randomly selected 48 data sets (15 % of data) are used for testing the network. The network is trained with Levenberg-Marquardt algorithm and gradient descent algorithm for hidden layer neurons ranging from 3

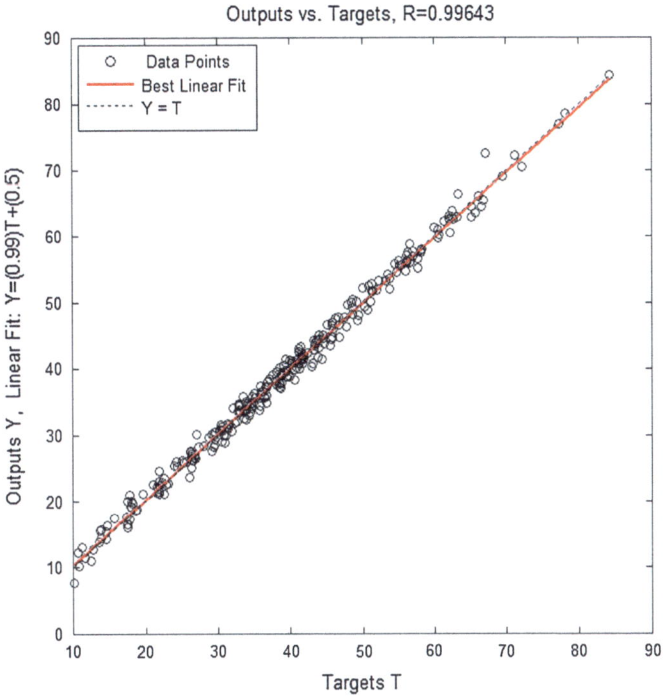

Fig. 5 Correlation plot for training data

to 20. The best performance is shown by network with Levenberg-Marquardt algorithm with 17 neurons in hidden layer. The Mean squared error for this network is 0.0009 Fig. 3 and time taken for training is very short i.e. 0.663 s. Regression coefficients for training, testing, and validation and over all are 0.99628, 0.99472, 0.9956 and 0.99577 respectively and are shown in Fig. 4. The correlation coefficient R for 272 sets of training data is 0.99643 as shown in Fig. 5. Correlation coefficients for individual outputs are 0.0.99643 for CO, 0.0.99485 for NO_X, 0.0.996015 for HC as shown in Fig. 6.

After training neural network is tested with 48 sets of unknown data and co-relation coefficient of prediction is 0.99322 as shown in Fig. 7.

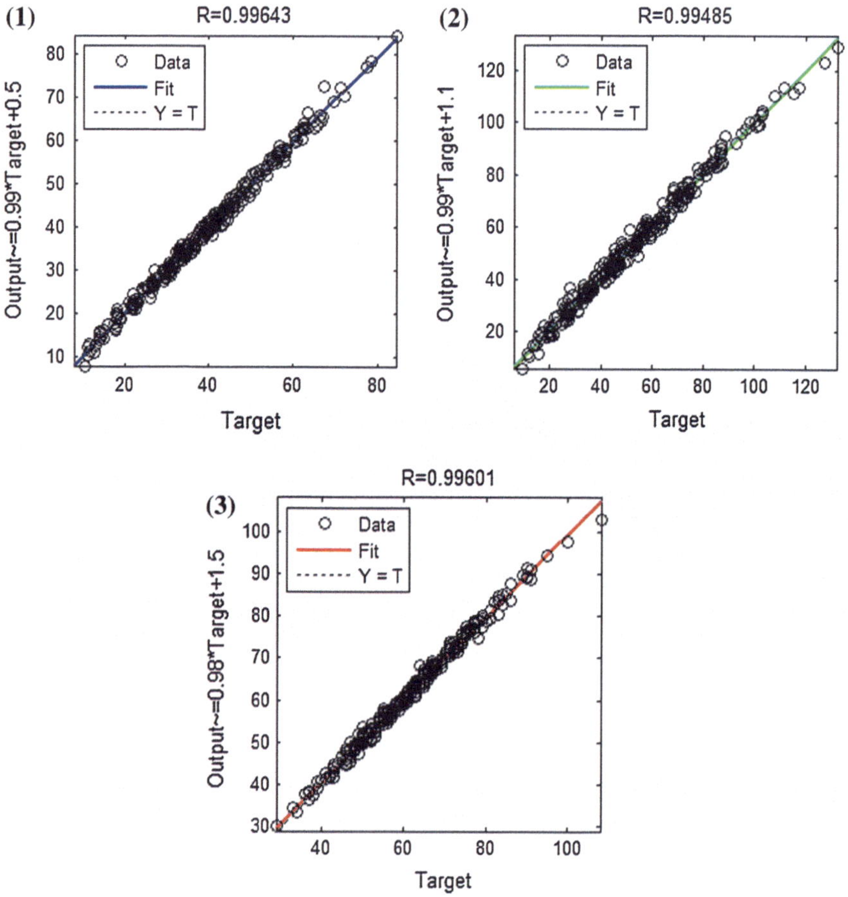

Fig. 6 Regression plots for 1. CO, 2. NO_X and 3. HC

Figure 8 shows variation of MSE with number of neurons in hidden layer. MSE does not show any significant fall after 17 neurons.

Figures 9, 10 and 11 represent experimental and predicted values of emission parameters. From the figures it is obvious that experimental and predicted values are in a very good agreement. From Table 3 it is evident that developed model is able to predict emission parameters with good accuracy.

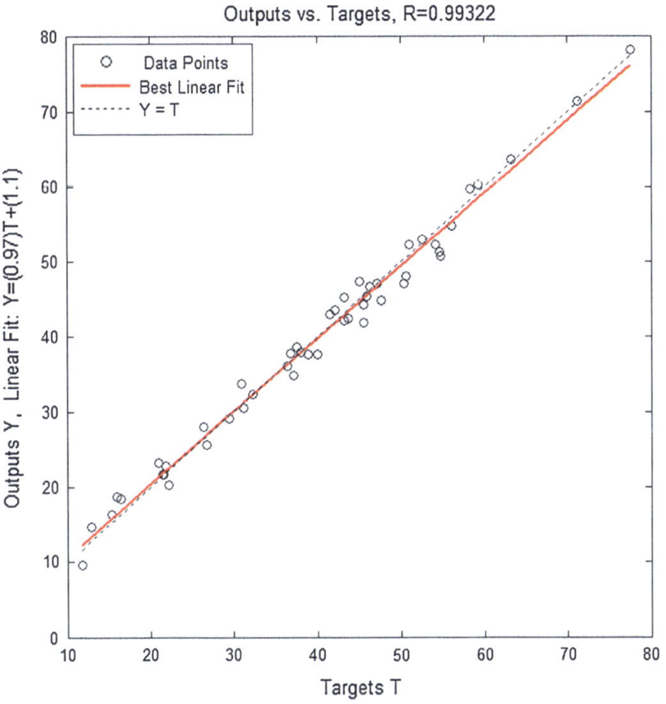

Fig. 7 Correlation plot for training data

Fig. 8 Variation of MSE with number of neurons in hidden layer

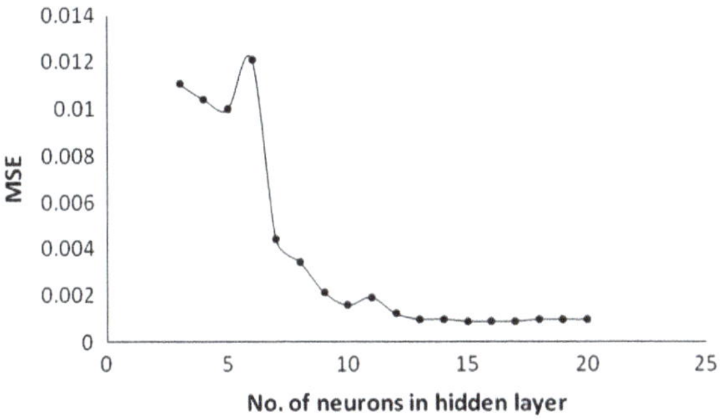

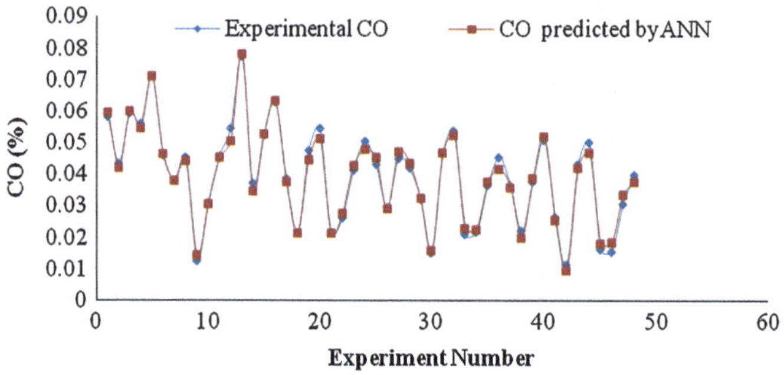

Fig. 9 Predicted and experimental values of CO

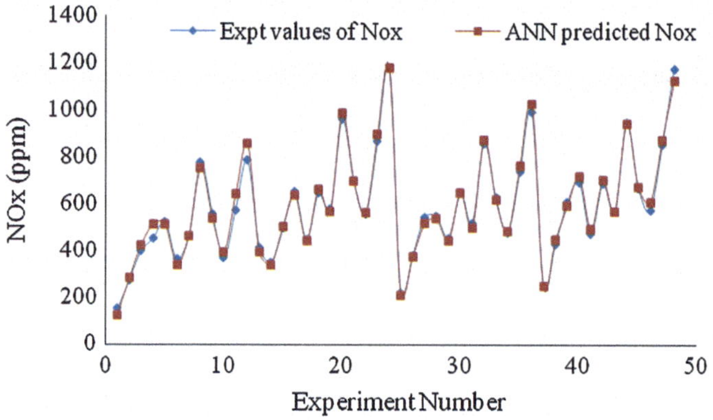

Fig. 10 Predicted and experimental values of NO_x

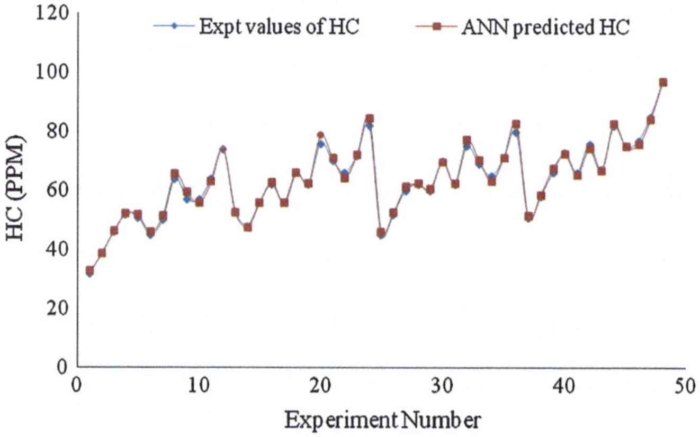

Fig. 11 Predicted and experimental values of HC

Table 3 Average percentage error for prediction for unseen data

Parameter	Average % error
CO	−0.16178
NO_X	−0.38814
HC	0.7459

5 Conclusions

In the present work a back propagation neural network is developed to predict CO, NO_X and HC for different compression ratios, injection timings, injection pressures and loads. A back propagation neural network is developed with Levenberg-Marquardt algorithm and gradient descent algorithm. For each algorithm tan-sigmoid transfer function and one hidden layer is considered and training is done by changing neurons in hidden layer from 3 to 20. It is observed that as number of neurons increase MSE value is decreasing and after certain optimum number of neurons is reached is remaining constant as shown in Fig. 6. Even though traingd (gradient descent) algorithm is performing equivalent to trainlm (Levenberg-Marquardt) algorithm traingd is consuming more time compared to train lm, for example network with 17 neurons in hidden layer completed training in 498 s with traingd whereas only 0.663 s with trainlm. Hence by comparing other performance parameters like MSE, regression coefficients, slope and intercepts of regression plots, correlation coefficients for prediction of output parameters trainlm algorithm with 17 neurons in hidden layer is declared to be the best of trained networks. From the values of correlation coefficients, MSE and regression coefficients it is strongly recommended the use of ANNs for diesel engine emissions since ANNs proved as an excellent tool for the prediction of emission parameters.

References

1. Zhen-tao LIU, Shao-mei FEI (2004) Study of CNG/diesel dual fuel engine's emissions by means of RBF neural network. J Zhejiang Univ SCI 5(8):960–965
2. Najafi G, Ghobadian B, Yusaf T, Rahimi H (2007) Combustion analysis of a CI engine performance using waste cooking biodiesel fuel with an artificial neural network aid. Am J Appl Sci 4(10):756–764
3. Hari prasad T, Dr. Hemachandra reddy K, Dr. Muralidhara rao M (2010) Performance and exhaust emissions analysis of diesel engine using methyl esters of fish oil with artificial neural network aid. IACSIT Int J Eng Technol 2(1):23–27
4. Manjunatha R, Badrinarayana P, Hemachandrareddy K (2010) Application of artificial neural networks for emission modelling of bio diesels for a C.I. Engine under varying operating conditions. CCSE Mod Appl Sci 4(3):77–89
5. Article in press: Taghavifar H, Taghavifar H, Mardani A, Mohebbi A, Khalilarya S, Jafarmadar S (2015) Appraisal of artificial neural networks to the emission analysis and prediction of CO_2, soot, and NOx of n-heptane fueled engine. J Clean Prod. http://dx.doi.org/10.1016/j.jclepro.2015.03.035. Elsevier

6. Roy S, Banerjee R, Bose PK (2014) Performance and exhaust emissions prediction of a CRDI assisted single cylinder diesel engine coupled with EGR using artificial neural network. Appl Energy 119:330–340 Elsevier
7. Roy S, Banerjee R, Das AK, Bose PK (2014) Development of an ANN based system identification tool to estimate the performance-emission characteristics of a CRDI assisted CNG dual fuel diesel engine. J Nat Gas Sci Eng 21:147–158. Elsevier
8. Rezaei J, Shahbakhti M, Bahri B, Aziz AA (2015) Performance prediction of HCCI engines with oxygenated fuels using artificial neural networks. Appl Energy 138:460–473. Elsevier
9. Mohammadhassania J, Dadvanda A, Khalilaryab Sh, Solimanpurb M (2015) Prediction and reduction of diesel engine emissions using a combined ANN–ACO method. Appl Soft Comput 34:139–150 Elsevier
10. Vinay Kumar D, Ravi Kumar P, Santosha Kumari M (2013) Prediction of performance and emissions of a biodiesel fueled lanthanum zirconate coated direct injection diesel engine using artificial neural networks. Procedia Eng 64:993–1002
11. Javed S, Satyanarayana Murthy YVV, Baig RU, Prasada Rao D (2015) Development of ANN model for prediction of performance and emission characteristics of hydrogen dual fuelled diesel engine with Jatropha Methyl Ester biodiesel blends. J Nat Gas Eng 26:549–557. Elsevier
12. Haykin S (1994) Neural networks, a comprehensive foundation. McMillian College Publishing Company, New York

Control System of Mobile Robotic Complex Based on Mini Tractor "Belarus 132H"

Makpal Zhartybayeva, Tamara Zhukabayeva and Ainur Zhumadillayeva

Abstract This article describes systems of navigation and communications of mobile robotic system (MRS), MRS manipulation modes, the most suitable mode solving assigned tasks. In this paper, we consider an example of designing robotic system on the basis of mini tractors Belarus 132 N chassis, which allows maximal use of mechanical components of series production, and would reduce the cost of the final product. The goal is to provide navigation and communication of mobile robotic system. Generalized block diagram of a multi-purpose mobile robotic complex, relationship diagram of main MRS components and object tracking were proposed.

Keywords Mechatronic systems · Mobile robotic systems · Adaptive systems · Mini-tractor belarus 132H

1 Introduction

Currently, researches in the field of robotics industry are highly challenging. One of the most popular designs of the robots are robotic systems for various purposes.

Researches were conducted in the field of control systems of mobile robotic systems (MRS) [1, 2, 3], driving system, sensor system (as localization systems [4, 5], visual systems etc.), systems of control, communication and navigation of robotic system, trajectory planning, control algorithms, etc.

M. Zhartybayeva (✉) · T. Zhukabayeva · A. Zhumadillayeva
Department of Computer Engineering, Faculty of Information Technologies, L.N. Gumilyov Eurasian National University, 010008 Astana, Kazakhstan
e-mail: makkenskii@mail.ru

T. Zhukabayeva
e-mail: tamara_kokenovna@mail.ru

A. Zhumadillayeva
e-mail: ay8222@mail.ru

K.R. Attele et al. (eds.), *Emerging Trends in Electrical, Communications and Information Technologies*, Lecture Notes in Electrical Engineering 394,
DOI 10.1007/978-981-10-1540-3_13

Many scientists have studied and investigated the various aspects of mobile robotics (design, engineering, modeling and control), among which the following works of scientists should be noted: Illa Reza Nurbaksh, head of the laboratory of urban robotics at the Institute of Robotics at Carnegie Mellon University (USA), which identified six major areas of priority research (• structure and robots chassis; • installed power; • electronics (sensors); • software (cloud and on-board); • connectivity; • monitor and control, without a breakthrough in which we can not achieve performance; [6] explores design issues and managing with two simple degrees of devices positioning [6]; (Bashir MY Nuri 2005) proposed new methods of modeling and control of mobile robots; (Jaroslav Hanzel 2011) studied computer-aided design of remote control mobile robotic systems via PC; (Ahmad A. Mahfouz 2012) who studies the modeling aspects of the motor dynamics in terms of output speed for mechatronics applications; (MBB Sharifian 2009) [7] examined the development and implementation of the system of motor speed PC-based DC-motor using PID; (Dada Nwe 2008) introduced a software implementation of obstacle detection for wheeled mobile robot, mathematical models and making control of mobile robots; (Gregor Klancar 2005) presented the new control design of nonholonomic mobile robot with differential drive; (Tao Gong 2003) considered architecture, characteristics and the principle of the mobile robot model;

Compactness, reliability, simple operation, standardized working load, and resistance to unfavorable external influences, versatility of useful board load, and other positive qualities of mobile robotic systems provided them with favorable conditions to be promoted in the industrial and domestic sectors [8]. Major companies in this field of the USA (iRobot, Adept—Technology Inc., etc.), Japan (FANUC, KAWASAKI, OTC DAIHEN etc.), Russia (Cubic Robotics, Copter Express, etc.) develop this industry actively, offering new opportunities for long-term improvements [9].

2 Prospects of Development of Technologies in the Field of Robotics

Application sphere of robotic systems and complexes is wide, but the main directions of mobile robotic systems development can be classified into three main groups:

- military and combat robots;
- emergency and research robots;
- robots, designed for domestic and industrial purposes.

It is important that most part of mobile robots are designed for working under conditions dangerous for human or emergency. Such conditions occur in the elimination of accident consequences, demining or working on other planets, during firefighting. Using mobile robots in such conditions helps to avoids casualties.

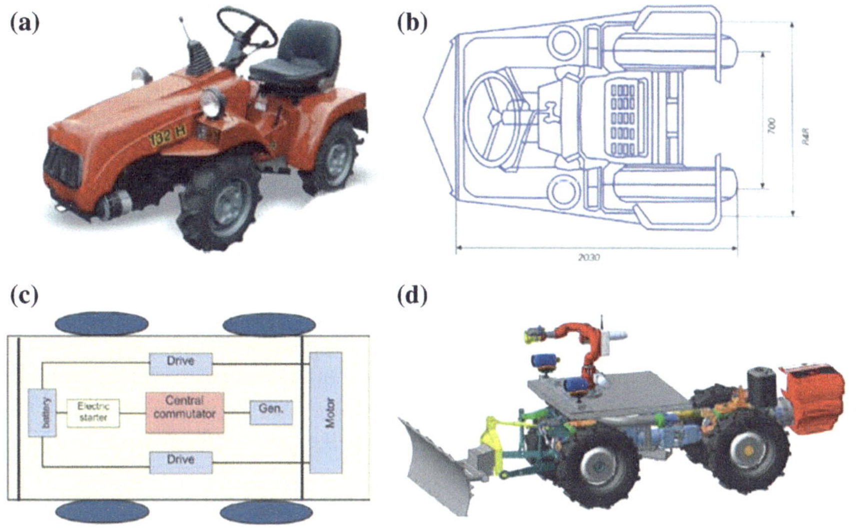

Fig. 1 **a** serial mini tractor Belarus—132 N; **b** scheme of loading and unloading of tractor; **c** mobile platform and a chain **d** mobile robotic complex based on mini tractor Belarus—132 N

Thus, the actual problem is the development of mobile robotic systems which have the possibility of independent movement and processing information about the object tracking. Created MRS with intelligent control system will ensure the effective performance of a number of tasks of the elimination.

To create a prototype robot car, it was decided to use a serial mini tractor Belarus —132 N with petrol engine Honda GX390 (132H/engine GX390/Honda/Indicated power 862 kW/mass 532 kg/2500 * 1000 * 2000 mm/Road (agrotechnical) clearance/mm 300/ specific fuel consumption at nominal rating) (Fig. 1) [3, 10].

2.1 Platform

Selecting wheeled chassis on the basis of mini tractor Belarus—132 N at a ratio of "price-quality" is the best, in our opinion. This chassis provides various kinds of twists and turns at the controls of Mobile Control Station. A special feature is an articulated frame, thereby achieving high maneuverability and flotation of machines, smaller turning radius [10].

Mini tractor will be equipped with additional devices which will give the versatility and effective application in such cases like explosion threat liquidation, poisoning, fire-fighting; in agriculture—for spraying pesticides etc. [11].

2.2 MRS Control System Designing

Manipulation of mobile robotic systems has two modes: human (operator) control and unmanned mode. The first mode requires continuous attention from the operator and the relatively small distance from the object that provide a lot of inconvenience. Navigation and communication is then carried by radio, providing a large range of motion, but which has a number of shortcomings: the dependence of the quality of radio communication on the application conditions, deterioration of the radio while driving, the probability of loss of data, low noise immunity, etc. These disadvantages can be avoided if the control by the operator will not be held at the level of movements, and at the level of goal. This approach requires from the robot independent decision-making on specific actions, and therefore there is a problem of obtaining and using knowledge of the underlying system of such a decision. One of the most effective options for building a knowledge base for the robot is teaching it on the basis of operator actions.

This mode requires accomplishment of a number of complex tasks. The control system should get information about the external world, objects and internal parameters of the robot as a system, components of which can be a variety of sensors, beacons, vision system, video system, etc. After processing this information, it generates control commands for the executive system, which includes means of propulsion and manipulators, as well as a variety of equipment. In the process of operation intelligent control system makes decisions about how to manage in order to fulfill the task. However, it uses the information stored in databases and knowledge of robot. Depending on the level of "intelligence" of the robot, the base may have different structure and store various information. If the control system is built on the principle of fuzzy inference, the database may contain information on the production rules and parameters of input and output variables [3].

After the analysis of intelligent systems, we decided that the control system will be based on fuzzy controllers that provide good adaptability of a mobile robotic system. Depending on the technical requirements and conditions of the problem fuzzy controllers can be realized as software and hardware [12].

Established mobile robot system should work in 2 modes, and carry out the following tasks [13]:

1. providing communication with the standard control system;
2. obstacles detection and ground map construction [14–18];
3. autonomous robot return in case of loosing communication or upon the operator command;
4. ensuring the adaptability of the system to different conditions

Consider the generalized structure of MRS (Fig. 2). The data about the object of exploration will be transmitted through the communication channels from video system, providing driving and video system of attached implements to the board computers. On-board computer, in turn, transmits the information via a wireless

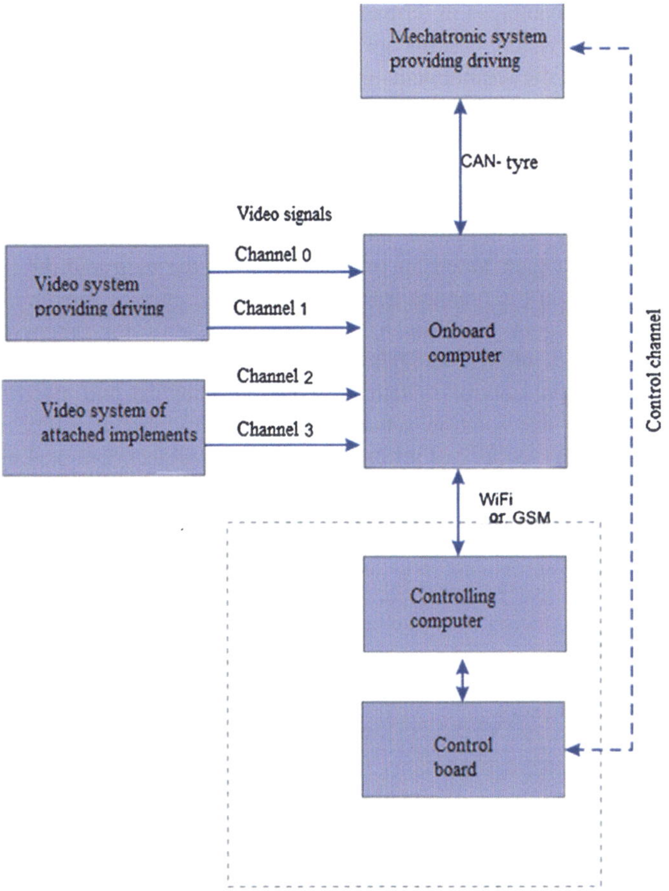

Fig. 2 The block diagram of a multi-purpose mobile robotic complex

interface (Wi-Fi) from the operator's position to the onboard processor, and vice versa [19].

Mechatronic system providing driving will control executive devices of MRS control system. It provides issuing commands to a PWM (pulse width modulation) that allows to control the speed of motor rotation. Module receives control commands via CAN-bus, often used in automotive equipment and which provides a single network for a variety of actuators and sensors. The system consists of a controller (Silicon Laboratories), electrical isolation and power keys that are controlled PWM (Fig. 2).

To provide navigation and communication mobile robotic system should perform the following tasks:

1. The formation of a more accurate image of the environment. This is done by means of laser range-finding devices and ultrasonic generators, which have a

number of disadvantages in terms of the perception of the data by the system. Today machine vision systems are more popular.

2. Development of the requirements for the target functions of algorithms of basic blocks and systems of MRS. The main ones are the navigation system on the basis of global (GPS, Glonass), network (WiFi, GPRS) and geomagnetic navigation systems. Determining the location coordinates of the mobile robotic system, in which peripheral beacons, GPS, ultrasonic emitters and receivers, mobile receivers and ultrasonic transducers and others will be active.
3. Planning motion path based on information from the map and the implementation of traffic on the planned trajectory in the presence of interference.
4. Providing intelligent machine visionbased on systems for processing stereo images of visible, infrared, and others spectrums, including with the use of range —finding devices; control systems of electromechanical and electrohydraulic with translation and rotation actuators of units providing mobility of robotic system, as well as security of information systems and technological operations; on the basis of the objectives, we present a diagram of the MRS main components relationship: video, object tracking and robot (Fig. 3).

The following systems can serve as a navigation system [14]:

1. Accelerometer;
2. Gyro systems;
3. Electronic compass
4. Range-finding sensors
5. Videosystems
6. GPS
7. Geomagnetic navigation systems

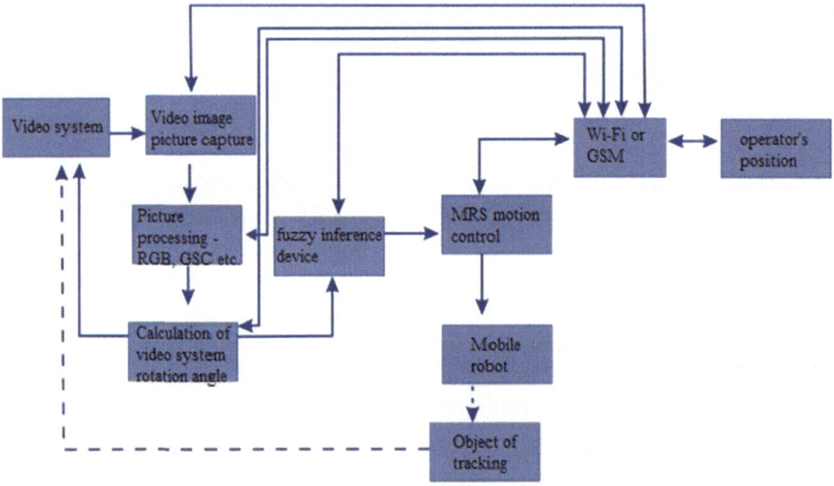

Fig. 3 Diagram of MRS major components and object tracking relationship

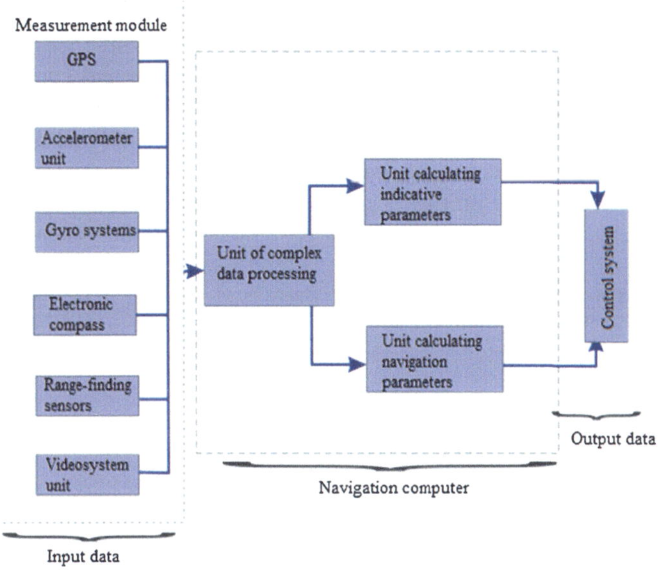

Fig. 4 Diagram of control system and measuring unit

The following systems can be used as communication systems: WiFi, ZigBee, VHF, IrDA, GSM etc. Depending on the distance of the operator's location for communication can be used: if the distance more than 1 km—GSM, if the distance is a small—Wi-Fi, as well as PC and laptop computers can be used as surveillance systems Each component of the navigation system provides the complex information processing block with input data, where parameters are distributed to evaluation units of indicative parameters and navigation parameters [20, 21]. For example, GPS provides information about the speed and position on site, an accelerometer—acceleration information in 3 planes. The gyroscope informs about the slope/turn angle, and electronic compass provides data about directions (orientation) in a horizontal plane, distance sensors are used to determine the distance to the obstacle/object (Fig. 3).

Good adaptability, the ability to continuously receive data about the object being in a safe place, intelligence of the system will provide developed mobile robot system not only with good popularization, but effective use in solving the problems of demining and elimination of consequences of man-made accidents (Fig. 4).

3 Conclusion

In this article, we presented basic specifications of our mobile robotic system. The proposed system of mobile robot based on the controlled methodology model will help to overcome the current difficulties of developing mobile robots. Highlighted features of our approach are simple, as well as proposed mobile platforms, model design that will provide ease of use and accessibility to potential users.

This article we described the system of MRS navigation and communication, assigned a task to provide navigation and communication for mobile robotic system. Mobile robotic systems manipulation modes were considered and the most suitable mode for the task was chosen.

In this article, we also proposed generalized structure of the MPS navigation system and a block diagram of a multi-purpose mobile robotic system.

Acknowledgments This work is supported by Kazakhstan Ministry of Education (Research Grant Scheme, contract № 47 from 12.02.2015).

References

1. Tatur M, Zhartybayeva M, Iskakov K, Babayev T, Pashayev A, Sabziev E (2015) Multifunctional mobile robot. In: The 5th international conference on control and optimization with industrial applications 27–29 Aug 2015. Baku, pp 432–435
2. Zhukabayeva T, Oralbekova, Z, Zhartybayeva M, Zhumadillayeva A, Adamova A (2015) Prospects of Development of Technologies in the Field of Robotics and the Stages of Design of Mobile Robotic Complex. In: IT convergence and security (ICITCS), 2015 5th international conference on, 24–27 Aug 2015. IEEE Catalog Number: CFP1591 W—ART, pp 1–4
3. Zhartybayeva M, Tatur M, Bairak M, Dadykin B, Mikhalkevich E (2015) Elaboration of management systems by robotic complexes materials of IV international scientific and practical internet-conference "Tendencies and prospects of development of science and education in the context of globalization" on 30–31 May 2015 Collection of Scientific Papers, 194 strmaterialy IV International Scientific and Practical Internet Conference "Tendencies and prospects of development of science and education the conditions of globalization" on 30–31 May 2015 Collection of scientific papers, p 194
4. Hanzel J, Jurišica L (2006) Comparisom of mapmaking methods for mobile robots. J Electr Eng 57(5):276–284
5. Hanzel J, Jurišica L (2004) mapmaking algorithms for mobile robot. selected topics in modelling and control, vol 4. Slovak University of Technology Press Bratislava, pp13–19
6. Deczky-Krdoss ES, Kiss BL (2004) Design and control of a 2-DOF positioning robot methods and models in automation and robotics, Miedzyzdroje, Poland
7. Mahfouz Ahmad A, Aly Ayman A, Salem Farhan A (2013) Mechatron Design Mobile Robot Syst. IJISA 5(3):23–3
8. Egerstedt M, Hu X, Stotsky A (2001) Control of mobile platforms using a virtual vehicle approach. IEEE Trans Autom Control **46**(aa):1777–1782
9. http://www.robo-hunter.com/news/kompanii-stroyashie-armiy-robotov-google
10. Instruction manual Mini tractor Belarus 132H http://avtovelomoto.by/instructions/minitractor-132-5-razdel.pdf
11. Dudek G, Jenkin M (2000) Computational principles of mobile robotics. Cambridge university press
12. Belevich AV, Lutsk VI, Odinets DN, Tatur MM (2013) Nguyen Trung Tin, Simulation control algorithms for automatic transmissions obespechpeniyu soft start transmission. Neurocomput Control Dyn Syst (2)
13. Kudryashov VB, Lapshov VS, Noskov VP, Rubcov IV Problems of robotization for military ground technics. Izvestiya SFedU. Engineering Sciences pp 42–57
14. Vőrős J (2006) On application of extended quadtrees. ATP J Plus 1:36–41

15. Vőrős J (2004) Quadtree-based path panning using potential fields. In: Proceedings 13th international workshop on robotics in Alpe-Adria-Danube region RAAD 04. June 1–6 2004. Brno University of Technology, pp 41–45
16. Vőrős J (2006) Quadtree-based representations of grid-printed data. Image Vis Comput 24:263–270
17. Duchoň F (2006) Map Building in mobile robotics. Elitech (10.5.2006, /CD/)
18. Duchoň F, Murár R (2006) Mobile robot modeling and control. ATP J Plus (2)
19. Barsukov AP (2005) Components and solutions to create robots and robotic systems, Directory Issue I:79
20. Zenkevich SL, Yushchenko A (2008) Management robots. Fundamentals of management manipulation robots. M.: Publishing House of the MSTU. NE Bauman, p 399
21. Telerik SB Concepts, models, algorithms and tools for adaptive technology of information management systems, thesis work, Kiev
22. Krasovsky AA (2008) Algorithmic foundations of optimal adaptive controllers new class // Automation and Remote Control. 2007, № 9. C. 104–116. Ladislav Jurišica, Roman Murár, Mobile robots and their subsystems, AT&P journal PLUS1 2008
23. Mohd Azizi Abdul Rahman1, Katsuhiro Mayama2, Takahiro Takasu2, Akira Yasuda2 and Makoto Mizukawa2 "Model-Driven Development of Intelligent Mobile Robot Using Systems Modeling Language (SysML)", Mobile Robots—Control Architectures, Bio-interfacing, Navigation, Multi Robot Motion Planning and Operator Training, pp 21–38

CA Based Design of Fault Detection Unit for Hierarchical Directories in Scalable CMPs

Supriti Mukherjee, Bhanu Pratap Singh, M. Chinnapureddy, Chandan Koley and Mamata Dalui

Abstract In a Chip Multiprocessors (CMPs) with large number of cores, directory size increases linearly with number of sharers. To over-come the shortcomings of flat directory structures, hierarchical directory structures are used. An insignificant fault in the sharer set representation of such a directory may introduce major inconsistency throughout the system. Therefore, the current work targets detection of any faulty recording in the sharer set representation of a hierarchical directory. The solution is developed around a special class of Cellular Automata (CA) with single length cycle attractors. The CA based design ensures low cost hardware implementation as well as high-speed operation.

Keywords Cache coherence · CMPs · Hierarchical directory · Fault detection

1 Introduction

The recent advancements in technology demand high degree of performance scaling and high speed computation. To meet these requirements, complex uniprocessors are getting replaced by multicore architecture in which multiple

---

S. Mukherjee · B.P. Singh · M. Chinnapureddy · C. Koley (✉) · M. Dalui
Department of Computer Science and Engineering, NIT, Durgapur 713209, India
e-mail: chandank375@gmail.com

S. Mukherjee
e-mail: supritimukherjee.mukherjee@gmail.com

B.P. Singh
e-mail: bhanu6715@gmail.com

M. Chinnapureddy
e-mail: cmchinnu4@gmail.com

M. Dalui
e-mail: mamata.06@gmail.com

K.R. Attele et al. (eds.), *Emerging Trends in Electrical, Communications and Information Technologies*, Lecture Notes in Electrical Engineering 394,
DOI 10.1007/978-981-10-1540-3_14

processor cores are integrated in a single chip leading to Chip Multiprocessors (CMPs).

In a typical CMPs system, L1 cache is the private cache of a processor core. L2 is shared among the cores. In a shared memory system, uncontrolled access to shared data by different processor cores may introduce inconsistencies in the data states. So, all the L1 caches are required to be kept coherent with the L2 cache, both from system performance and power efficiency point of view.

CMPs, implementing directory-based protocol, maintains a directory between the private and shared caches. Directory based coherence protocols can scale well for large number of processors in shared-memory based systems. It takes care of all cached copies of a shared data. However, the directory size grows linearly with increase in number of cores, making the implementation area inefficient.

In hierarchical directory organizations, each level of hierarchy maintains information about their lower levels. So, energy and area grow logarithmically with increasing core count. Scalable Coherence Directory (SCD) introduced in [1], is an efficient hierarchical directory structure that scales to thousands of cores efficiently keeping exact sharer representation. But, fault-free recording of sharer information between the various levels of SCD is highly required for its proper functioning. This motivates us to design a fault detection unit for detecting faults in the sharer set of SCD. The proposed design is modeled around a special class of Cellular Automata (CA), having single length cycle attractor(s). The inherent modular and scalable structure of CA ensures high scalability to the design.

2 CA Preliminaries

The n-cell Cellular Automaton (CA), an autonomous finite state machine (Fig. 1a), evolves with time and its state transition diagram consists of cyclic and non-cyclic states (Fig. 1b, '0' is the cyclic state, called attractor). The combination of present states of a CA-cell is the Min Term of a 3-variable S_{i-1}^t; S_i^t; S_{i+1}^t switching function and is referred to as the Rule Min Term (RMT). The 4-cell null-boundary 3-neighborhood (2-state per cell) CA <248, 248, 248, 248> of Fig. 1(b) is an

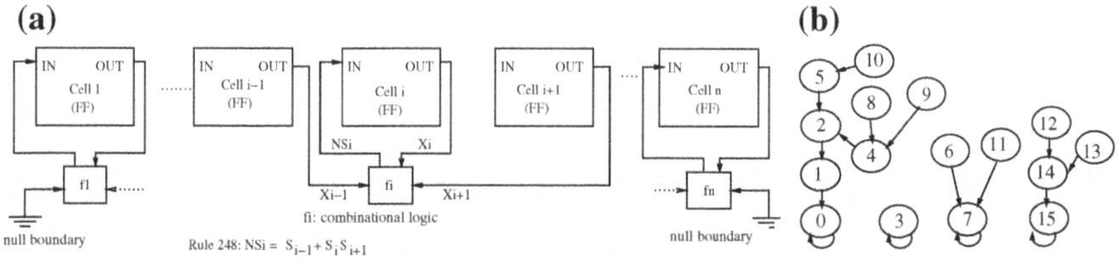

Fig. 1 An n-cell CA and state diagram of rule 248. **a** An n-cell CA. **b** State transition diagram of 4-cell irreversible CA <248, 248, 248, 248>

MACA (single length cycle multiple attractor CA) having attractors (state '0', '3', '7' and '15') of single length cycle. The next state logic (f_i) of each CA cell follows $NS_i = S_{i-1} + S_i . S_{i+1}$ (defined as rule 248).

3 Overview

In CMPs implementing flat directory, the directory size grows linearly with the increase in number of cores. To overcome the limitations of flat directory structure for large scale CMPs, hierarchical directory structure is used. Scalable Coherence Directory (SCD) is an efficient hierarchical directory structure. The SCD uses a multi-tag format and is represented using hierarchical bit-vectors.

Figure 2 shows an instance of SCD for 1024-sharer directory, where 32-bit root-bit vector shows the subset of sharers that share the line and the corresponding bits are set to 1. Each 32-bit leaf-bit vectors track individual sharers and the corresponding bits are set to 1. The first tag in a root-bit/leaf-bit vector represents the cache line address, second tag represents the type of vector ('10' for root-bit vector and '11' for leaf-bit vector). Third tag in the root bit vector gives the coherence status whereas in leaf-bit vector it gives particular leaf number. Fourth tag gives the subset of sharers in root-bit vector and in leaf-bit vector it represents individual sharers, that is, the presence bits. Each sharer performing read operation to a cache line, say B, are represented in the leaf-bit vectors by setting corresponding (presence) bits to 1. For a write operation by processor P_i on a cache line B, invalidations are sent to all sharers. Thus all the presence bits in the leaf-bit vectors are set to '0' except the one corresponding to processor P_i and in the root bit vector all presence bits corresponding to the subsets of sharers are set to '0' except the one to which processor P_i belongs. For example, let us consider a system of four cluster each having four processors. Now, if the current operation is a write by processor P_i, which is the 1st processor of cluster 2. Then the correct root-bit vector entry is '0100' and leaf-bit vector entry is '1000'. However, due to some fault in the system, a leaf-bit or root-bit vector entry for the presence bits may encounter a faulty recording. For example, more than one presence bits in the leaf-bit vector or in root-bit vector is still remaining '1' which indicates a faulty recording of sharers. Such a faulty recording may cause directory indirection as well data loss which in turn may cause huge system power consumption. Thus an explicit verification

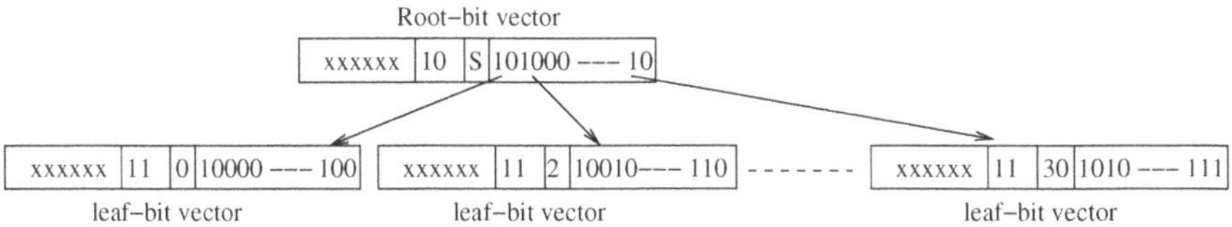

Fig. 2 Root-bit vector and leaf-bit vector of SCD

logic/fault detection unit is required to detect the presence of any faulty recording in SCD's sharer set. Conventional solutions for fault detection are not suitable for highly scalable directory structure like SCD. Therefore, we have considered a Cellular Automata (CA) based test logic for detecting faults in SCD's sharer set. On each write operation, the root-bit and leaf-bit vectors are updated. Both, root-bit and leaf-bit vectors are fed as input to the test logic for detecting the presence of faults. The CA based design of test logic is described next.

4 Proposed Solution

The proposed CA based test logic takes the root-bit vector and leaf-bit vector to detect the presence of any faulty recording in SCD's sharer set. It requires synthesis of a CA which will correctly respond to faulty ('F') and non-faulty ('NF') cases, so that the decision on 'NF' or 'F' can be taken at speed. This requires an n-cell CA for n number of sharers represented in the bit-vector of SCD. Now, at each write operation, once the n-bit root-bit and leaf-bit vectors are updated with sharer information, these vectors are fed as input to the fault detection (test) unit. An n-cell CA at the fault detection unit is configured with suitable rules and is run for $t(= n-1)$-time steps considering the presence bit in root-bit/leaf-bit vector as the initial seed. The CA settles to an attractor (final) state. The CA reaches to an attractor, say X_1, in the non-faulty case and it reaches to a different attractor, say X_2, in the presence of a fault. Thus by checking the attractor, decision on presence of fault can be taken. However, sensing of attractor is reduced to a single bit, also called the check bit. The state '0'('1') of a particular cell (corresponding to the check bit) designates the absence or presence of single or multiple fault(s) in the sharer set of SCD.

The proposed test logic first checks the root-bit vector. If there is any fault, it indicates the presence of fault. If it finds no fault in the root-bit vector, it then tests the leaf-bit vectors. If any of the leaf-bit vectors is found to be faulty, it indicates fault in the sharer set of SCD.

The CA based design of fault detection unit demands the design of a CA in which a single bit of the attractors will distinguish faulty and non-faulty cases. So, to fulfill the above requirement we require single length cycle attractor CA having at least two single length cycle attractors, one of which corresponds to faulty case and the other corresponds to non-faulty case.

For realizing the CA based test logic, we need to select appropriate CA rule (R_i) for the ith CA cell. The CA <248, 248, 248, 252> has five single length cycle attractors (Fig. 3a). The states 8, 4, 2, and 1 belonging to the attractor basin-1, and state 0 belonging to attractor basin-0, correspond to a non-faulty ('NF') recording. All other states belonging to attractor basins-3, 7 and 15, correspond to faulty ('F') cases. Now, these two classes of attractors are distinguishable at the 2nd LSB. Thus the check bit here is the 2nd LSB of attractor. It is '0' for the attractor 0 (0000) and attractor 1 (0001) and '1' for the attractors 3 (0011), 7 (0111) and 15 (1111).

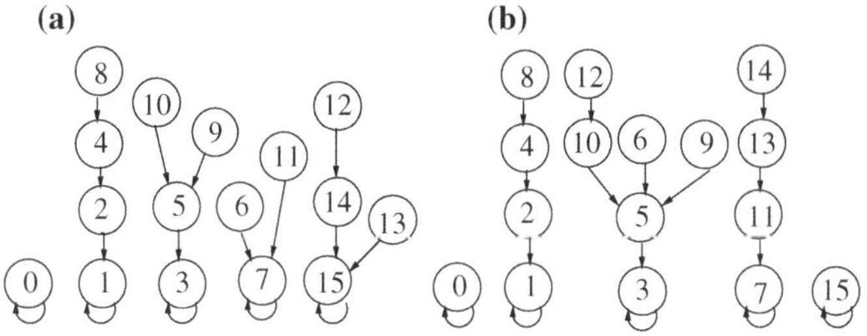

Fig. 3 State transition diagram. **a** (CA <248, 248, 248, 252>). **b** (CA <56, 184, 184, 118>)

A bit vector 0000 is considered as non-faulty, as it represents the initial state of the sharer set when there is no sharer of a block.

Observation 1: The uniform CA <248; 248; 248; 248> settles to the attractors 0, 3, 7, 15 and the uniform CA <252; 252; 252; 252> settles to the attractors 0, 1, 3, 7, 15. Attractors 0, 3, 7, 15 is common for both the uniform CA, However, 1 is an additional attractor for the uniform CA of 252.

Theorem 1 *The CA rule-vectors which are rule 248 dominated, i.e. maximum number of cells are configured with rule 248 (3 out of 4) settles to attractors same as that of uniform CA with rule 248 (i.e. attractors 0, 3, 7, 15). On the other hand, the CA rule vectors which are rule 252 dominated, settles to attractors as that of uniform CA with 252 (i.e. attractors 0, 1, 3, 7, 15).*

Theorem 2 *The CA rule vectors which settle to attractors 0, 3, 7, 15, are configured with rule 248 as the last cell rule.*

Theorem 3 *The CA rule vectors which settle to attractors 0, 1, 3, 7, 15, are configured with 252 as the last cell rule.*

Observation 2: For the CA rule vectors where two cells are configured with rule 248 and two with rule 252, the rule vectors having 248 as the last cell rule settles to the attractors 0, 3, 7, 15 and the rule vectors having 252 as the last cell rule settles to the attractors 0, 1, 3, 7, 15.

Extensive experimentation has resulted in a variety of candidate rules for the different CA cells as listed in Table 1. We can synthesize alternative CA rule vectors by choosing any one of the rule from the column 1 for the first cell, any combinations of the candidate rules from column 2 for the intermediate cells and

Table 1 Rules for cells

Rules for		
First cell	Intermediate cells	Last cell
56, 40, 24, 120, 184, 8	184, 248	118, 116, 126, 86, 246, 124
104, 168, 88, 152, 248, 72		84, 244, 94, 254, 214, 92
136, 232, 216, 200		252, 212, 222, 220

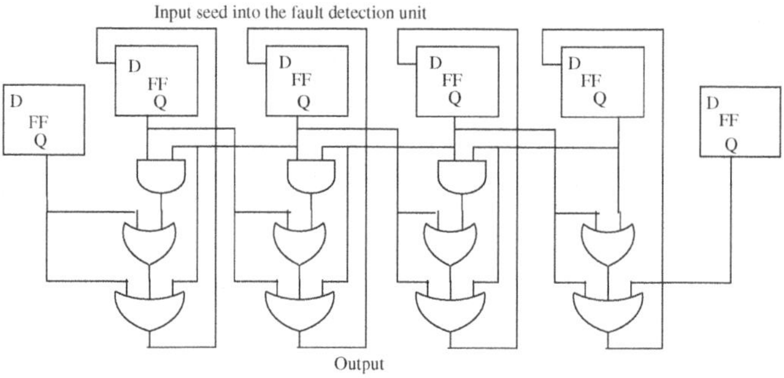

Fig. 4 Hardware realization

one candidate from the column 3 as the last cell rule. For example, for synthesizing a 4-cell CA, we choose rule 56 for 1st cell, rule 184 for the two intermediate cells and rule 118 for the last cell and the resulting CA <56, 184, 184, 118> is shown in Fig. 3b. Another alternative CA is shown in Fig. 3a.

The hardware realization of the CA based fault detection unit is shown in Fig. 4. It is designed by considering the CA <248, 248, 248, 252> at the test unit. The hardware overhead of the design is reported below.

Hardware Overhead: The researchers in [2] proposed a solution to detect fault in the coherence states of a data block. Table 2 reports the gate counts (FFs, 2 input ANDs, 2 input ORs, 3 input ORs and 2 input NANDs) and gate area requirements of our proposed solution and the solution in [2] for CMPs with 16 to 256 cores. The area is computed following the units specified in [3].

Simulation Results: The MACA based solution is experimented in Multi2sim [4], an open source simulator for multicore architecture. The architectural parameters of L1 and L2 caches used for the simulation are shown below.

L1 cache: size = 64 KB, associativity = 4, page replacement policy = LRU, block size = 64 byte, latency = 2 cycles.

L2 cache: size = 2 MB, associativity = 4, page replacement policy = LRU, block size = 64 byte, latency = 20 cycles.

Table 2 Hardware requirements for MACA based solution

No. of cores	No. of FFs	No. of ANDs	No. of ORs		Area (units)	
		(2 i/p)	(2 i/p)	(3 i/p)	MACA based solution	Solution proposed in [2]
16	18	15	16	16	271904	47824016
32	34	31	32	32	539168	78737552
64	66	63	64	64	1073696	154313872
128	130	127	128	128	2142752	298784912
256	258	255	256	256	4280864	591498384

Table 3 Simulation results for parsec benchmark suite

		Execution time (s)		
S. no.	Program	Without verification logic	With verification logic	Fault detection (%)
1	Bodytrack	1.01	92.35	100
2	Canneal	0.88	35.07	100
3	Ferret	0.27	1.42	100
4	Fluidanimate	0.81	45.78	100
5	Freqmine	0.71	25.29	100
6	Swaptions	0.76	36.87	100
7	X264	0.28	2.77	100
8	Blachscholes	0.25	1.83	100

The application programs in Parsec benchmark suit are used as the standard workloads and the simulation results are tabulated in Table 3 which indicates 100 % fault coverage in all cases.

5 Conclusion

The current work proposes a CA based effective solution for detecting fault(s) in the sharer set representation of SCD. The regular, modular and cascadable structure of CA ensures high scalability to the design. The simple and efficient realization of the solution ensures 100 % fault detection.

References

1. Sanchez D, Kozyrakis C (2012) SCD: a scalable coherence directory with flexible sharer set encoding. In: IEEE 18th international symposium on high-performance computer architecture
2. Wang H, Baldawa S, Sangireddy R (2008) Dynamic error detection for dependable cache coherency in multicore architecture. In: VLSI conference
3. Santovichu EM, Singh KJ, Lavagno L, Moon C, Saladanha A, Savoj H, Stephen PR, Murgai R, Brayton R, Sangiovanni-Vincentelli AL (1992) Sis: a system for sequential circuit Synthesis, Technical report, UCB/ERL M92/41, Electronic Research Laboratory
4. Ubal R, Jang B, Mistry P, Schaa D, Kaeli D (2012) Multi2Sim: a simulation framework for CPU-GPU computing. In: International conference on parallel architectures and compilation techniques

A New Stratified Immune Based Approach for Clustering High Dimensional Categorical Data

G. Surya Narayana, D. Vasumathi and K. Prasanna

Abstract With development in Database Technology, many existent real world applications contain outsized volumes of categorical data, which are playing an important role in data analysis and effective decision making. However, the clustering algorithms are deliberated for numerical data only, for the reason that of their similarity of measures. There is an enormous work carried on clustering categorical data with predefined similarity measure explicitly defined over categorical data. However, intricate problem with real world domain is that the feature in the data may depend on some hidden and transonic perspective, which is explicitly not in the given form of predictive features. So this poses a covenant with categorical data competently and proficiently. In this paper, a stratified immune based approach is proposed for clustering categorical data CAIS, is proposed with new similarity measure to minimize distance function. CAIS adopts an immunology based approach for effective discovery of clusters over categorical data. It selects frequently subsist nomadic feature as representative object and perform grouping into clusters with new affinity measure. CAIS is scaled to large number of attributes to minimize miscluster rate in the datasets. The extensive empirical analysis on CAIS shows that the proposed approach attains better mining efficiency on various categorical datasets and outperforms with Expectation Maximization (EM) in different settings.

Keywords Categorical data · Immune based clustering · High dimensional data · Expectation maximization

---

G. Surya Narayana (✉) · K. Prasanna
Department of CSE, AITS, Rajampet, India
e-mail: surya.aits@gmail.com

K. Prasanna
e-mail: prasanna.k642@gmail.com

D. Vasumathi
Department of CSE, JNTU, Hyderabad, India
e-mail: rochan44@gmail.com

K.R. Attele et al. (eds.), *Emerging Trends in Electrical, Communications and Information Technologies*, Lecture Notes in Electrical Engineering 394,
DOI 10.1007/978-981-10-1540-3_15

1 Introduction

In the real world the developments in information and communication technology has introduced large amount data for database analysts. The real world environment extracting knowledge is deepening with transonic characteristics, which are not represented explicitly in the form of predictive features. However, extraction of knowledge over categorical domain has not been extensively discussed [1] since categorical attributes exist in real data with peripatetic concepts. How to covenant with the categorical data competently and proficiently is the core topic in clustering categorical data [2, 3].

Clustering is mainly focused to discover dense and sparse domain in the data by partitioning the dataset into disjoint groups called. Clusters are useful for sympathetic and enhanced understanding [4, 5]. The clustering problem is defined as an objective function which minimizes the similarities among objects in the same group while maximizes the dissimilar with objects in another group according to the predefined similarity measures [3].

At present two types of data domains are available in the real life applications. The numerical domain is continuous or persistent, and can be interminable. On the other hand, categorical data domain takes only discrete values in a finite number. An intrinsic feature of categorical data is that it will have a diminutive number of attribute values. In the literature, STIRR [6], CACTUS [7], ROCK [8] and Squeezer [9] are the few clustering algorithms designed to covenant with categorical data values. STIRR is used for categorical data. It is iterated until a fixed point of assigning basins for each cluster. In ROCK it adopts an agglomerative hierarchical clustering algorithm, to minimize a cost function defined as the number of links between samples. It also uses nearest neighbor search strategy to group common samples in the dataset. In CACTUS, it provides a fresh view of clusters with categorical data values. CACTUS uses a vector space to compose data objects into clusters. Several heuristic methods were proposed on clustering categorical data. The main memory utilization and cluster accuracy are heavily affecting because of the database size and large number of attributes. The previously stated algorithms can get entirely fair clustering impacts with numerical data. This deficiency is adequate to accept with the aid of researchers, because of the intricacy in similarity measurement in categorical values.

To do this, a fast stratified immune based algorithm clustering algorithm called CAIS for clustering categorical data is described. The vital logic behind CAIS is to determine affinity threshold measures over the representative antibodies identification (RAI) instead of calculating among data objects. So in a single scan of the dataset the affinity information is plotted over matrix which easily fits in main memory, and clusters can be constructed efficiently, which results in significant improvement over the performance than the earlier reported algorithms. The framework of CAIS is organized as

1. Calculating Representative Antibody Identification for centroids in the dataset.
2. Adopting an immune based algorithm CAIS for clustering categorical data.

3. Extending CAIS to discover miscluster rate in datasets consisting of a large number of attributes.
4. Performing extensive experimental study of CAIS, evaluating and comparing its performance with earlier reported algorithms on synthetic and real datasets.

The rest of the paper is prepared as follows. Section 2 defines basic definitions, and related work used in our proposed work. Section 3 presents the framework of CAIS. Section 4 describes the empirical evaluation of the proposed work and Sect. 5 will present conclusion.

2 Basic Preliminaries

In this section, a formal description of the proposed work is characterized. Section 2.1 defines the basic definitions and notations and Sect. 2.2. Defines problem descriptions associated with categorical data is presented.

2.1 Definitions

Before introducing CAIS, some imperative definitions and notations are used in the proposed work is given in the following section.

Let A_1, A_2, … A_m, be 'm' attributes describing a high dimensional data space 'D_p'.

Definition 1 Cluster is a collection of data objects [7] which satisfies

(a) Data object are similar to each other with in the same cluster.
(b) Data objects from diverse clusters which are dissimilar from each other.

Definition 2: Categorical Attributes Let A_1, A_2, … A_m be m attributes describing a space, D_p and $D(A_i)$, the domains of the attributes. A domain high dimensional data space of $D_p(A_i)$ contains finite and unordered then it is defined as categorical e.g., for any α, β of D_p (A_i), if $\alpha = \beta$ or $\alpha \neq \beta$ then D_p (A_i), is entitled as a categorical data.

Definition 3: Distance of Categorical values Let a_1, a_2 indicates two objects and a_2 belongs to the clusters C_i. The distance between categorical value [10] is denoted by $\Phi(a_1, a_2)$ from a_1 to a_2 is defined as below:

$$\Phi(a_1, a_2) = \sum_{i=1}^{m} \varnothing(a1, i{:}\, a2, i);$$

Where

$$\varnothing(a1, i, a2, i) = \begin{cases} 1 - f_r(a_{j,i} = a_{1,i}/c_i), & (a_{1,i} = a_{2,i}) \\ 1, & (a_{1,i} \neq a_{2,i}) \end{cases}$$

Definition 4 Let a1, a2 $\in D_p$, the similarity ℓ^j(a1, a2) between a1 and a2 with respect to Aj(j $\neq$ i) is [7] defined as follows.

$$\ell^j(a1, a2) = |\{x \in D_p : \Phi(a_1, x) > 0 \text{ and } \Phi(a_2, x) > 0\}|$$

2.2 *Problem Statement*

Consider D is a dataset with N instances and let A = $\{A_1, A_2 \ldots A_m\}$. where A_j is the jth categorical data attribute $1 \leq j \leq m$. A = {Ai}, where i = 1 to N, in which each instance $A_i = \{A_1, A_2 \ldots A_m\}$ has m features. Grouping these instances into partitions Ᵽ with a predefined number of subsets ‘k’ of clusters i.e. $Ᵽ_j = \{Ᵽ_j^1, Ᵽ_j^2, \ldots, Ᵽ_j^N\}$ and D_p is into several continuous subsets $Ś^t$ where $Ś^t$ = N. The main objective is to perform clustering on these dataset and consider the affinities between $Ś^t$ and $Ś^{t+1}$.

3 Stratified Immune Based Clustering on Categorical Data: CAIS

As described in previous section, choosing k value for number of initial clusters was not done properly in the earlier reported algorithms. The existing algorithms will get stuck in producing cluster accuracy. This problem restricts the usage clustering in many applications. Earlier maximum work was presented in focusing clustering over categorical data with predefined assumption about initial clusters. The existing algorithms still suffer with problem of representative clusters and accuracy.

Now this section, CAIS the two-phase clustering algorithm is depicted. The motivational idea behind CAIS is that an affinity of the complete dataset is adequate to compute a set of “antibodies” clusters which can be collaborated to discover the set of clusters. CAIS consists of two phases: *Representative Antibody identification (RAI)* and *clustering with similarity measures.* In the RAI phase, the antibody information from the dataset is evaluated as a set of representative antibodies. In the clustering phase, using the compute RAI information, a set new affinity measures are evaluated to discover a set of entrant clusters. The CAIS process is illustrated with a hypothetical example used throughout the paper and presents the successive phases in the algorithm.

3.1 Representative Antibody Identification (RAI)

The basic initiative behind RAI is to characterize the distribution of the attribute values called representative antibodies as a cluster. To discover RAI measures of each antibody, in a cluster it is evaluated as:

1. If the frequency of an antibody is high in the cluster then RAI is important.
2. If the antibody is appears in this cluster rather than in other cluster.

The RAI of antibody is defined as attribute name as follows:

$$\mathrm{R}(\mathrm{Ci}, \mathrm{I}_{ir}) = \frac{|I_{ir}|}{m_i} * f(I_r)$$

$$f(I_r) = 1 - \frac{-1}{\log k} * \sum_{i=1}^{k} P(I_{ir}).\log(P(I_r))$$

and

$$P(I_r) = \frac{|I_{ir}|}{|I_{zr}|}$$

Where m_i = number of data points I_{ir} = representative value of C_i $|I_{ir}|$ = number of occurrences of I_{ir}

$\mathrm{R}(\mathrm{C_i}, \mathrm{I_r})$ stand for the RAI of node I, in cluster Ci. $f(I_r)$ the frequency. The procedure of identifying RAI nodes is explained with the following example shown in Fig. 1.

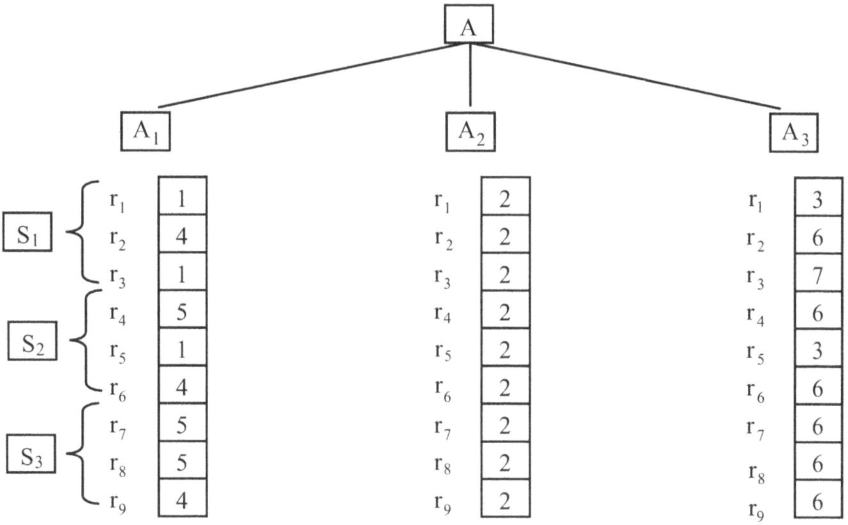

Fig. 1 Example dataset

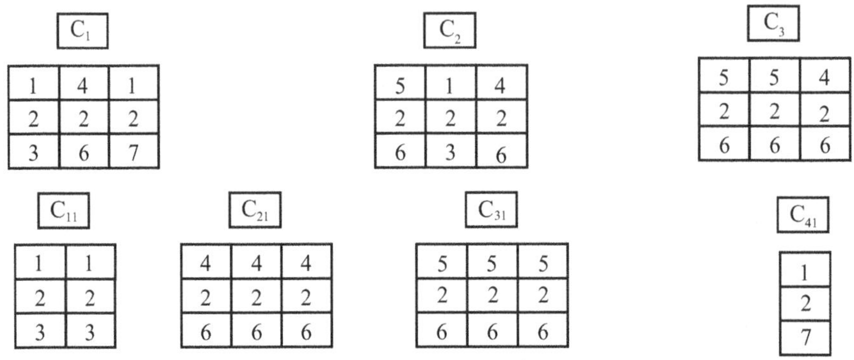

Fig. 2 The RAI of clusters

Figure 1 shows the data set with 9 data points, with three attributes and subset size = 3. The representative value of I_{ir} in C_i, and the second concept, frequency function $f(I_{ir})$ and $R(C_i, I_{ir})$ both in range of [0, 1]. The first subset S1 is clustered as shown in C_1, which further explored to two clusters, C_{11} and C_{12}, as shown in Fig. 2. The calculated RAI of C_1 is 0.66. Same procedure is applied on C_2 and C_3. The calculated RAI of C_2 is 1 and C_3 is 1. For the second iteration C_1 and C_2 is split into C_{11} and C_{21}. The C_2 and C_3 is split into C_{31} and C_{41} as shown in Fig. 2.

Consider the dataset in Fig. 2. Cluster C_{11} contains three data points. The object A1 = 1 occurs 2 times in C_{11} and does not occur in C_{21}. The frequency function f (Ci, I_{ir}) = 1. Therefore the RAI of A1 in C_{11} is 0.66 and in Cluster C_{21}, it is 0. The frequency of A2 = 2 is 0.029and the RAI of A2 in C_{21} is 0.66, in C_{31} is 0.66 and in C_{41} is 0.029. Finally the RAI of all clusters is given in the following Table 1. From the Fig. 2 the cluster C_{41} has lowest RAI on A1, A2 and A3. Therefore Cluster C_{41} is dose not considered for future iterations. Finally the dataset produces only 3 clusters with high RAI values.

Table 1 Performance of CAIS on synthetic dataset

Dataset name	No of attributes	No of antibodies (RAI)	Min Affinity threshold	No of instances = 100			
				Precision	Recall	F-Measure	Accuracy
Synthetic1	100	37	0.2	83.1	80.1	77.8	80.0
Synthetic2	200	38	0.2	69.0	69.7	66.6	69
Synthetic3	300	48	0.2	80.7	81.0	79.1	80
Synthetic4	400	47	0.2	78.6	77.0	75.8	77

3.2 Clustering with Similarity Measure

The basic reason to adopt immune system, it will easily groups cells into clusters improve the resistance in human body. Similarly to improve the accuracy of the clustering is need to distribute the data objects into discrete and disjoint clusters. To do this, in our proposed approach uses modified immune methodology with new similarity measure to define the affinities among the data objects. The procedure of discovering candidate clusters over categorical data is given as follows.

Step1 input high dimensional dataset 'X' of size 'N' such that each data object as an antigen in the dataset

Step2 Generate the 'K' initial antibodies using Representative Antibody Identification (RAI)

Step3 Identification of antigens using the affinity value between antigen and antibodies. For each categorical attribute in the given dataset, the affinity values are constructed as

$$A_{ij}(d) = \sqrt{1 - \frac{(x_{id} - x_{jd})^2}{\|x_i - x_j\|^2}}$$

Let a Database 'D' with 'N' instances and 'd' antibodies. Let X_i, X_j be the two instances of same antibody. If $(x_{id} - x_{jd})^2$ is dominates $\|x_i - x_j\|^2$ then the affinity $A_{ij}(d)$ is close to '0' and having both X_i and X_j are similar. Sort the affinity values according to the similarity measure.

Step4 Cluster each antigen to an antibody

Repeat step 3 and step4 and update the affinity matrix according to occurring frequencies of each antigen value in its corresponding cluster.

Step5 for each cluster

Step 5.1 the antigen value, evolved antibodies to get a new antibody So that the sum of affinity between antibody and antigen is maximum The new antibody can be evolved as follows. let A1 and A2 denote the two antibodies and A2 belongs to cluster C_i, then the distance function is defined as follows: $\Phi(a_1, a_2) = \sum_{i=1}^{m} \varnothing(a1, i, a2, i)$.

Step 5.2 for each affinity pair, choose the highest affinity value as best antibody and eliminate others whose Affinity Threshold (AT) value is less than pre supplied threshold

Step 5.3 calculate miscluster rate. It is formulated as follows:

miscluster rate = *Nm/N*

Nm denotes the number of categorical objects and *N* denotes the size of the data set.

Step6 Check for stop condition

4 Performance Evaluation

In this section, the detailed evaluation of scalability and accuracy of CAIS on synthetic and real datasets is described. The proposed approach is evaluated based on elapsed time and miscluster rate and also different performance measures are used to evaluate the accuracy of the clustering process. The evaluations are mainly focused on performance measures of the proposed algorithm. Experiments were conducted on a synthetic dataset and a benchmark dataset from UCI machine learning repository.

4.1 CAIS Performance Evaluation on Synthetic Dataset

In this section, we present CAIS evaluation on synthetic dataset. The synthetic datasets were generated using the clustering data generator. The detailed evaluation on synthetic dataset is shown in the following Table 1.

From above Table 1 it is observed that the F-measure and accuracy values are high when number of antibodies is low. As the number of RAI are increased the f-measure and accuracy are increased gradually. The Fig. 3 shows performance on a dataset with 50,000 samples. The time values are measured in milliseconds. From the Fig. 3 it is observed that CAIS reduces to half of the elapsed time with EM.

The Table 2 shows the impact of RAI on number of clusters and dimensionality on a synthetic dataset of size 50,000 samples. Synthetic dataset with varied

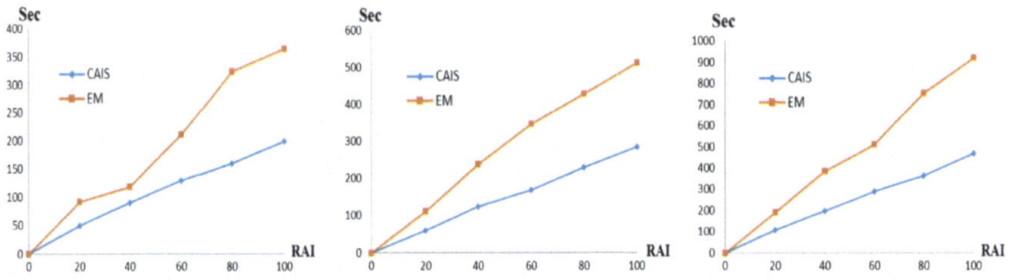

Fig. 3 Elapsed time for discovering k = 20 with varying RAI

Table 2 Elapsed time for discovering number of clusters (k) with fixed dimensionality **(n = 20)**

No of antibodies (RAI)	Time for n = 20 in ms	Time for n = 30 in ms	Time for n = 50 in ms
0	2	2	4
20	54	69	104
40	89	117	186
60	110	164	270
80	160	215	340
100	194	281	415

dimensionality and varying clusters are shown in above Table 2. The elapsed time values are measured in seconds. The execution of CAIS is still faster than EM[11] regardless of how many RAI are identified using CAIS.

4.2 CAIS Performance of Vote Dataset

In this section, we discuss CAIS on vote dataset taken from UCI. The results from vote show that CAIS finds intuitively meaningful clusters from dataset. Table 3 depicts the performance on vote dataset. From the table it is observed that clusters increases the F-measure and accuracy values are reduced when number of clusters increases. This is because of selection of efficient and scalable RAI objects and affinity threshold. The rate of miscluster increases slightly with the increase of number of clusters and RAI objects.

Table 4 shows the execution time of CAIS&EM on vote dataset. From the Table observations, it is noticed that CAIS takes very less time in discovering number of clusters compared to EM. Hence EM algorithm takes more time to cluster categorical data in two phases. During Expectation phase and Maximization phase it is also predicted that an average CAIS takes 50 % less time compared to EM.

Table 3 Performance on vote dataset

No of instances	No of RAI	Affinity threshold	K	Precision	Recall	F-measure	Accuracy	Miscluster rate
3944	232	0.4	3	97.7	94	95.8	94.9	0.045
			4	97	95.9	96.4	95.63	0.045
			5	98	91.8	94.8	93.7931	0.05
			6	96.9	94.4	95.6	94.7126	0.051
			8	97.6	93.3	95.4	94.4882	0.048
			10	97.3	92.9	95.0	94.023	0.053
			15	98	91	94.4	93.333	0.053
			20	97.2	91	94.0	92.8736	0.06

Table 4 Elapsed time on vote

No of clusters	Elapsed time on CAIS in sec	Elapsed time on EM in sec
3	0.22	2
4	0.19	2
5	0.20	4
6	0.23	5
8	0.17	7
10	0.19	8
15	0.19	11
20	0.17	14

5 Conclusion

In this, the proposed a stratified immune based unsupervised clustering of categorical data based on a new similarity measure and representative antibody identification (RAI). For each attribute or antigen, the representative object is selected as distribution of frequency function defined over the attributes and on the basis of adoption of representative clusters method and the knowledge of the true number of clusters. The proposed CAIS is compared with EM in several settings. The empirical analysis on CAIS shows that it significantly outperforms with EM algorithm. The major strength of the proposed work is it evaluates the miscluster rate over the clustering on categorical data. From our experiments it is observed that the miscluster rate is in between 0.05–0.07 up to 20 clusters. The elapsed time is greatly reduced when it is scaled with larger size attributes and number of clusters and outperforms better as compared with EM algorithm.

References

1. Nasraoui O, Soliman M, Saka E, Badia A, Germain R (2008) A web usage mining framework for mining evolving user profiles in dynamic web sites. IEEE Trans Knowl Data Eng 20 (2):202–215
2. Anderberg MR (1973) Cluster analysis for applications. Academic Press
3. Han J, Kamber M (2001) Data mining: concepts and techniques. Morgan Kaufmann Publishers, San Francisco
4. Jain A, Dubes R (1998) Algorithms for clustering data. Prentice Hall
5. Jain AK, Murty MN, Flynn PJ (1999) Data clustering: a review. ACM Computing Surveys
6. Gibson D, Jon K, Prabhakar R (1998) Clustering categorical data: an approach based on dynamical systems. In: Proceedings of the 24th international conference on very large databases, pp 311–323 (and Squeezer [9])
7. Ganti V, Gehrke J, Ramakrishnan R (1999) CACTUS-clustering categorical data using summaries. In: Proceedings of 1999 international conference on knowledge discovery and data mining, pp 73–83
8. Guha S, Rastogi R, Shim K (1999) ROCK: a robust clustering algorithm for categorical attributes. In: Proceedings of 1999 international conference on data engineering, pp 512–521
9. He Z, Xu X, Deng S (2002) Squeezer: An efficient algorithm for clustering categorical data. J Comput Sci Technol 5:611–624
10. Zhendong P, Jiafu T, Mu L (2006) An immune-based clustering algorithm for large data sets with categorical values. World J Model Simul 2(1):28–35
11. Dempster AP, Laird NM, Rubin DB (1996) Maximum likelihood from incomplete data via the EM algorithm. J R Stat Soc

Multiclass SVM Classifier with Named Entity Recognition for Scheduling Workflows in Cloud

Jyothi Bellary and E. Keshava Reddy

Abstract Present Internet based applications are generating huge data due to demands of various scenarios. This gives several challenges in analyzing data and performances of the cloud based systems. In order to represent the named entity recognition problem which is the challenging task for the cloud environment, we present a new model called as multi class SVM (Support Vector Machine) for workflow scheduling in cloud. This workflow scheduling provides a framework for scheduling the entity identification with multiclass SVM classifier. The algorithm for the scheduling of resources in cloud called as improved allocation, which continuously and vigorously reallocates multiple types of named entities to the cloud resources to fulfill the cost and performance requirements. Experimental results shown that the system can handle resources effectively, time and cost is optimized.

1 Introduction

The word Named Entity, now extensively used in NLP (Natural Language Processing), was used for the 6th Message Understanding Conference (MUC-6). At that period, MUC was focused on named entity projects where it is designed for company actions and protection related actions are recovered from the semantic relations, such as information document articles. In essential with this people recognized that the essentiality of acknowledging details like titles of individuals, locations and company titles and number expression like time, date, percent and money expression. Determining sources and acknowledging the writing is essential sub projects of details recovery known as Named Entity Recognition (NER), can

J. Bellary (✉)
A Department of Computer Science & Engineering,
Aditya College of Engineering, Madanapalle, India
e-mail: jyothibellary@gmail.com

E.K. Reddy
Department of Mathematics, JNTUA College of Engineering,
JNTUA, Ananthapuramu, India

K.R. Attele et al. (eds.), *Emerging Trends in Electrical, Communications and Information Technologies*, Lecture Notes in Electrical Engineering 394,
DOI 10.1007/978-981-10-1540-3_16

label the data automatically with high accuracy. The computer needs to know how to recognize a small text with having semantic principles in order to make a correct prediction. Thus making an exact relation between the entities is a typical task for a natural language understanding applications. In this paper we are focused on recognizing entities with specified techniques.

A Monitored learning strategy using Support Vector Devices and high perspective features is used to recognize the named entities. Single-Class and Multi-Class category efficiency and outcomes are mentioned. The strategy used removes prior language information such as part-of-speech or noun term labeling thereby enabling for its usefulness across 'languages'. No domain-specific information is involved. The initial outcomes are much like those acquired using more complicated techniques despite problems experienced due to storage and computational power restrictions. The Multi-Class $F\beta = 1$ ranking for the right border recognition is 79.3 %, for the left border recognition is 69.9 % and the ranking for the semantic blend category is 63.5 %. In the above mentioned approach the memory and the computational power is limited so we didn't achieved expected results then here we are presenting a new system for scheduling the resources and computational power called as work flow scheduling which can be performed in the cloud environment.

There has been various types of workflow scheduling algorithms exists in cloud computing system. Most of them can be applied in the cloud environment with suitable verifications. The main advantage of job scheduling algorithm is to achieve a high performance computing and the best system throughput. Traditional job scheduling algorithms are not able to provide scheduling in the cloud environments. Scheduling process in cloud can be generalized into three stages namely, (1) Resource discovering and filtering: Datacenter Broker discovers the resources present in the network system and collects status information related to them. (2) Resource selection: Target resource is selected based on certain parameters of task and resource. This is deciding stage. (3) Task submission: Task is submitted to resource selected.

Workflow scheduling deals with the automation of procedures whereby the data and files are transferred between the participants based on the different set of rules to fulfill the overall goal. A workflow management system executes defines and manages the computing resources, which should have the capability to deal with the dynamic applications and process the multiple tasks to the distributed resources. A workflow is poised by connecting numerous tasks according to their dependencies.

We considered a Multi class SVM (Support Vector Machine) based prediction model to design the data intensive tasks. Most likely we formulate the scheduling problem in cloud and proposed a new algorithm called as Improved Allocation, to extend the allocation resources capability in cloud environment. The experimental results claimed that the algorithm performs better utilization accordance with resource utilization and power compared to previous techniques which is employed.

The respite of this paper is well thought-out as follows: Sect. 2 provides an introduction to the structural design of the work-flow management system. Section 3 provides the suggested strategy for improved allocation. Section 4 analyzes our methods for a real-world datasets for evaluating relation identification. We evaluate our strategy with simulation setup in Sect. 5, and lastly we determine in Sect. 6.

2 System Model

This section deals with the framework for multi-class SVM classifier with work flow scheduling.

The workflow system is depicted in Fig. 1. The top layer of the work flow contains the user applications to define the tasks to the middle ware which is present in the architecture. The application specific activities are performed by the user applications. It gives the input values to the named entity recognition.

2.1 Named Entity Recognition

In the NER we use the Conditional Random Fields (CFR) to perform the classification in our framework. These CRFs are undirected graphical models trained to increase the probability of a label sequences given to the corresponding input sequences. Let I, $I = i_1 \ldots i_N$ be the input sequence, and L, $L = l_1 \ldots l_N$ be the label sequence for the input sequence. The probability for the L to I is given as:

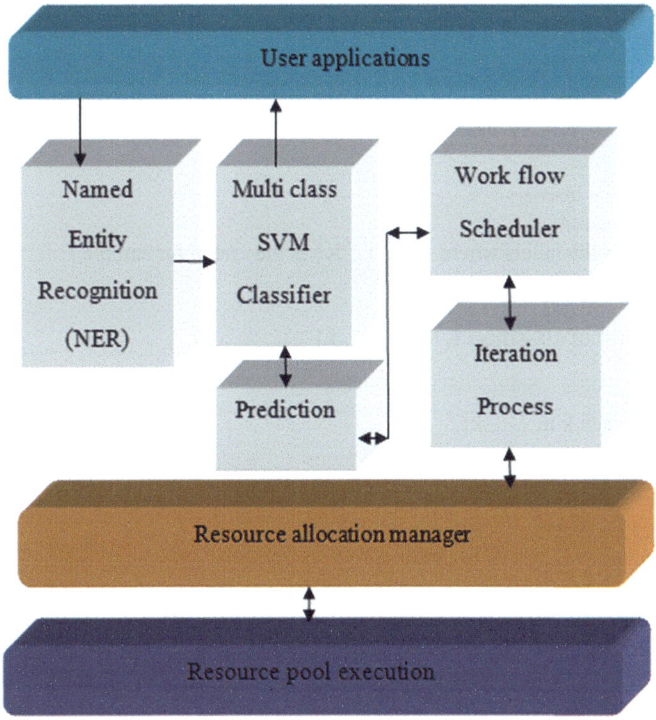

Fig. 1 Model for MSVM workflow scheduling in cloud

$$p(L|I) = \frac{1}{Z(I)} \exp\left(\sum_{n=1}^{N} \sum_{k} \lambda_k f_k(l_{n-1}, l_n, I, n) \right) \tag{1}$$

Where normalization term is coined as Z (I), fk is a feature function; it takes a binary value, and k is a leaning weight associated with the feature fk. The log likelihood parameter is given as M which can be represented as

$$M = \sum_x \log p(L_x | I_x) - \sum_k \frac{\lambda_k^2}{2\sigma_k^2} \tag{2}$$

Where smoothening parameter for f_k is defined as. The penalty term, used for regularization, basically imposes a prior distribution on the parameters.

2.2 Multi Class SVM Classifier

The Multi class SVM classifier classifies the instances into two or more binary classes. This classifier uses one-Vs-all strategy to distinguish that class form all other classes. Prediction is performed on binary classifiers and choosing the prediction with high confidence rate.

The training algorithm for the binary classifier is given in Algorithm 1.

Algorithm: 1 Binary classifier Training Algorithm

Input: C is a learner
I is a sample
L is the labels where $L_n \in \{1...K\}$ is the label for sample I

Output:
The list of classifiers f_k for K {1…k}

Procedure:
For each k in {1…k}

Define a new label vector $L'_n = 1$ where $L_i = k, 0$ (or -1)
Else where
Apply C to I, L' to obtain f_k.

Making decisions from all classifiers to the unseen sample I and predicting label L, the corresponding highest confidence is given as

$$L^* = \arg\max_{k \in 1..k} p(L|I)$$

The prediction layer collects the confidence score and gives the results to the work flow scheduler in cloud environment. The work flow scheduler connects to the scheduling process for the execution of applications.

2.3 Workflow Scheduler

This layer deals with actual scheduling, predictive analysis, iteration control, and dispatching of workflow tasks. The work flow scheduler to process the complex analytical workflows extends the workflow engine by adding the adaptive computational methods. The task dispatcher dispatches the tasks which are received from the work flow engine to the concerned resources selected by the scheduler. In each step the task execution time, status and VM type will be send back to the system for further analysis. Prediction model trains the data set for each step to execute the data processing tasks. Hence the system from the each step will be trained better.

The resource allocation manager contacts with the physical infrastructure which has facilities like adaptive resource allocation and best resource allocation to meet the user cost and time requirements. By looking the performance and characteristics of tasks the system can adjust its behavior accordingly and improves the resource allocation periodically. The job of the scheduler in the workflow is to assign all the workflow tasks to cloud resources. The resource pool execution layer contains the virtual machines which executes the tasks which are allocated to them. With this architecture, we plan to take an optimal schedule to achieve the minimum economic cost and completion time. The following section deal with improved allocation for mapping the tasks in distributed heterogeneous resources which was discussed earlier.

3 Improved Allocation

We proposed a method that extends the current ability of the workflow management. It is designed for the workflow analytical applications to in which analytical applications are periodically repeated. In order to improve the performance and financial cost for such applications, the workflow engine analyzes historical data, produces efficiency forecast for each process, and maximizes accordingly. Hence the performance of running the analytical applications can be consistently improved by adjusting system configurations to achieve the user requirements. To run the information in cloud, we believe there is on demand source provisioning and various types of VMs have different efficiency and prices. The work-flow scheduler decides the appropriate amount of resources and designates projects to appropriate resources to meet up with user specifications. Clouds add extra complexness to the work-flow scheduler because the efficiency duration of projects clearly differs when implemented on different resources. Therefore, the cloud work-flow scheduler needs to spend projects and to find an appropriate bargain between budget expenses and efficiency speedup. Our work-flow management system provides a new interesting procedure and new arranging criteria to help users spend appropriate resources and complete their programs with less completion time and economic cost.

Algorithm: 2 Workflow scheduling by improving VMs and combining Tasks (Improved Allocation algorithm)

Input: Assigning tasks S to the workflow system
Output: VM allocation and Task distribution
Step 1: Identify the time span T of applications and for some applications the time span is constant.
If time span T is not assigned then
While S_{max} -> new task do
For all V_i -> VMs sorted by type do
Step 2: Calculate Utility function U_{K+1} with the equation 3
If $U_{K+1} \leq 0$ then
Improve the VM for S_{max} to new type V_i
End if
End for
End while
$T = S_{max}$
End if

In the algorithm 2 how the resources will be allocated to each task can be calculated using an improved version of allocation.

This algorithm contains two steps. In step 1 identifies the time span of each application which is allocated to the work flow system. For some applications the time span is constant. If the time span is not assigned to any application allocate the max time span to the new task and allocate the respected VM based on their Smax value.

In step 2 calculate the utility function UK + 1 with the following equation.

$$U_{k+1} = (K-1).T^{'}.q_k + T^{'}.q_{k+1} - K.q_k.T \tag{3}$$

Where qk is the price of VM Vk, Vi is the improved type of VM Vi, K is the number of tasks, T is the time span of the longest task before improving the VM type Vi to Vi + 1, T' is the time span of the longest tasks after improving. In our case, minor utility indicates the additional benefit or price that a consumer originates from buying an improved unit of reasoning service. U $\leq$ 0 indicates there is no price rise, and we can securely update the VM for the lengthiest process consistently until U > 0.

4 Experimental Results Evaluation

The assessment of entity relation extraction relies on the characteristics of the technique used (supervised or unsupervised) and the type of dataset used. First, we temporarily explain some of the datasets that are available for entity relation extraction process. Here we are selected several data sets namely from MUC (Message Understanding Conference). The datasets listed are 1999 IE, MUC-2, MUC-3, MUC-6, CoNill-03, Met2 and Cora. Our algorithm is tested over different real world data sets. The data set size and dimensions are of variable. The parameters precession, recall and f-score in English language is given in Table 1.

These data sets can be classified and submitted to the work flow scheduling algorithm which is called as improved allocation.

4.1 Evaluating in Supervised Process

In the supervised process establishing, relation extraction is indicated as a category task and hence, metrics like Precision, Recall and F-score are used for efficiency assessment. These metrics are described as follows:

$$\text{Precision}\, P = X/Y \quad (4)$$

$$\text{Recall}\, R = X/Z \quad (5)$$

$$F - \text{score} = 2PR/P + R \quad (6)$$

Here X is the number of correctly extracted entity relations; Y is the Total number of extracted entity relations, Z is the actual number of extracted entity relations.

Table 1 Performance results of multi class SVM algorithm with different datasets

English	Precision (%)	Recall (%)	F-score
1999 IE	88.99	88.54	88.76 ± 0.7
MUC-2	84.29	85.50	84.89 ± 0.9
MUC-3	82.02	81.39	81.70 ± 0.9
MUC-6	83.87	63.71	72.41 ± 1.3
CoNill-03	76.33	80.17	78.20 ± 1.0
Met2	75.20	59.35	66.34 ± 1.3
Cora	85.93	86.21	86.07 ± 0.8

4.2 Evaluating in Semi-supervised Process

In the lack of labeled test data, analyzing semi-supervised process is a little bit different procedure though the actual analytics remain the same (Precision, Recall and F-score). Semi-supervised process of relation extraction is typically used on considerable quantities of information resulting often in the finding of a huge variety of new patterns and relations. Therefore, getting an actual precision and recall is challenging. Since the real variety of entity relations are challenging to obtain from considerable quantities of information, it is challenging to calculate recall to evaluate semi-supervised process.

5 Experimental Setup

The simulation results setup has carried in a simulator named Cloudsim. Cloudsim is a java based work environment in which it is used for cloud application development. The improved allocation algorithm has set up in a simulation environment. The details of experimental process and the simulation results have been discussed below.

5.1 Simulation Setup

By taking the requests which is generated by the multiclass SVM classifier by using prediction function the tasks which has been divided we termed these tasks as jobs in cloud terminology.

The software which is used for to implement workflow scheduling is cloudsim V3.0.2.

(a) Virtual Machine
    Virtual Machine is a piece of software which is layered on the hardware that executes programs on the software. The configuration of Virtual Machine to perform the workflow scheduling is given in Table 2.
(b) Cloudlet
    Cloudlet will work as input job where we can load the tasks, it contains encapsulated information. Each and every job contains a cloud ID.

To experimentally implement the work flow scheduling by using improved allocation algorithm we are going to use both FCFS scheduling policy and Round Robin scheduling policy. Here we are using the constant VMs count which is taken as 50, 100 and 150. The experimental results for FCFS scheduling are shown in Table 3.

Table 2 Configuration of VM

Configuration	VM
RAM	512
No. of processors	1
MIPS	250
Band width	1000
Storage space	10000 MB
Processor of type	Xen

Table 3 FCFS workflow scheduling using improved allocation

No. of cloudlets	No. of VMs	Execution time using proposed approach (sec)	Average response time (ms)
200	50	286.21	317.40
	100	246.79	247.15
	150	201.16	220.76
400	50	529.19	415.32
	100	465.76	355.25
	150	379.60	317.50
800	50	1091.47	625.77
	100	961.59	513.24
	150	843.28	435.14
1500	50	1918.93	1172.65
	100	1714.67	1081.48
	150	1587.26	974.67

FCFS scheduling is a first come first serve basis in the mean time the task which is arrived first can be allocated to the VM. The process continuous until the tasks has been completed. The priority of the task is not taken in to consideration.

In Fig. 2 it shows the comparison graph of execution time and no. of cloudlets which is participated in the workflow scheduling. It defines that execution time is increased automatically when the no of cloud lets increases. The average response time is also shown in Fig. 3. It can variable when we are using random VMs.

The experimental results of Round Robin scheduling policy are shown in Table 4. The Round Robin is similar to FCFS, but it has some time quantum for each and every task if the task is not finished within the time the process acts like FCFS and it swaps to the waiting queue.

In Fig. 4 it shows that the comparison of execution time in round robin with the no. of cloudlets. Here the round robin acts as efficient scheduling policy when compared to FCFS. The response time of round robin is smaller when compared to FCFS it is shown in Fig. 5.

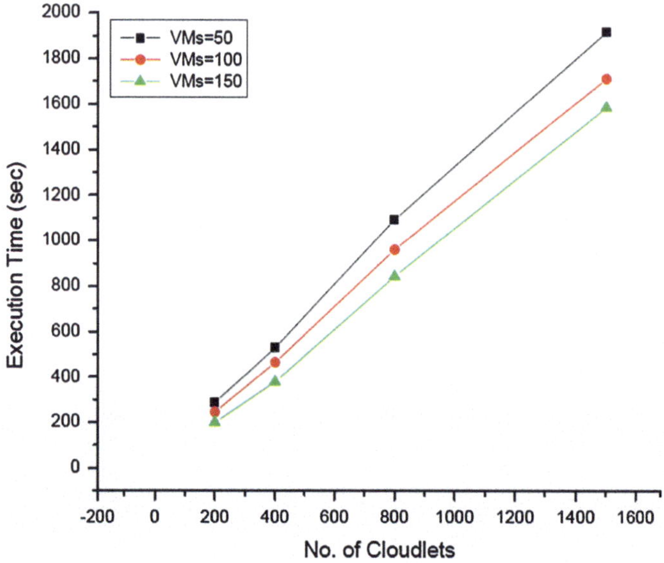

Fig. 2 Comparison of execution time with No. of cloudlets in FCFS

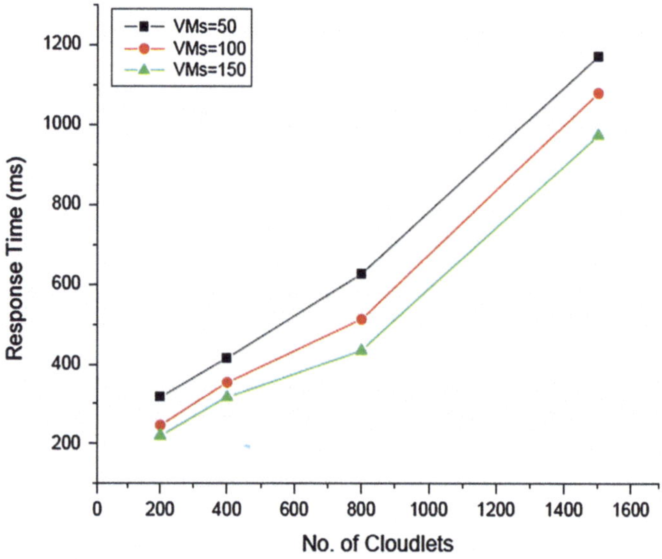

Fig. 3 Average response time in FCFS

Table 4 Round Robin workflow scheduling using improved allocation

No. of cloudlets	No. of VMs	Execution time using proposed approach (sec)	Average response time (ms)
200	50	284.78	291.54
	100	234.50	265.43
	150	194.32	250.64
400	50	528.40	367.23
	100	435.76	327.58
	150	345.27	312.54
800	50	1090.34	541.92
	100	923.40	513.86
	150	821.49	489.30
1500	50	1917.78	996.45
	100	1634.26	873.28
	150	1435.31	825.71

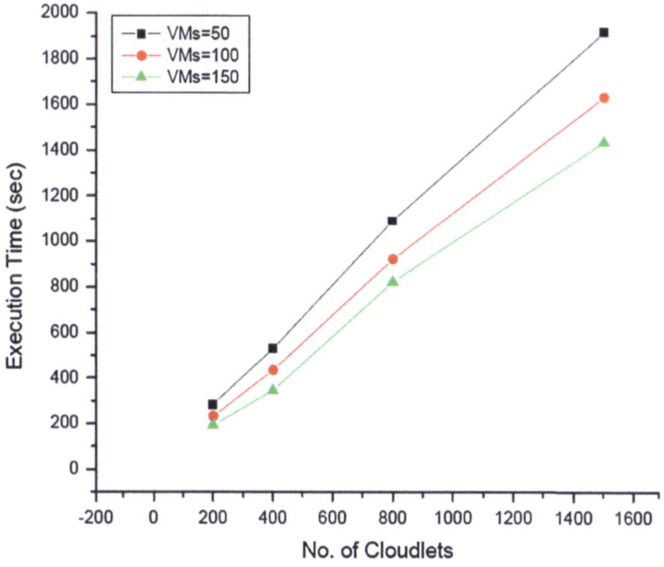

Fig. 4 Comparison of execution time with No. of cloudlets in Round Robin

The previous techniques which were employed for the NER had taken a maximum execution time of about 17 days to execute the tasks in i3 and i5 processors and 2–4 GB of RAM. But the workflow scheduling with improved allocation had shown that the given tasks can be executed in seconds.

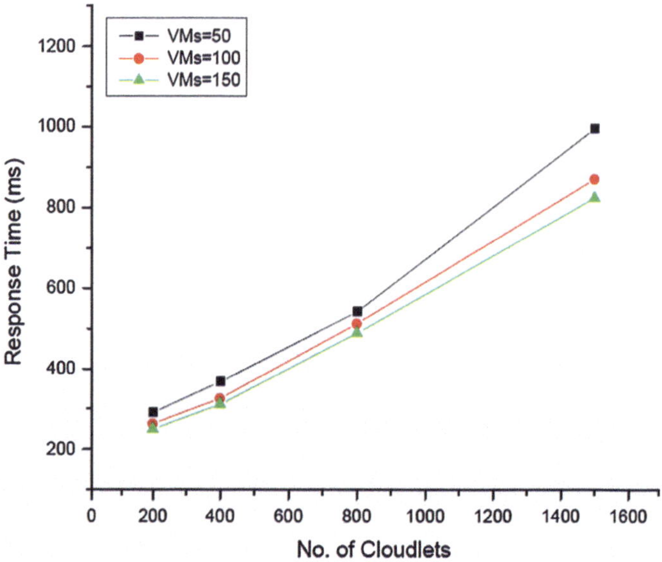

Fig. 5 Average response time in Round Robin

6 Conclusion

In this paper we are addressed the interesting and new problem called named entity recognition through resource allocation in cloud. The technique which is employed to detect the named entities called as multiclass SVM classifier which is combined with workflow scheduling in cloud achieved better results. The time and cost of establishing resources is reduced through VM allocation by applying FCFS scheduling and Round Robin scheduling.

Experimental results indicate that the multiclass classifier performed well in identifying named entities in different datasets. The proposed algorithm is efficient and effective for solving the problems of practical size. In order to improve the performance the improved allocation algorithm analyzes the record of each run, generates the predictive model performance of each task and optimizes the economic cost and performance accordingly.

References

1. Aiolli F, Sperduti A (2002a) An efficient SMO-like algorithm for multiclass SVM. In: Proceedings of IEEE workshop on neural networks for signal processing, pp 297–306
2. Allwein E, Schapire R, Singer Y (2000) Reducing multiclass to binary: a unifying approach for margin classifiers. J Mach Learn Res

3. Banko M, Cafarella MJ, Soderland S, Broadhead M, Etzioni O (2007) Open information extraction from the web. In: IJCAI '07: proceedings of the 20th international joint conference on artificial intelligence. Hyderabad, India
4. Babu LD, Krishna PV (2014) An execution environment oriented approach for scheduling dependent tasks of cloud computing workflows. Int J Cloud Comput 3(2):209–224
5. Bikel DM, Schwartz RL, Weischedel RM (1999) An algorithm that learns what's in a name. Mach Learn 34:211–231
6. Brin S (1998) Extracting patterns and relations from the world wide web. In: WebDB workshop at 6th international conference on extending database technology, EDBT '98
7. Buyya R, Pandey S, Vecchiola C (2009) Cloudbus toolkit for market-oriented cloud computing. CloudCom
8. Culotta A, McCallum A, Betz J (2006) Integrating probabilistic extraction models and data mining to discover relations and patterns in text. In: Proceedings of the main conference on Human language technology conference of the North American chapter of the association of computational linguistics New York. association for computational linguistics, New York, pp 296–303
9. Deelman E et al (2003) Mapping abstract complex workflows onto grid environments. J Grid Comput, LNCSD9 1:25–39. ISSN 1570-7873
10. Dhinesh Babu LD, Venkata Krishna P (2013) Versatile time-cost algorithm(VTCA) for scheduling non-preemptive tasks of time critical workflows in cloud computing systems. Int J Commun Netw Distrib Syst 11(4):390–411 (Inderscience Publishers)
11. Dhinesh Babu LD, Gunasekaran A, Venkata Krishna P (2014) A decision based pre-emptive fair scheduling strategy to process cloud computing work-flows for sustainable enterprise management. Int J Bus Inf Syst 16(4):409–430 (Inderscience Publishers)
12. Finkel JR, Grenager T, Manning C (2005) Incorporating non-local information into information extraction systems by gibbs sampling. In: ACL '05: Proceedings of the 43rd annual meeting on association for computational linguistics, Morristown, NJ, USA: Association for computational linguistics, pp 363–370
13. Grishman R, Sundheim B (1996). Message understanding conference—6: A brief history. In: Proceedings of the 16th conference on computational linguistics, pp 466–471
14. Jyothi B, Kesava Reddy E (2012). named entity recognition using support vector machines and high dimensional features-scalability issues. In: Presented conference at international conference on emerging trends in electrical, electronics and communication technologies, ICECIT
15. McDonald R, Pereira F, Kulick S, Winters S, Jin Y, White P (2005) Simple algorithms for complex relation extraction with applications to biomedical ie. In: ACL '05: proceedings of the 43rd annual meeting on association for computational linguistics. Ann Arbor, Michigan, pp 491–498
16. Nagavaram A et al (2011) A cloud-based dynamic workflow for massspectrometry data analysis. eScience, 2011
17. Nguyen D P, Matsuo Y, Ishizuka M (2007) Subtree mining for relation extraction from Wikipedia. In: Human language technologies 2007: the conference of the North American chapter of the association for computational linguistics; Companion Volume, Short Papers. Association for computational linguistics, Rochester, New York, pp 125–128
18. Rajasekhara Babu M, Venkata Krishna P, Khalid M (1997) A framework for power estimation and reduction in multi-core architectures using basic block approach. Int J Commun Netw Distrib Syst 10(1):40–51 (Inderscience Publishers, Netherlands); Baldonado M, Chang C-CK, Gravano L, Paepcke A (1997) The stanford digital library metadata architecture. Int J Digit Libr 1:108–121
19. Reddy TSK, Krishna PV, Reddy PC (2014) Power aware framework for scheduling tasks in grid based workflows. Int J Commun Netw Distrib Syst (Inderscience Publishers (accepted))

20. Wang L, Duan R, Li X, Lu S, Hung T, Calheiros R, Buyya R (2013) An Iterative Optimization Framework for Adaptive Workflow Management in Computational Clouds. In: 12th IEEE international conference on trust, security and privacy in computing and communications. IEEE computer society, 2013, pp 1049–1056. doi:10.1109/TrustCom.2013.128
21. Zhao S, Grishman R (2005) Extracting relations with integrated information using kernel methods. In: Proceedings of the 43rd annual meeting on association for computational linguistics, pp 419–426

Spatial Data Analysis Using Various Tree Classifiers Ensembled With AdaBoost Approach

S. Palaniappan, T.V. Rajinikanth and A. Govardhan

Abstract The Spatial Data is growing very fast but the available statistical techniques are not sufficient to analyze. The existing Spatial Data Mining Techniques also has certain limitations. The size and complexity of the data sets are posing challenges to the research community. In order to overcome these it is required to do deep study on the suitability of the existing Machine Learning Techniques apart from that check for the suitability of hybrid machine learning techniques. In our paper Classifier Ensembling Technique called AdaBoost Approach was applied on the Spatial Data set for rigorous Analysis. The AdaBoost Technique combines multiple weak classifiers into a single Strong Classifier. It is used in conjunction with many machine learning classifier algorithms in order to boost up their performances. In this connection various Tree Classifier Techniques like J48, Random Forest, BF Tree, F Tree, REP Tree, Random Tree, Simple Cart etc., were considered and applied on the Spatial Data set considered and did the comparative study in terms of various performance metric values both in terms of Numerically and Visually and finally made effective conclusions out of that study. This paper also states that ensemble methods perform in better way than any individual classifier.

Keywords Spatial data set · Spatial data mining techniques · Machine learning techniques · Classifier ensembling technique · AdaBoost approach

---

S. Palaniappan (✉)
Department of CSE, Saveetha University, Chennai, India
e-mail: s.palani@gmail.com

T.V. Rajinikanth
Department of CSE, SNIST, Hyderabad, India

A. Govardhan
Department of SIT, JNTUH, Kukatpally 500085, Hyderabad, India

K.R. Attele et al. (eds.), *Emerging Trends in Electrical, Communications and Information Technologies*, Lecture Notes in Electrical Engineering 394,
DOI 10.1007/978-981-10-1540-3_17

1 Introduction

Spatial Data [1] has spatial coordinates and topology and it can be analyzed by accessing, through GIS. **Spatial analysis** [2] is a collection of techniques for analysis of spatial data. The **Analysis** of Spatial data set depends on the locations of the objects. The Software used for **spatial analysis** depends on two factors namely spatial locations and its attributes. Spatial Data Mining Techniques involves Classification, Clustering and Association Techniques. Spatial Analysis or Spatial Statistics [3, 4] are the techniques used to study properties of topological, geometrical or geographical objects. It is used to analyze the spatial data. The type of this analysis includes Spatial data Analysis, Spatial Auto correlation, Interpolation, Regression, Interaction and Simulation and Modeling. Spatial Analysis turns raw data into potential information. It is a set of techniques for analyzing spatial data. It is the analysis of Geographical data sets. Shape files are the vector data and composed of a few required files (a).shp—the shape format file (b).shx—the shape index file (c).dbf—the attribute file. **Machine Learning** [5] is a technique that provides computers an ability to learn without being explicitly programmed. It makes development of computer programs so as to train themselves to grow and change whenever they are subjected to new data. Classification is a supervised learning technique.

2 Literature Survey

AdaBoost [6] is short form of Adaptive Boosting a popular boosting machine learning meta-algorithm technique which helps you combine multiple "weak classifiers" into a single "strong classifier". A weak classifier is simply a classifier that performs poorly, but performs better than random guessing. Its major advantage is it can be used in association with other types of classifier algorithms to improve their performance. The output of weak learners is combined into a weighted sum and results a boosted classifier. It is sensitive to noisy data and outliers. AdaBoost algorithm is used along with decision tree learning, information received at every stage is fed into the tree growing algorithm.

In this section, we will investigate the impact of AdaBoost as classifier ensembles on classification accuracy. Consequences of choosing different base classifier are monitored. In our case we used different type of decision tree algorithm, such as DecisionStump, J48, ADTree, LADTree and BFTree. Later on, comparisons of results of measuring the performance of classifiers are presented.

2.1 Ensembling Method

Ensemble methods [7] are used to construct a collection of classifiers. These are used to classify new data points by considering their weighted vote of predictions. The original ensemble method uses Bayesian averaging. But the present algorithms combine error correcting output coding, Bagging and boosting. E**nsemble methods** [8] use multiple learning algorithms to get better predictive performance than could be obtained from any of the constituent learning algorithms.

2.2 J48 Classifier

J48 algorithm [9] is an advanced version of C4.5 algorithm. The output of this J48 algorithm results a Decision tree. J48 algorithm [10] output results a Decision Tree.

2.3 Random Forest

Random forest [11] is an ensemble learning method used for classification, regression and other tasks.

2.4 BF Tree

In BF Tree the node in which split leads to maximum reduction of impurity across all other nodes is called best node.

2.5 F Tree

F Tree algorithm is used for building ‘Functional trees. It has logistic regression functions at the inner nodes and/or leaves.

2.6 J48 Consolidated

It is used for constructing a C4.5 consolidated tree. In this a single tree is built based on a set of sub samples. New options are added to the J48 class to set the Re-sampling Method (RM) for the generation of samples to be used in the consolidation process.

2.7 REP Tree

It is a Fast decision tree classifer. It constructs a decision/regression tree using information gain. It prunes the tree by the method reduced-error pruning.

2.8 Random Tree

It is used for constructing a tree with K randomly chosen attributes at each node.

2.9 Simple Cart

It implements minimal cost-complexity pruning. It deals with missing values. It makes use of "fractional instances" method.

3 User Classifier

User Classifier classifies interactively through visual means by constructing a scatter graph of the data.

3.1 Decision Stump

This classifier is uses a decision stump for building tree.

3.2 Extra Tree

This Class is used for generating a single Extra-Tree. It is used with the Random Committee Meta classifier to generate an Extra-Trees forest for classification or regression. This classifier requires all predictors to be numeric. Missing values are not allowed. Instance weights are taken into account for this classifier.

3.3 Hoeffding Tree

It is an incremental decision tree induction algorithm. It has capability of learning from massive data streams. Even a small sample is sufficient to choose an optimal splitting attribute and is supported mathematically by the Hoeffding bound. The features of Hoeffding Trees are not shared by other incremental decision tree learners and it gives very good performance.

4 Proposed Approach

The proposed approach is implemented by considering the Crime data set which is fetched from RRR+-Tree data structure. The fetched data set is pre processed using various data cleaning techniques. It is then subjected to different Decision Tree Classification techniques like J48, J48 Consolidated, Hoeffding Tree, Extra Tree, Decision Stump, User Classifier, Simple Cart, Random Tree, Rep Tree, F Tree, BF Tree, and Random Forest. After that in combination with ensembling methods like AdaBoost is applied in combination with these Tree Classifiers and their performance were found. Performance comparisons were made among these classifiers with that of Classifiers without the influence of ensembling method AdaBoost. It is found that their performances were enhanced.

5 Implementation of Proposed Approach

The proposed approach is implemented by considering the Crime data set which is fetched from R-Tree data structure [12, 13]. The fetched data set is pre-processed using various data cleaning techniques for removal of Noisy, Inconsistency, Incomplete and removal of Outliers etc. The data set was converted into the required format in which it has 66 attributes indicates type of crime and 6 districts as 6 instances. The 66 Attributes are namely Crime, BSI, TSV, CFOP, BR, MP, ANFC, T, AFA, PT/LP, OTC, GT, I, TITT, GA, NSEA, TWSV, GI, DSO, TAI, DD, TOM, H, P, PT, PTI, FFJ, HR, ATL, PCS, TANI, UPII, SL, BB, S, BV, VG, RB, M, FP, RU, BNCS, TVWO, SPNC, v, DAD, RPS, WS, CR, FITO, PDP, BC, COR, PCV, AOIS, LP, OAA, PV, HRID, RSAP, FOB, FD, FIR, SB, DAC, CW. The original data set is then divided into two data sets namely Training Data set and Test Data set with ratio 70: 30. Initially Training data set is subjected to different Decision Tree Classification techniques like J48, J48 Consolidated, Hoeffding Tree, Extra Tree, Decision Stump, User Classifier, Simple Cart, Random Tree, Rep Tree, F Tree, BF Tree, AD Tree, Multi Classifier and Random Forest. After that Test data set was subjected for these classifiers and performances were calculated. After that same procedure is conducted in combination with ensembling methods like AdaBoost is applied in combination with these Tree Claassifiers and their performance were found. Performance comparisons were made among these classifiers with that of Classifiers without the influence of ensembling method AdaBoost. It is found that their performances were enhanced.

6 Results and Analysis

The Pre- processed Trained and Test data sets are then subjected to 12 various classification techniques namely Random Forest, J48, J48 Consolidated, F Tree, REP Tree and Random Tree, Simple Cart, User Classifier, BF Tree, Decision Stump, Extra Tree, Hoeffding Tree and the results were shown in Table 1. In the same way again the tree Classifiers namely Random Forest, J48, J48 Consolidated, F Tree, Rep Tree and Random Tree, Simple Cart, User Classifier, BF Tree, Decision Stump, Extra Tree, Hoeffding Tree were applied in combination with ensembling method AdaBoost and the results were shown in Table 2. The 9 parameters taken for comparison are Time taken to build model for Train and Test Data sets, Kappa Statistic, % Correctly Classified Instances, Precision, Recall, F-measure, Mean Absolute error, Relative absolute error and Root Mean square error. It is found from the Table 1 that Random forest and Hoeffding Tree classifiers are best suitable for Crime data sets for classification Analysis. In that also although the parameters like Precision, Recall and F-measure values are same but the other parameters like Mean Absolute Error, Relative absolute error, Root Mean Square error are not same for both the algorithms. When these Error parameters are considered the Hoeffding Tree classifiers is proved to be better than Random forest classifier. Again the same was observed from Table 2 when the 12 classifiers were considered in combination with ensembling method AdaBoost same results were observed i.e. Random forest and Hoeffding Tree classifiers are best suitable for Crime data sets for classification Analysis. The advantage with ensembling method AdaBoost is found only in terms of reduction of time taken to build the Training Model.

$$\text{Accuracy} = \frac{\text{TP} + \text{TN}}{(\text{TP} + \text{FN} + \text{FP} + \text{TN})} \times 100 \quad (1)$$

$$\text{Specificity} = \frac{\text{TN}}{(\text{TN} + \text{FP})} \times 100 \quad (2)$$

$$\text{Sensitivity} = \frac{\text{TP}}{(\text{TP} + \text{FN})} \times 100 \quad (3)$$

The comparisons were made in between the algorithms Random forest and Hoeffding Tree classifiers and found that these classifiers are suitable for the Crime data set. The 3 performance parameters considered for comparison were Accuracy, Specificity and Sensitivity and whose formulae is shown above as Eqs. (1), (2) and (3). Among all these classifiers it is found that Hoeffding Tree classifiers is best suitable particularly for Crime Spatial Data set with or with out ensembling Ada-Boost method. The performance in terms of time complexity improved with ensembling AdaBoost method. Figure 1 shows the Comparison Graph between

Table 1 Performance comparisons of various classifiers without ensembling AdaBoost

Without ensembling AdaBoost	Random forest	J48	J48 Consoli dated	FT	REP tree	Random tree	Simple cart	User classifier	BF tree	Decision stump	Extra tree	Hoeffding tree
Train	0.09	0.03	0.13	0.03	0.02	0	0.03	10.97	0.02	0	0	0.09
Test	0	0	0	0	0	0	0	0	0	0	0	0
Kappa statistic	1	0.2	0.2	0	0	0.4	0	0	0	0	0	1
% of correctly classified Instances	100	33.3333	33.3333	16.6667	16.6667	50	16.6667	16.6667	16.6667	16.6667	16.6667	100
Precision	1	0.125	0.125	0.028	0.028	0.306	0.028	0.028	0.028	0.028	0.033	1
Recall	1	0.333	0.333	0.167	0.167	0.5	0.167	0.167	0.167	0.167	0.167	1
F-measure	1	0.178	0.178	0.048	0.048	0.361	0.048	0.048	0.048	0.048	0.056	1
Mean absolute error	0.1978	0.2407	0.2407	0.2778	0.2778	0.1667	0.2778	0.2778	0.2778	0.2778	0.2778	0
Relative absolute error	71.2	86.6667	86.6667	100	100	60	100	100	100	100	100	0
Root mean square error	0.2758	0.36	0.36	0.527	0.3727	0.4082	0.3727	0.3727	0.3727	0.4082	0.527	0

Table 2 Performance comparisons of various classifiers without ensembling AdaBoost

Ensembling AdaBoost	Random forest	J48	J48 consoli dated	FT	REP tree	Random tree	Simple cart	User classifierr	BF tree	Decision stump	Extra tree	Hoeffding tree
Train	0.06	0	0.14	0.03	0.02	0	0.03	11.66	0.03	0	0.02	0.05
Test	0	0	0	0	0	0	0.02	0	0	0	0	0
Kappa statistic	1	0.2	0.2	0	0	0.2	0	0	0	0	0	1
% of correctly Classified Instances	100	33.3333	33.3333	16.6667	16.6667	33.3333	16.6667	16.6667	16.6667	16.6667	16.6667	100
Precision	1	0.125	0.125	0.028	0.028	0.208	0.028	0.028	0.028	0.028	0.028	1
Recall	1	0.333	0.333	0.167	0.167	0.333	0.167	0.167	0.167	0.16 7	0.167	1
F-measure	1	0.178	0.178	0.048	0.048	0.233	0.048	0.048	0.048	0.048	0.048	1
Mean absolute error	0.1994	0.2407	0.2407	0.2778	0.2778	0.2222	0.2778	0.2778	0.2778	0.2778	0.2778	0
Relative absolute error	71.8	86.6667	86.6667	100	100	80	100	100	100	100	100	0
Root mean square error	0.2733	0.36	0.36	0.527	0.3727	0.4714	0.3849	0.3849	0.3849	0.4082	0.527	0

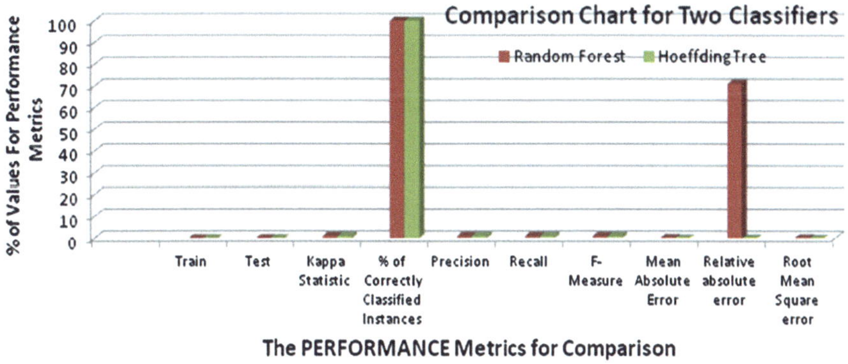

Fig. 1 Comparison graph between two classifiers random forest versus Hoeffding tree

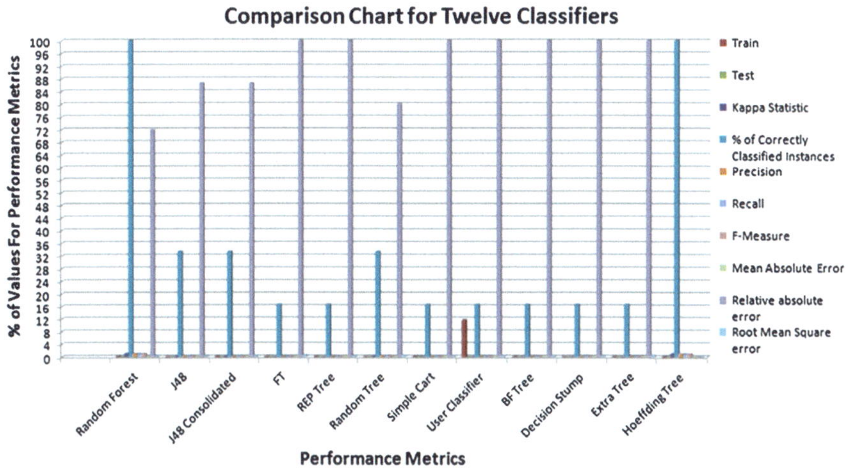

Fig. 2 Comparison graph of twelve classifiers when ensembling AdaBoost is considered

Two Classifiers Random Forest versus Hoeffding Tree and the Fig. 2 shows Comparison Graph of Twelve classifies when ensembling AdaBoost is considered.

7 Conclusions

The AdaBoost Algorithm is used with many different classifiers. It improves classification accuracy and is commonly used in many areas. It is simple to implement and not prone to over-fitting. The considered Spatial data set was subjected to 12 various classification techniques namely Random Forest, J48, J48

Consolidated, F Tree, REP Tree, Random Tree, Simple Cart, User Classifier, BF Tree, Decision Stump, Extra Tree, Hoeffding Tree. Initially all these 12 classifiers were applied without ensembling method AdaBoost and later with ensembling method AdaBoost was applied. There is a gain observed in terms of performance in reduction of time to build the trained model. It is found that two tree Classifiers Random Forest and Hoeffding Tree are suitable in classifying Spatial Crime data set effectively. Between these two classifiers Hoeffding Tree proved to be good than the Random Forest because it has non-zero error values for Men Absolute error, Relative absolute error, Root Mean Square error. The mean absolute error [14] is one of a number of ways of comparing forecasts with their eventual outcomes. The absolute error is the absolute value of the difference between the forecasted value and the actual value. MAE tells us how big of an error we can expect from the forecast on average. The **Mean Absolute Error** (MAE) is the quantity used to measure how close forecasts or predictions are to the eventual outcomes. The larger the difference between RMSE and MAE the more inconsistent the error size. The Classifier Hoeffding Tree classifier has zero value so it proves to be more efficient.

References

1. http://www.webopedia.com/TERM/S/spatial_data.html
2. https://apps.carleton.edu/collab/spatial_analysis/SpatialAnalysis/
3. https://en.wikipedia.org/wiki/Spatial_analysis
4. http://dusk.geo.orst.edu/gis/Chapter14_notes.pdf
5. http://whatis.techtarget.com/definition/machine-learning
6. https://en.wikipedia.org/wiki/AdaBoost
7. Dietterich TG Oregon State University, Corvallis, Oregon, USA, Ensemble methods in machine learning. http://www.cs.orst.edu/~tgd
8. https://en.wikipedia.org/wiki/Ensemble_learning
9. Palaniappan S, Rajinikanth TV, Govardhan A Enhancement of effective spatial data analysis using R
10. Kumar Y, Sahoo G Analysis of Bayes, neural network and tree classifier of classification technique in data mining using WEKA
11. https://en.wikipedia.org/wiki/Random_forest
12. Palaniappan S, Rajinikanth TV, Govardhan (2015) A RRR+Tree: rough set theory-based reduced R+tree to indexing and retrieval of spatial data. Aust. J Basic Appl Sci 9 (23):482–494. © 2015 AENSI Publisher. ISSN:1991-8178
13. Guttman A (1984) R-trees: a dynamic index structure for spatial searching. In: Proceedings of the 1984 ACM SIGMOD international conference on management of data (SIGMOD), Boston, MA, vol 14(2), pp 47–57
14. https://en.wikipedia.org/wiki/Mean_absolute_error
15. https://en.wikipedia.org/wiki/Multiclass_classification
16. https://en.wikipedia.org/wiki/Alternating_decision_tree

Critical Analysis of Congestion Control for the Future Networks

G.N. Vivekananda and P. Chenna Reddy

Abstract The Internet has always done a remarkable job at surprising people. Technology oscillates, and some envisioned architectures of the future fail while others thrive. From the origination of peer-to-peer computing, it became easier task for communication and information transfer. But there exists still few hysterical issues like congestion, and security. Congestion occurs when one part of subnet becomes overloaded. The means of moderating the traffic in the networks is referred as congestion control. This paper critically analyses various issues behind congestion in various networks and various techniques to avoid and prevent from congestion and to efficiently control the congestion in the networks. This paper essentially focuses on the future networks posing various challenges and exposed issues regarding congestion control.

1 Introduction

Increase in the network traffic bursts to sufficiently large, temporary congestion may occur, causing delay, jitter, and loss. If this Congestion is not controlled applicably it leads to lasting congestion [1]. The increase in various mobile networks equipment's such as laptops, tablets, handsets causes a major traffic generator, lots of mobile data, death of 2G due to raise of small screen, offloading from Cellular, and The Internet of Things for these reasons devices deals with the end user content and applications that are not supported by former groups of mobile devices. These trends are all expected to have far reaching implications. More advanced cloud

G.N. Vivekananda (✉)
Research Scholar, Department of Computer Science and Engineering
JNTUA College of Engineering, Pulivendula, India
e-mail: vivekanandagn@gmail.com

P.C. Reddy
Department of Computer Science and Engineering JNTUA College of Engineering, Pulivendula, India
e-mail: pcreddy1@rediffmail.com

K.R. Attele et al. (eds.), *Emerging Trends in Electrical, Communications and Information Technologies*, Lecture Notes in Electrical Engineering 394,
DOI 10.1007/978-981-10-1540-3_18

applications will require increasingly powerful networks as demand grows for faster upload speeds as well demand for improvements to latency. Congestion must be controlled to avoid performance degradation and network break down that occur when network storage overflows and when packets are missing [2]. Congestion control is invoked as network reaction that refers to mechanisms as network provisioning, admission control, traffic throttling, traffic aware routing and load shedding. The bulk of web traffic uses the TCP as a congestion control protocol. TCP was successful dealing congestion in timely periods of the Internet. These theories use allocating fairly using max-min and proportional fair allocation techniques. Various metrics and parameters are used to specify how to control congestion. Various metrics like loss, delay, bitrates, bandwidth, lossy links, fairness features that improves the performance and sender, receiver issues also acts as platform for the congestion [3].

From Fig. 1a and b, from the surveys of cisco [4], it is expected to be 37 basic phones traffic can be generated by a single smart phone. The count increases to 94 feature phones in case when a tablet is used. Similarly, when a laptop is used it generates heavy burst of traffic that equals to 119 feature phones. By 2019 global IP traffic will be equal to 37 M DVDs/hour, mobile traffic will be ten times than in 2014, besides total traffic will be three times larger than 2014.

Primal algorithms for congestion control consider TCP algorithms and Dual algorithms for congestion control consider queue management algorithms [5, 6]. Various issues that repeatedly lead to the network congestion are Heterogeneity of networks and data, offensive Stability and fairness, Network support based on performance and robustness, information acquisition, corruption loss, flow start-up, sender and receiver issues, and malfunctioning devices.

Various internet standards and drafts are aimed as mechanisms to control the congestion. Different RFC's like RFC 793, RFC 2001, RFC 2581, RFC 2914, RFC 3426, RFC 3649 and RFC 5681 exploits various considerations for jamming in variable networks.

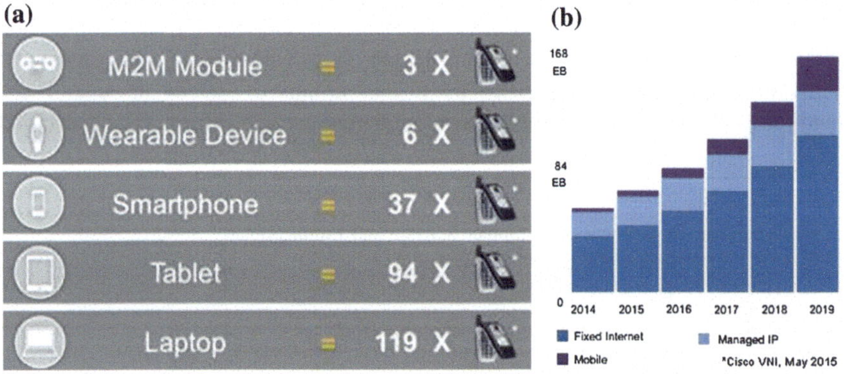

Fig. 1 Cisco traffic forecast 2014–2019. **a** Device traffic. **b** Monthly basic mobile phone data traffic

Wired Networks Congestion control:
The classic algorithms that help for congestion management in TCP are slow-start and congestion avoidance that is used by TCP Tahoe, TCP Reno uses fast crash recovery and fast retransmits [6]. These are interrelated with one another, i.e., once the connection is recognized between client and server slow start algorithm starts initially sending a segment and once it receives the acknowledgement of successful transmission it doubles the segments every time transmission got succeeded. At certain point when window size is huge, packets perhaps dropped and sender may use congestion avoidance algorithm. Congestion avoidance slows the broadcast rate in case of congestion and in this retransmission timer plays a crucial task. Window is reset by one segment, when timeout indicates the congestion. This automatically sets the sender into slow start state, and if the duplicate acknowledgements are indicated, fast crash recovery and fast retransmit algorithms are enforced. In fast recovery, when continuously sender is receiving more duplicate acknowledgements, segment transmission occurs without waiting for termination of retransmission timer. In the state of fast recovery, sender enters into congestion avoidance mode rather than slow start.

Ad hoc Networks Congestion control:
Presently, wireless networks are been used extent everywhere. There is no doubt that these may lead the future. TCP mechanisms are not appropriate here, due to the time varying nature of wireless channel and interference that occurs because of other paths or nodes that may result issues regarding out of delivery, delay and packet loss [7]. It is well-known that a network may perform extremely unwell when congestion control is absent. Controlling the congestion in wireless networks is one of the difficult tasks due to its highly dynamic nature. Most congestion control mechanisms are designed with the assumption of a chain of links and routers (or switches) stretching between two end systems [8]. There are no proper links in multi-hop wireless networks, because each wireless transmission does not affect the congestion only at its intended receiver but to all nearby receivers as well. Thus efficient congestion control mechanism is needed that monitors and react to congestion not only on the path taken by the packets, but also around it [9]. This becomes tough condition to handle in multicast applications. Layered technique may fail while supporting congestion here due to the mobility of nodes and lack of different concerns like reliability, and modularity that are absent in MANETs. Thus different approaches like Abstraction layer approach and Cross-layer approaches came into existence [10].

2 Literature Review

In [11], a scenario is shown where congestion in TCP can't be regulated that offers inflexible conditions in multi-hop wireless networks. In terms of shared medium, hop-by-hop congestion mechanisms are discussed. In the scheme that is recommended, inactive medium and backpressure occurs modestly while establishing the source nodes. Error detection and correction schemes are employed for TCP and UDP data streams to lessen the overhead and to change the medium directives for guaranteeing faster conditions. Simulations are studied and conducted in depth on both software and hardware to determine performance and outcomes by test bed results and implementing protocol.

In [12], author's highlight how the diversified data is gathered from wide varieties of sensors from the WMSNs. For the maintenance of Quality of Service (QoS) credibility, that has consistent and impartial features various transport protocols are used in case of multimedia applications. To control the congestion WMSNs are used. A WMS network employs many resources for packet broadcasting which carries the confidential data, which has different levels of important information. WMSNs are also considered for the nodes that are being dropped from the real-time traffic. To take immediate actions to prevent the data loss, the network traffic must support low latency and high consistency. Therefore alike wired, different services of the WMSNs also plays a significant role in congestion control. Provision variation and congestion control techniques are suggested in WMSNs for a priority based control technique. In this situation, input traffic flow is examined considering the changeover in High impact real time traffic with slight significances non-real time traffic.

In [13], in general TCP which is an conventional protocol was deliberate for wired networks thinking that congestion in wireless network causes packet loss, however many data packets are lost in the ad hoc networks happens due to the superior Bit Error Rate (BER), nodes, mobility etc. Congestion control can't be applied to ad hoc networks using TCP, from its issues of general TCP mechanisms that may cause network performance degradation. Authors examine the factors that affect TCP performance in ad hoc networks, and afford several improved congestion control approaches, and compare them.

3 Critical Analysis and Open Issues

Stability, Scalability, and Deploy ability are the basic features of a network. A network must be in a stable state for its better performance gains. Congestion control is one of the essential objectives of the networks. The stableness must be proven analytically leastwise in the synchronous RTT and fluid traffic model case, where other cases are simulated often. As a network is used by different users, this mechanism should be scalable. It must be instantly deployable in the Internet, which means that it needs to be fair and friendly to TCP. There are various issues of congestion control as follows

- The main aim of congestion controlling algorithms must be to recover from the congestion reducing the queue delay interval and loss. And this can be achieved by keeping queues short results in low loss and low delay

- Utilizing as much as possible existing capacity can lead to high throughput, but the question of wastage of bandwidth in case of limited transfers must be analyzed.
- UDP and its wide applications are not ideal enough to support congestion control as TCP. It is uncertain to develop UDP for the benefits of just single application, thus UDP must be reviewed in case of wireless networks for better real time activities.
- Performance estimation is a problematic task in the networks, as it may shuffle between static and dynamic. Thus we should create measures such that it includes (a) low loss and low delay and (b) high capacity, and assigning a priority for the first two factors. It is a smart thought to achieve these two objectives. Because, this gives a clear idea to choose the regulation schemes in UDP based applications, by measuring suitable control mechanism that does a better job.
- SCTP offers various conceptions such as Multi-homing by which seamless mobility for wireless networks can be offered and multi streaming is used for increasing the stream delivery, thereby increasing throughput and data rate of the transmissions. The concurrent multipath feature in SCTP makes it much more efficient for Ad hoc networks [14].
- SCTP can perform well in wireless networks for resolving congestion concerns and enables reliable transfer that makes use of both TCP reliability and UDP streaming, but until what extent this can be used and what are various mechanisms, algorithms, parameters that are required to calculate exact SCTP performance in Ad hoc networks [15]. The following table, Table 1 gives in detail status of the SCTP in wireless networks.

Many approaches are being considered continuously for controlling the congestion, like adaptation layer approach and Cross-layer approach, which deviates from the general layered methodology. In Adaptation layer approach, it intends to alleviate the difficulty of choosing and tuning a transport service by hiding transport layer details from applications. In cross-layer approach, by fusion of the layers communication can happen among the layers, which is the main idea to acquire the improved network performance.

Table 1 Comparison of Various Features of MANET with TCP, UDP and SCTP

MANET features	TCP	UDP	SCTP
Reliability	Yes	No	Yes
Flow control	Yes	No	Yes
Congestion control	Yes	No	Yes
Multi streaming	No	No	Yes
Multi homing	No	No	Yes
Check of reachability	No	No	Yes
Security against SYN attacks	No	No	Yes

On the social front, these techniques for congestion control using Cross-layer approach for transmission of various heterogeneous data using SCTP protocol in Ad hoc networks can be greatly scalable and can improvise the performance of future networks that are wireless.

4 Conclusion and Future Work

Congestion control in the MANETs has different types of issues ranging from different levels. A perfect congestion control technique for MANETs is not available until now, as the tradeoffs in wireless networks are not certain. When we consider different environments, we face many problems with heterogeneous, hybrid environments and their various modes of operation. There exist different problems related to fairness, and pricing too. There is a fundamental design problem in ad hoc networks, when we consider the congestion control. As it happens between the layers, it is essential to deal with the inner network and thus should be placed in between network, transport and application layers. However, no such cross-layer design strategy is established until now to build a successful solution for controlling congestion. Congestion control in Ad hoc networks while heterogeneous data transmission and by using the cross-layer strategy and various features of SCTP protocol can be food for thought for the better survival of wireless and future internet.

References

1. Jacobson V, Karels MJ (1998) Congestion avoidance and control, symposium proceedings on communications architectures and protocols SIGCOMM '88 **18**(4):314–329
2. Jain R (1990) Congestion control in computer networks: issues and trends. IEEE Netw Mag 24–30
3. Floyd S, Jacobson V (1993) Random early detection gateways for congestion avoidance. IEEE/ACM Trans Netw **1**(4):397–413
4. Cisco visual networking index: global mobile data traffic forecast update 2014–2019. http://ciscovni.com/forecast-widget/advanced.html
5. Alexander L (2006) Stolyar: Greedy primal-dual algorithm for dynamic source allocation in complex networks. Queueing Syst 54(3):203–220
6. Allman M, Paxson V, Stevens W (1999) TCP congestion control, internet draft. http://www.ietf.org/rfc/rfc2581.txt
7. Floyd S, Fall K (1999) Promoting the use of end-to-end congestion control in the internet. IEEE/ACM Trans Netw 7(4):458–472
8. Chen L et al (2005) Joint congestion control and media access control design for ad hoc wireless networks. In: INFOCOM, 24th annual joint conference of the IEEE computer and communications societies, pp 2212–2222
9. Rakocevic V Congestion control for multimedia applications in the wireless internet 1–23
10. Shakkottai S, Rappaport TS, Karlsson PC (2003) Cross-layer design for wireless networks. IEEE Commun Mag 41:74–80

11. Scheuermann B, Lochert C, Mauve M (2007) Implicit hop-by-hop congestion control in wireless multi-hop networks. Sci Direct Ad Hoc Netw 6:260–286
12. HosseinYaghmaee M, Adjeroh D (2009) Priority-based rate control for service differentiation and congestion Control in wireless multimedia sensor networks. Sci Direct Comput Netw 53:1798–1811
13. Teng YP, Wang HZ, Mei Jing, and Lian ZZ A study of improved approaches for tcp congestion control in ad hoc networks In: Proceedings of science direct IWIEE procedia engineering, vol 29, pp 1270–1275
14. Stewart R et al (2000) Stream control transmission protocol, internet draft. http://www.ietf.org/rfc/rfc2960.txt
15. Stewart R (ed) (2007) Stream control transmission protocol, internet draft. https://tools.ietf.org/rfc/rfc4960.txt

Part II
Trends in Image and Speech Processing, VLSI and Embedded Systems, Communications Technology

An Intelligent Frame Work System for Finger Touch Association on Planar Surfaces

S. Asif Hussain, M.N. Giri Prasad and Chandrashekar Ramaiah

Abstract In this work an intelligent projection framework system (IPS) is proposed, which empowers uncovered finger touch association on normal planar surfaces (e.g., dividers, tables), with one and only standard camera and one projector. The test of uncovered finger touch recognition is recouping the touching data just from the 2-D picture caught by the camera. In our system, the graphical client interface (GUI) catch is anticipated at first glance and is twisted by the finger when clicking it, and there is a huge positive relationship between the button's distortion and the finger's stature to the surface. Hence, we propose a novel, quick, and strong calculation, which exploits the catch's mutilation to identify the touch activity. The current consoles utilized keys construct console for writing in light of the PC. These consoles are chipping away at the mechanical push standard. In any case, for the little gadgets like cellular telephones and tablets it is difficult to convey enormous console with them. The touch screen based consoles accessible in such gadgets are extremely badly designed to compose in light of the fact that the measure of individual's finger is enormous and the extent of the keys on the touch screen is little. So writing chip away at the little gadgets is not helpful and on PC our fingers get torment in the wake of doing long time writing work as a result of mechanical vibration of the keys.

S. Asif Hussain (✉) · C. Ramaiah
Department of ECE, Middle East College, Muscat, Sultanate of Oman
e-mail: sah.ssk@gmail.com; shussain@mec.edu.om

C. Ramaiah
e-mail: chandrashekar@mec.edu.om

M.N. Giri Prasad
Department of ECE, JNTUA, Ananthapuramu, A.P., India
e-mail: mahendragiri1960@gmail.com

K.R. Attele et al. (eds.), *Emerging Trends in Electrical, Communications and Information Technologies*, Lecture Notes in Electrical Engineering 394,
DOI 10.1007/978-981-10-1540-3_19

1 Introduction

Mobile devices (e.g., mobile phones, pads) with significant computational power and capabilities have been a part of our daily life. Benefiting from the small size of these devices, they are easy to carry. However, the screen real estate of today's Mobile devices is limited by their small sizes. This greatly diminishes their usability, functionality, and comfort. A pico-projector can be used to significantly increase the limited screen size of the mobile devices. With the development of the projection technology, we believe that embedded projectors in the mobile phones will be very common in the future, and people will enjoy a way of displaying digital contents on everyday surfaces (Fig. 1).

Meanwhile, the interactions (e.g., touch, gesture) on the projected display are thought to be appealing. To achieve the touch interaction, the biggest challenge lies in how to determine whether the fingers touch the projected surface or not. Most of the researchers in this area use multi cameras or a depth camera to obtain the relative position between the fingertip and the projected surface [2, 3].

2 Related Work

Touch identification on a projection screen is a vital issue in the range of human PC cooperation. Bio-acoustic detecting cluster incorporated with armband to recognize touch activity on the skin applies an accelerometer to distinguish the increasing speed produced by the finger click. Both systems require extra hardware and are delicate to environment impacts. Scope of data give two or more cameras are extraordinarily enhances the exactness of touch detection. One camera is utilized for to track the finger direction another is put parallel to the surface to identify the

Fig. 1 Hardware prototype of IPS, cited in Ref. [1]

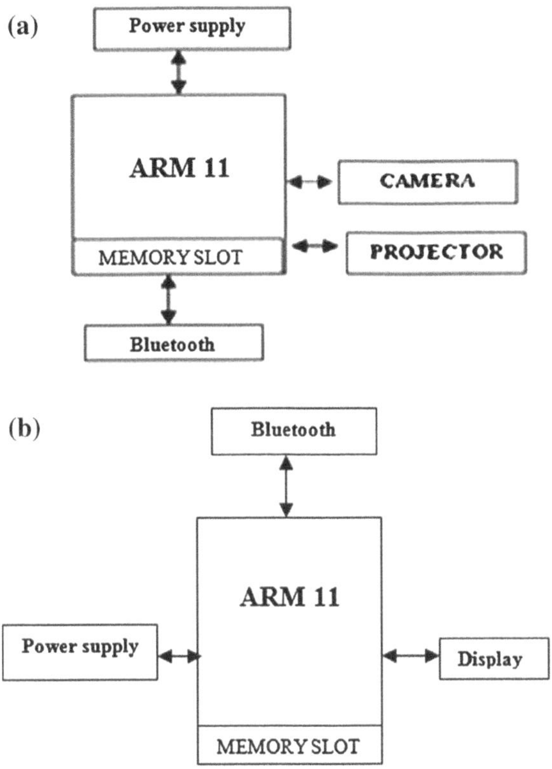

Fig. 2 System framework. **a** Transmitter section, **b** receiver section, cited in Ref. [1]

whether the finger touches the surface. The camera is put in the side of the LCD display. Their optical pivot is parallel to the screen to distinguish the touch occasions and focus the position of touch on the screen. The shadow of the finger can be utilized to perceive touch action where Gaussian blend model is used to recognize shadow. The extraction of the fingertip is simple, accurate and hearty however assistant Equipment is required in the framework. Profundity detecting Camera has turned out to be exceptionally famous in identifying touch. The location of a catch's bending can be did utilizing edge identification. In the PC vision and picture handling developed innovation (Fig. 2) [4].

Our framework is planned by utilizing ARM 32-bit miniaturized scale controller which bolsters distinctive elements and calculations for the improvement of car frameworks. We are anticipating a GUI on surface by projector and camera for catching GUI, The camera will catch the spots where client put his finger and the development of the finger. The camera catch pictures are broke down by the calculations and projects present in the ARM microcontroller [5].

Grayscale

Grayscale is a scope of shades of dim without obvious shading. The darkest conceivable shade is dark, which is the aggregate nonattendance of transmitted or reflected light (Fig. 3).

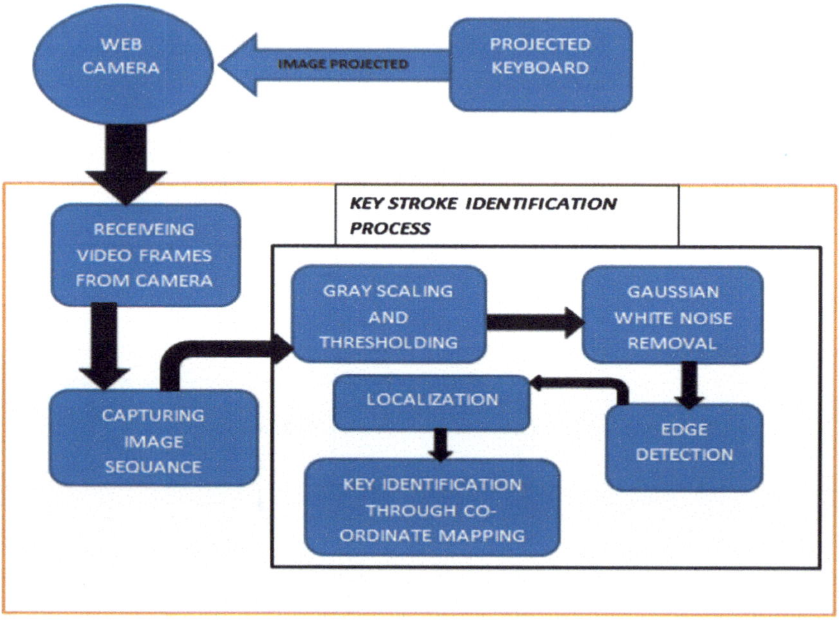

Fig. 3 Key stroke identification process, cited in Ref. [1]

Gaussian Noise

Primary wellsprings of Gaussian clamor in computerized pictures emerge amid obtaining e.g. sensor commotion brought on by poor enlightenment and/or high temperature, and/or transmission e.g. electronic circuit commotion.

Edge Detection

Edge data in a picture is found by taking a gander at the relationship a pixel has with its neighborhoods. In the event that a pixel's dark level quality is like those around it, there is presumably not an edge by then [6].

Localization

Limitation are non-viewable pathway, hub choice criteria for restriction in vitality compelled system, planning the sensor hub to enhance the tradeoff between confinement execution and vitality utilization.

3 Hardware Implementation

Raspberry Pi Board

The Raspberry Pi is a MasterCard measured single-board PC created in the UK by the Raspberry Pi Foundation with the aim of advancing the educating of fundamental software engineering in schools (Fig. 4).

Fig. 4 Raspberry pi board, cited in Ref. [1]

The Raspberry Pi has a Broadcom BCM2835 framework on a chip (SoC), which incorporates an ARM1176JZF-S 700 MHz processor, Video Core IV GPU, and was initially delivered with 256 megabytes of RAM, later moved up to 512 MB.

A. TFT Display Unit

TFT remains for Thin Film Transistor, and is a kind of innovation used to enhance the picture nature of a LCD. Every pixel on a TFT-LCD has its own transistor on the glass itself, which offers more control over the pictures.

B. Camera

The depicted parts for equipment are utilized to make a fitting model that could coordinate the prerequisite of programming configuration. Positions of the three primary parts are kept at an altered stature and edge such that the camera can catch the entire console image.

4 Software Requirements

A. Linux Operating System

Linux or GNU/Linux is a free and open source software operating framework for PCs. The working framework is a gathering of the essential guidelines that tell the electronic parts of the PC what to do and how to function. Free and open source programming (FOSS) implies that everybody has the flexibility to utilize it.

B. Qt for Embedded Linux

Qt is a cross-stage application structure that is broadly utilized for creating application programming with a graphical client interface (GUI) (in which cases Qt is named a widget toolbox), furthermore utilized for growing non-GUI projects such as command-line devices and comforts for servers [7].

5 Experimental Results

Bare-Finger System

The detection of the finger's height, which is shown in the form of the button's offset, is accurate. The projector–camera system is usually used in the measurement field with the help of structured light technique (Fig. 5).

Display Unit

A display is a computer output surface and projecting mechanism that shows text and often graphic images to the computer user. The display is usually considered to include the screen or projection surface and the device that produces the information on the screen (Fig. 6).

Virtual Keyboard

In our application, IPS projects the keyboard on the table and supports the way to directly type on it (Fig. 7). For example, when we want to send a message to somebody on the left side of the screen, we click his or her portrait. Then the keyboard appears and touch interaction is detected by the distortion of the key.

Fig. 5 Bare-finger system

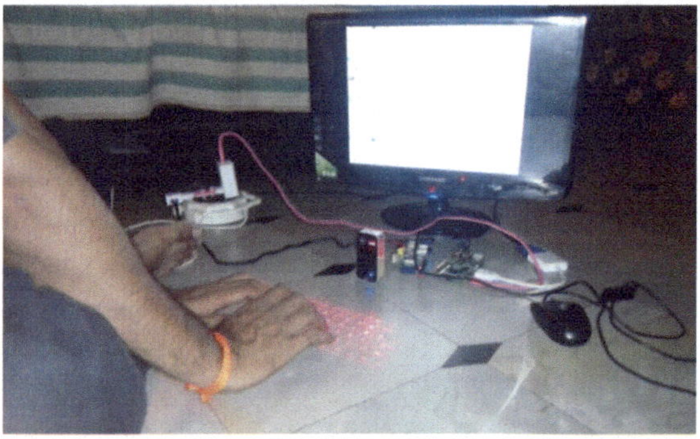

Fig. 6 Display unit

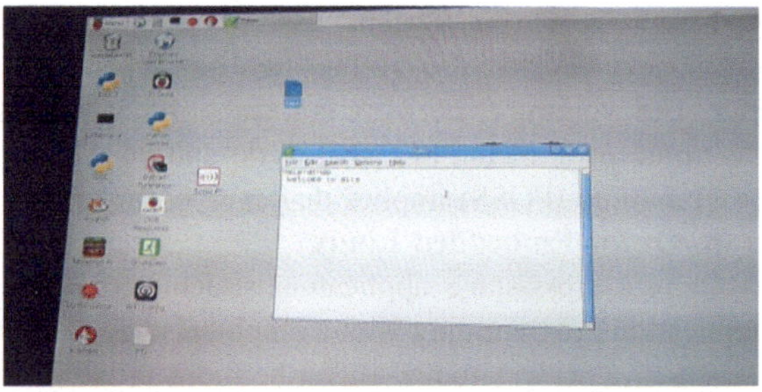

Fig. 7 Virtual keyboard

6 Advantages

The most vital point of preference of this framework is to make the framework autonomous of on-board handling which utilizes the majority of the battery power and in addition the independency of every module makes it simpler for move up to new element. The remote component makes it valuable to work from a far off area with for all intents and purposes controlling the portable PC or PC.

7 Conclusion

The project "An Interactive Projection System to Enable Bare-Finger Touch Interaction" has been effectively composed and tried. Vicinity of each module has been contemplated out and set deliberately in this way adding to the best meeting expectations of the unit. Furthermore, utilizing very propelled ARM board and with the assistance of developing innovation the task has been effectively actualized.

References

1. Asif Hussain S, Veera Pratap N (2015) An interactive projection system to enable bare-finger touch interaction. IJIRCCE 3(8)
2. Khalilbeigi M, Lissermann R, Mühlhäuser M (2011) Xpaaand: interaction techniques for rollable displays. In: Proceedings of ACM CHI, pp 2729–2732
3. Cowan LG, Li KA (2011) ShadowPuppets: supporting collocated interaction with mobile projector phones using hand shadows. In: Proceedings of ACM CHI, pp 2707–2716
4. Wilson D (2010) Using a depth camera as a touch sensor. In: Proceedings of ACM ITS, pp 69–72
5. Harrison C, Benko H, Wilson AD (2011) OmniTouch: wearable multitouch interaction everywhere. In: Proceedings of ACM UIST, pp 441–450
6. Harrison C, Tan D, Morris D (2010) Skinput: appropriating the body as an input surface. In: Proceedings of ACM CHI, pp 453–462
7. Optoma Ltd. http://www.optomausa.com/products/ search/pk301

Robust Invisible Watermarking for Image Authentication

Priyanka R. Kulkarni, Altaaf O. Mulani and P.B. Mane

Abstract This paper proposes implementation of digital image watermarking using 3-level discrete wavelet transform. In this experimentation, digital image watermarking algorithm uses discrete wavelet transform (DWT) for decomposing cover image and watermark image. A defined algorithm does not change any information of the cover image. Information obtained from low frequency DWT coefficient of cover image and the watermark image is being used in order to get watermark embedding. Watermark extraction has been simply done by wavelet decomposition of watermarked image and cover image.

Keywords Digital image watermarking · Discrete wavelet transform (DWT) · Scaling factor · PSNR · MSE etc.

1 Introduction

Exchange of digital contents such as images, text, audio, video etc. have been widely increased due to internet. Because of this, copying and manipulation of these digital contents have become easier. To protect data against this illegal manipulation watermarking is the solution. Different theories for digital image watermarking are presented by different authors. The surveyed literature on digital image watermarking is as follows: Subramanyam et al. [1] have suggested the technique for robust watermarking of encrypted and compressed images. The proposed watermarking algorithm in this paper gives watermarking of JPEG 2000 compressed and encrypted images. This is advantageous in case of watermarking of compressed and encrypted images. A method of invisible image watermarking has

P.R. Kulkarni (✉) · A.O. Mulani
Department of Electronics and Telecommunication,
SKN Sinhgad College of Engineering, Korti, Pandharpur, India
e-mail: priyanka.kul28@gmail.com

P.B. Mane
AISSMS's Institute of Information Technology, Pune, India

K.R. Attele et al. (eds.), *Emerging Trends in Electrical, Communications and Information Technologies*, Lecture Notes in Electrical Engineering 394,
DOI 10.1007/978-981-10-1540-3_20

been suggested by Prof. Dixit et al. [2]. Proposed algorithm uses variable scaling factor and combination of Discrete Wavelet Transform (DWT) and singular value decomposition (SVD). Kundu et al. [3] have presented a new method to improve robustness. This paper uses singular value of watermark image and singular value of 3rd-level DWT approximation matrix of original image are embedded. The genetic algorithm is used to optimize the scaling factor with which the watermark is embedded to host image. Under various attacks watermark robustness of watermarking has been defined. Li et al. [4] give watermarking of medical images. The studied method uses Arnold transform and Discrete Wavelet Transform. The watermarking image is scrambled by Arnold transform to enhance its privacy.

Bhargava et al. [5] have presented digital image authentication using digital image watermarking. For authentication, watermarking of image is carried out using Discrete Wavelet Transform. Discrete Wavelet Transform based digital image watermarking is done by Tianming et al. [6]. In this article, an algorithm based on DWT coefficients is summarized also it presents basic procedure of watermark embedding and watermark extraction. Jabade et al. [7] give comparison of different techniques used for digital image watermarking. This paper elaborates suitability of wavelet transform for image watermarking, wavelet transform based image watermarking process, classification and analysis of wavelet based watermarking techniques. Shi et al. [8] have suggested digital watermarking algorithm based on singular value decomposition and Discrete Wavelet Transform. For optimizing the image Arnold scrambling algorithm is used. By combining the characteristics of singular value decomposition and processing the singular decomposition of wavelet transform image enhances the digital watermark invisibility and robustness effectively. Lai et al. [9] presents image watermarking technique to satisfy perceptibility and robustness. This is the paper that presents about the robustness scheme. This defines a hybrid technique based on Discrete Wavelet Transform and singular value decomposition. A new technique for increasing robustness of image watermarking is presented by Liu et al. [10]. This paper suggests robust watermarking scheme for copyright protection. Proposed method is based on Discrete Wavelet Transform and it achieves watermark embedding by taking difference values of original image and reference image to overcome weak robustness problem of embedding watermark. In this proposal we use original image and watermark image to perform image watermarking. To perform watermark embedding and extraction both images are processed in wavelet domain. After that mean square error (MSE) and peak signal to noise ratio (PSNR) is calculated.

Rest of the paper is organized as follows: Sect. 2 provides brief idea about proposed system. It is followed by Sect. 3 which gives results and analysis of this experimentation. Section 4 summarizes conclusions of this study.

2 Proposed System and Methodology

Proposed system uses the images from database having different sizes such as 512 × 512, 256 × 256, 128 × 128, and 64 × 64. Preprocessing i.e. conversion of RGB image to gray, resizing is done. Block diagram of proposed system is shown below: Basically, proposed system has two steps. First step is watermark embedding and next is to watermark extraction. In watermark embedding, cover image is selected from database. Then it is preprocessed. Watermark image is subsequently selected and preprocessed. Both the images are then decomposed using discrete wavelet transform to get LL, LH, HL and HH sub-bands up-to third level. These sub-bands obtained, are nothing but the approximation, horizontal, vertical and diagonal coefficients. LL sub-band obtained by 3 level DWT decomposition of both the images are added along with scaling factor and watermark embedding process is done. Inverse DWT is taken and watermarked image is obtained. In the next part, extraction of embedded watermark is done. In watermark extraction process, input image is nothing but the watermarked image. This image

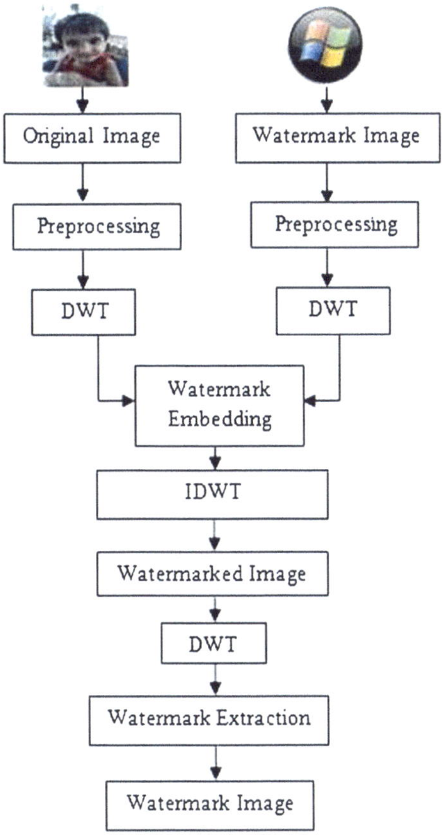

Fig. 1 Block diagram of proposed system

is again decomposed by using 3 level DWT and sub-bands are obtained. Approximation coefficient i.e. LL sub-band of watermarked image and LL sub-band of original image are subtracted to get watermark image (Fig. 1).

3 Experimental Results and Discussion

Proposed system has graphical user interface (GUI) which allows selection of input and displays output. GUI has been created by using MATLAB 2007b software which is as shown below in Fig. 2. Firstly, cover image of size 512 × 512 from the database is selected using GUI as shown in the Fig. 3. Then watermark image of size 128 × 128 is selected from the database as shown in the Fig. 4. After selection of the images, 3 level discrete wavelet transform (DWT) is applied to the cover image and LL sub-band is displayed as shown in Fig. 5. In the next part 3 level discrete wavelet transform (DWT) is applied to the watermark image and LL sub-band is displayed as shown in Fig. 6. After that actual embedding of both the images is done and 3 level Inverse DWT is obtained which gives watermarked image as shown in Fig. 7. In the next part, watermark extraction is done, which is as shown in the Fig. 8. In the last part, Mean square Error and PSNR is calculated using GUI itself. A number of experiments on different original gray scale images of size 512 × 512 and watermark gray scale images of different sizes from database have been obtained.

Performance of system has been analyzed using parameters such as PSNR and Mean square Error. Peak signal to noise ratio (PSNR) is used as metrics for measure of quality of watermark image. PSNR is measured in decibels (dB).

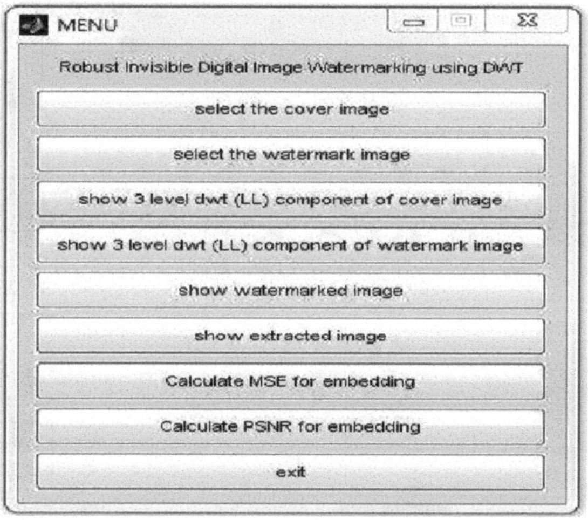

Fig. 2 Created GUI of proposed system

Fig. 3 Cover image selection

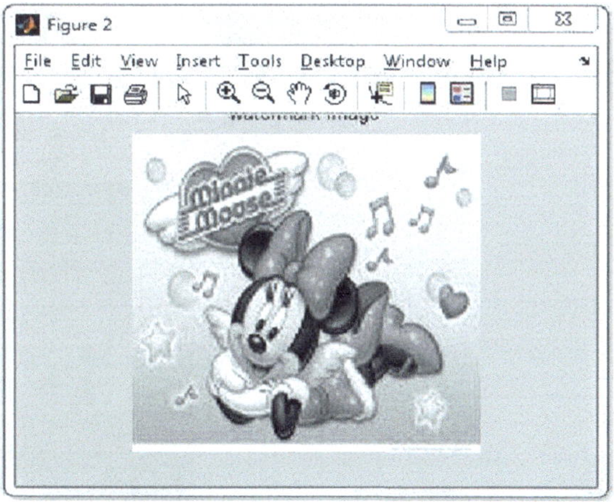

Fig. 4 Watermark image selection

Fig. 5 LL sub-band of cover image

Fig. 6 LL sub-band of watermark image

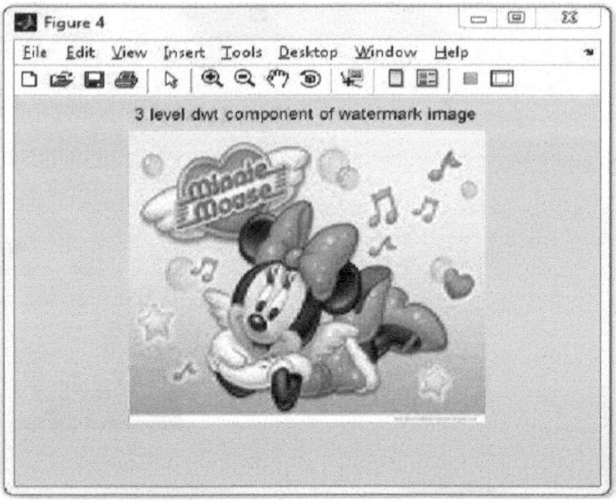

Fig. 7 Watermarked image

$$\text{PSNR} = 10\log_{10}\frac{(\text{Max})^2}{\text{MSE}}\,(\text{dB})\ \text{and}\ \text{MSE} = \frac{\sum_{i=1}^{M}\sum_{j=1}^{M}\left[f(i,j) - f'(i,j)]^2\right]}{M \times N}$$

For gray scale images, MAX value is 255. Where, Mean square Error can be calculated by using above equation. In this case, f is original image and f' is watermarked image.

Proposed system is based on the Discrete Wavelet Transform which gives better efficiency and accuracy. System is able to embed and extract watermark successfully from used database. If we observe other methods, Peak Signal to Noise Ratio

Fig. 8 Extracted watermark

Table 1 Experimental results

Original image	Watermark image	Extracted image	MSE	PSNR (dB)
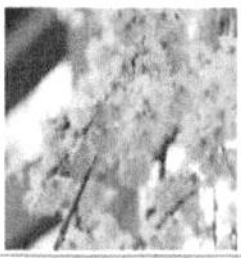			8.7274e-005	93.5113
			2.2844e-004	84.8559
			1.6414e-004	88.7030

is less whereas, but for this system we have obtained higher PSNR values. For different attacks system shows better performance, this suggests about robustness of the system.

Several experimental results are tabulated using MATLAB 2007b platform in which cover image of size 512 × 512 has been selected and watermark embedded is of various sizes such as 256 × 256, 12 8 × 128 and 64 × 64. Mean square error and PSNR values are calculated as shown in Table 1.

4 Conclusion

In this paper we have explained implementation of invisible digital image watermarking system using third level DWT with the help of MATLAB 2007b platform. If we look for implemented system, one can go for invisible watermarking of images. Accuracy of system is good. Watermark embedding and extraction ifs fairly simply with this system. Quality of image remains same even after watermark embedding and extraction process. Here, we have achieved watermarking system using self-generated database so we have not compared our results with results of other researchers.

References

1. Subramanyam AV, Emmanuel S, Kankanhalli MS (2012) Robust watermarking of compressed and encrypted JPEG2000 images. IEEE Trans Multimed 14(3)
2. Dixit MM, Kulkarni PK, Somasagar PS, Angadi VC (2012) Variable scaling factor based invisible image watermarking using hybrid DWT–svd compression-decompression technique. In: IEEE students' conference on electrical, electronics and computer science
3. Kundu S, Kumar S, Chander K (2012) Efficient genetic algorithm based image watermarking using DWT-SVD techniques. In: International conference on computing sciences
4. Li J, Dong C, Huang M, Bai Y, Zhang H (2012) The medical images watermarking using DWT and Arnold. IEEE
5. Bhargava N, Sharma M, Singh A, Garhwal, Mathuria M (2012) Digital image. SKP Engineering College, Tiruvannamalai, TN, India 21–22, pp 185–189
6. Tianming G, Yanjie W (2011) DWT-based digital image watermarking algorithm. In: IEEE the tenth international conference on electronic measurement and instruments, ICEMI, pp 163–166
7. Jabade VS, Gengaje SR (2011) Literature review of wavelet based digital image watermarking techniques. Int J Comput Appl (0975–8887) 31(1):28–31
8. Shi F, Shi Y, Lai L (2011) Optimization on digital watermarking algorithm based on SVD-DWT. In: IEEE international conference on granular computing
9. Lai C-C, Tsai C-C (2010) Digital image watermarking using discrete wavelet transform and singular value decomposition. IEEE Trans Instrum Meas 59(11):3060–3063
10. Liu J-L, Lou D-C, Chang M-C, Tso H-K (2010) A robust watermarking scheme using self-reference image. Comput Standards Interf Sciencedirect 356–367

Automatic Digital Modulation Recognition System Using Feature Extraction

H.L. Punith Kumar and Lakshmi Shrinivasan

Abstract Automatic modulation recognition is the vital part in the advanced communication system used for both military and civil applications. In this paper a new methodology is proposed for distinguishing five digital modulation schemes (ASK-2, ASK-4, FSK, BPSK and QPSK). The algorithm extracts the features from the received signal and they are tested against preset thresholds to determine the modulation type of received signal. The simulations are done using MATLAB 2013 and results show that the system has an average recognition rate of 99.6 % at SNR as low as 4 dB.

1 Introduction

The wireless networks are nearly at the edge of their capacity. But there is an increasing demand for wireless services as more and more consumers are entering the wireless networks. With advancements in connected cars, smart grids, machine to machine communication and domestic installations such as home health monitoring system in the field of civil applications resulting in the increased demand for wireless networks. The military applications such as electronic warfare, electronic surveillance, radio spectrum management, threat analysis, interference identification calls for increased demand on wireless network services.

But studies also show that allotted spectrums are not being used at all times efficiently [1]. Hence there is a need for effective spectral management. As a solution for this, intelligent and flexible radios called software defined radios are developed. Cognitive radios, adaptive radios and intelligent radios are advanced radio systems which improve the utilization of allotted spectral bandwidth efficiently.

H.L. Punith Kumar (✉) · L. Shrinivasan
Department of ECE, MSRIT, Bangalore, India
e-mail: puntihkumar067@gmail.com

L. Shrinivasan
e-mail: Lakshmi.s@msrit.edu

K.R. Attele et al. (eds.), *Emerging Trends in Electrical, Communications and Information Technologies*, Lecture Notes in Electrical Engineering 394,
DOI 10.1007/978-981-10-1540-3_21

The modern radio system includes spectrum sensing methodologies at the transmitter side and AMC at the receiver side. The intelligent radio system adopts different modulation techniques at the transmitter side based on spectrum sensing and channel condition. The receiver includes automatic modulation recognition system which identifies the type of modulation adopted in the received signal and there by enables the corresponding demodulator for recovering the data being sent from transmitter. Automatic modulation recognition of a communication signal plays a vital role in the performance of the whole advanced communication radios. As a result the demand for high efficient signal processing algorithms for automatic modulation recognition system is increased.

2 Literature Survey

The Automatic modulation classification (AMC) systems are developed for both analog and digital modulation types. There are mainly 2 approaches used in digital AMC and are: Maximum likely hood approach and Feature based approach.

Likelihood algorithms that make a decision based on the comparison of a likelihood ratio with a predefined threshold to minimize false decision probability. The advantage of this method is that its performance is usually optimal. The disadvantage is computational complexity and it is not robust [2]. In likelihood approach AMC is considered as composite hypothesis testing problem under likelihood based approach and is mainly depends on the PDF (probability distribution function) derived from the observed waveform. The PDF is derived such that it contains all information necessary for classification of the modulated signal.

Feature based method is most widely used approach and has an advantage of ease of implementation and most robust compared to likelihood based approaches. But are non-optimal, however if FB based systems are designed carefully by choosing the proper set of features and appropriate thresholds, near optimal results can be obtained [3]. Feature based algorithm mainly consists of subsystem which extracts the features from the signal and other subsystem which makes the decision based on the derived features.

The algorithm proposed [4] uses Lempel-Ziv complexity and designed to recognize modulation signals like FSK-2, FSK-4, PSK-2 and PSK-4. The results have been presented at 10 and 20 dB SNR only. A method proposed in [5] which extracts the features of signal and then decision theoretic approach is used to identify modulation type. The sample results have been shown only for 15 dB SNR. The method proposed in [3] used feature extraction and decision theoretic approach and results have shown a better recognition rate of 98.6 % at 4 dB.

3 Algorithm Implementation

Modulation recognition system involves feature extraction section which is designed for extracting the features from the signal followed by decision making section which identifies the modulation format of the signal and generates the necessary control signals for configurable demodulator.

The feature set used in our algorithm includes following parameters.

3.1 Maximum Value of Power Spectral Density

It is the Peak value of the power spectral density of the normalized centered instantaneous amplitude of the intercepted signal segment, and is defined [2] by

$$\gamma_{\max} = \frac{max|DFT(a_{cn}(i))|^2}{N} \tag{1}$$

where N is the No. of samples taken; a_{cn} (i) = (a(i)/m) − 1; in which a(i) is the instantaneous amplitude and m is the mean.

$\gamma_{\max}$ feature can express the character of signal's envelope and can be used to differentiate between the modulation scheme which carry amplitude information and those that do not. The method of obtaining instantaneous amplitude of a signal is an interesting task, which is done with the help of Hilbert transform. The instantaneous amplitude obtained for ASK4 and FSK with added noise is shown in Fig. 1. It can be seen from the Fig. 1 that the instantaneous amplitude follows the envelope of the signal. $\gamma_{\max}$ value is more for ASK4 signal, because ASK4 has more variation in its envelope when compared to that of FSK signal.

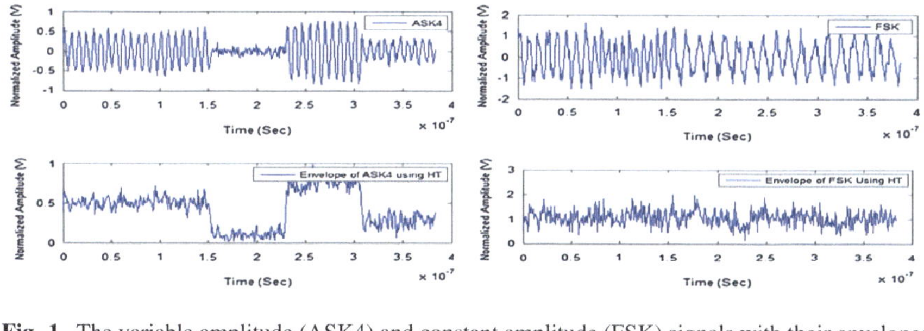

Fig. 1 The variable amplitude (ASK4) and constant amplitude (FSK) signals with their envelopes using HT

3.2 4th Moment of Mean (μ_4)

It is the statistical parameter and is the 4th power of difference between instantaneous value of signal and the mean value of signal over a period. This feature is the numerator of the statistical parameter called kurtosis. This feature is used in the proposed algorithm for distinguishing between ASK2 and ASK4 as well as for PSK and FSK signal identification.

3.3 Feature Extraction by Spectrum Analysis

The third feature is extracted from spectrum analysis methodology and is used for differentiating BPSK and QPSK and is found easier and effective when compared to all other methods of estimation including feature extraction and other statistical parameter. The basic idea for opting spectrum analysis method is that when we square the BPSK signal, it loses its phase information since it includes ±180° phase shifts, in contrast QPSK signal has phase shifts which is multiple of ±45° or ±90°, there by it preserves its phase information even after squaring the signal that can be easily observed in the spectrum analysis [5]. This fact can be clearly understood by observing their spectrums as shown in Figs. 2 and 3.

After the above cited features are extracted, the decision making algorithm is developed to classify the modulation type of the received signal. Every 2000 samples of the incoming data are given for feature extraction section which extracts the necessary features and feeds the result to the decision making algorithm. The flow chart of the decision making algorithm is given in Fig. 4.

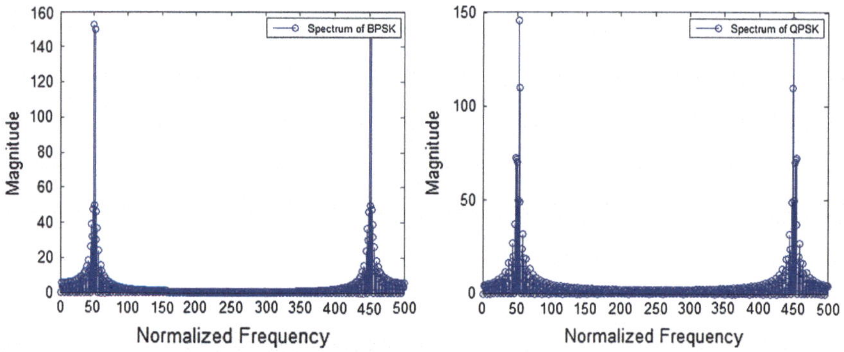

Fig. 2 Spectrum of BPSK and QPSK signals before squaring

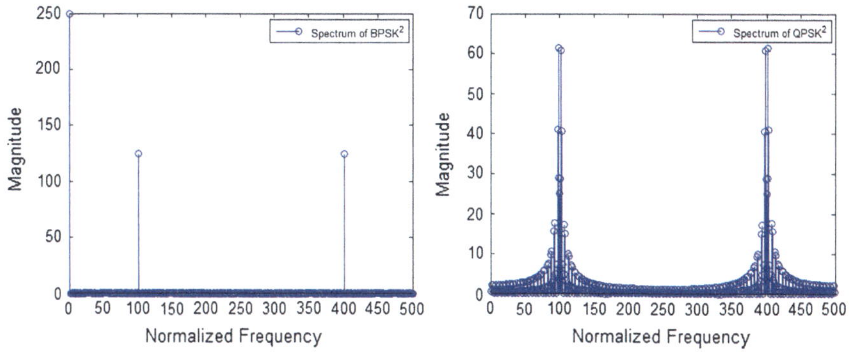

Fig. 3 Spectrum of BPSK and QPSK signals after squaring

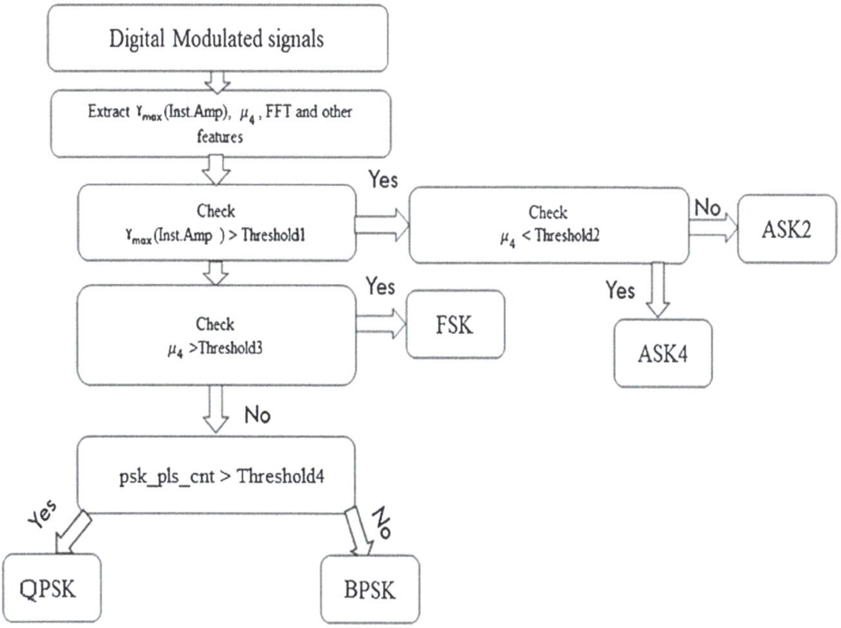

Fig. 4 Flow chart of decision making algorithm

4 Result and Discussion

The plot of recognition rate of individual modulation type and the average recognition rate of the AMC algorithm developed with varying SNR is given in the Fig. 5.

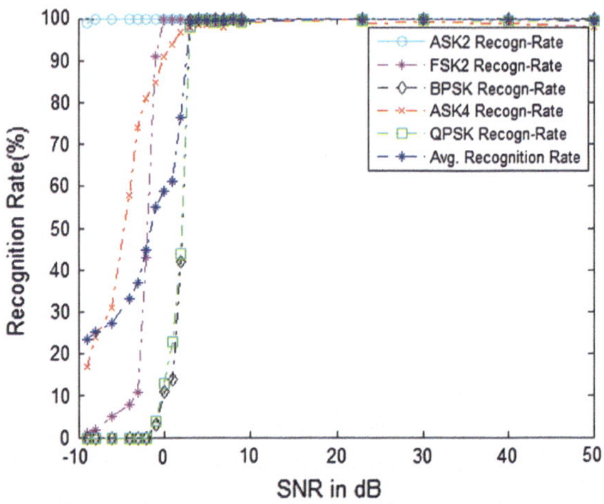

Fig. 5 The plot of recognition rate of the AMR system with different SNRs

The sudden decrease in the recognition rate of the modulated signals when SNR of the signals falls below 3 dB can be seen from the graph shown in Fig. 5. The recognition rate obtained for different modulated signals in the transition region of Fig. 5 are given in Table 1.

The results obtained at 10 dB SNR for Automatic modulation classifiers used in [2, 6, 7] are given in Table 2. The difference between proposed methodology and methods used in [2, 6] is the type of features extracted from the signals. In addition, a method proposed in [2, 6] uses ANN for decision making. The method proposed in [7] uses wavelet features and SVM classifiers.

The average recognition rates obtained in existing methodologies [2, 6, 7] are 99.5 %, 97.12 % and 95.1 % respectively at 10 dB SNR. The result obtained in proposed methodology has an average recognition rate of 99.6 % at 4 dB SNR and is shown in Table 1. Hence performance of the proposed methodology has an advantage of improved performance even at low level of SNR (4 dB) compared to the methodologies used in [2, 6, 7].

Table 1 Recognition rate of AMR system

Proposed method	2 dB (%)	3 dB (%)	4 dB (%)	5 dB & >5 dB (%)
ASK2	99.3	~100	~100	~100
FSK2	99.4	99.5	~100	~100
BPSK	42	98.7	~100	~100
ASK4	97	98	98.5	~100
QPSK	44	98	99.8	~100
Overall recognition rate	76.3	98.8	99.6	~100

Table 2 Results obtained by existing methodologies [2]

	Modulation type	Correct recognition rate (%)
Method used in [6] with SNR = 10 dB	ASK2	97.0
	ASK4	99.8
	FSK2	92.5
	BPSK	100
	QPSK	96.3
Average recognition rate		97.12
Method used in [7] with SNR = 10 dB	ASK2	100
	ASK4	95.5
	FSK2	95.75
	BPSK	100
	QPSK	84.25
Average recognition rate		95.1
Method used in [2] with SNR = 10 dB	ASK2	98.9
	ASK4	99.2
	FSK2	99.8
	BPSK	99.8
	QPSK	99.7
Average recognition rate		99.48

5 Conclusion

Automatic modulation recognition is a vital part in the modern digital modulation recognition system. The performance of the AMR section in advanced communication system affects the overall system performance. The Proposed methodology enhances the AMR system performance, even at lower level of SNRs compared to existing methodologies by the use of features extracted from the received signal. However, the system performance can be further improved by the use of adopting threshold technology for finding the thresholds for extracted features used in decision making section.

References

1. Subhedar M, Birajdar G (2011) Spectrum sensing techniques in cognitive radio networks: a survey. Int J Next-Gener Netw (IJNGN) 3(2)
2. Popoola JJ (2014) Automatic recognition of both inter and intra classes of digital modulated signals using artificial neural network. J Eng Sci Technol 9(2):273–285
3. Punith Kumar HL, Lakshmi S (2015) Automatic digital modulation recognition using minimum feature extraction. In: 2015 2nd international conference on "Computing for sustainable global development", 11th–13th March 2015

4. Ronghua Z (2007) A new key features extraction for automatic modulation recognition. WiCom
5. Subbarao MV, Khasim NS, Jagadeesh T, Sastry MHH (2013) A novel technique for automatic modulation classification and time-frequency analysis of digitally modulated signals. IJSIPR 6(2)
6. Azzouz EE, Nandi AK (1997) Automatic modulation recognition—II. J Franklin Inst 334 (2):275–305
7. Park C-S, Choi J-H, Nah S-P, Jang W (2008) Automatic modulation recognition of digital signals using wavelet features and SVM. In: Proceedings of 10th international conference on advanced communication technology, pp 387–390

A Secure Route Discovery Protocol for AODV Based Mobile Adhoc Networks Using Hyperelliptic Curve Cryptography

P. Vijayakumar, R. Rajashree and P. Sandhya

Abstract A Mobile Adhoc Network (MANET) consists of numerous number of nodes, neighbor nodes are connected by means of radio links. Each node is constrained with limited power, bandwidth and scalability. Many researchers lay their effort to develop various techniques for providing a secure route in MANET. As a result, an Elliptic Curve Cryptography (ECC) technique is used to develop a secure shortest algorithm for routing the packet from source node to destination node, since ECC is famous due to its shorter key length and less computational overhead for encryption and decryption operation. ECC is more suitable for power constrained devices. But ECC based secure routing protocol requires a large number of Route Request (RREQ) packet in Adhoc on Demand Distance Vector (AODV) protocol and consumes more power. In order to reduce the number of RREQ packet and power consumption, a proposed secure energy efficient routing protocol is implemented. Finally, discuss the improvement of the proposed approach with the existing systems using GloMoSim software tool.

Keywords AODV protocol · Elliptic curve cryptography · Hyperelliptic curve cryptography · Mobile adhoc network

---

P. Vijayakumar (✉)
School of Electronics Engineering, VIT University, Chennai Campus, Chennai 600127, Tamilnadu, India
e-mail: vijaya.kumar@vit.ac.in

R. Rajashree
GKM College of Engineering and Technology, Chennai 600063, Tamilnadu, India
e-mail: rajashree.ece@gmail.com

P. Sandhya
Aksheyaa College of Engineering, Chennai, India
e-mail: sandynakshatra@gmail.com

K.R. Attele et al. (eds.), *Emerging Trends in Electrical, Communications and Information Technologies*, Lecture Notes in Electrical Engineering 394,
DOI 10.1007/978-981-10-1540-3_22

1 Introduction

A Mobile Adhoc Network (MANET) is an infrastructure less, self-configured, power constrained wireless network. Each client is provided with a wireless radio link which makes network free to move in any direction any place. So, it changes its link to other devices frequently. Due to limited battery power, wireless nodes are all vanished and damaged once it dissipates power tends to link failure. Hence reducing power consumption becomes very important in MANET routing. In order to reduce power consumption for MANET, the energy efficient broadcasting algorithm has to be proposed. In addition to this, security is the major issues in routing of MANET [1]. Hyperelliptic Curve Cryptography (HECC) solves the security issues in many networks. The research on HECC highlighted on finding effective methods to select secure Hyperelliptic Curves, fast operations on the Jacobian and implementation of HECC for use in practical applications to enhance the network security [2–8]. This section gives the introduction about MANET and HECC. Section 2 details about secure energy efficient routing protocol. Section 3 shows the simulated results and finally conclusion of the paper.

2 Proposed Secure Energy Efficient Routing Protocol (SEER)

The existing AODV based Secure Routing Protocol (SRP) [9] is based on Elliptic Curve Cryptography (ECC) which requires 160 bit key size to encrypt the data leads to increased power consumption. The traditional system also requires more RREQ packet to discover the shortest path to deliver the packet. In order to reduce the power consumption as well as to maintain the power level of the node. The expansion of ECC known as HECC over genus 2 curve based encryption technique is used to route the packet from source to destination which reduces the energy consumption as well as to maintain the power level of the node. This maximizes lifetime of the network and also provides the secure routing of packets from source to destination. The proposed Secure Energy Efficient Routing protocol (SEER) is especially designed for AODV routing protocol based MANET. The proposed protocol selects a secure route which has more battery power. Addition of new POWER field and finding relative mobility factor of each node with respect to neighbor node will provide additional feature to the proposed protocol. SEER protocol does not require a new route discovery process to find a new path in case of any link failure. Data packets are transmitted through the shortest path in a more reliable manner. Because of back up routes available in source node, there will be no link breakage in the network. It also provides the authenticated routing of packets from source to destination using HECC. It also reduces the number of RREQ packet to find the shortest path in AODV protocol.

2.1 Certificate Distribution Centre

During the route discovery process, SEER makes use of HEC having genus g = 2 of polynomial having degree 2g + 1 value is used to generate the certificates to provide data integrity, mutual authentication, and non-repudiation. Certificate Distribution Centre (CDC) generates a cryptographic certificate and distribute to the entire node in the network. SEER requires trusted certificate T from the CDC (server), whose public key (PUK_A) is known to the entire user in the network. These cryptographic certificates consists of IP address of the user A (IPA_A), Public key of user A (PUK_A), Time Stamp (TS), Expiry Date (ED) and Public key of the Trusted Server (PUK_T) are used to authenticate each and every node in the network. During the exchange of routing messages, HECC based key exchange algorithm is used to substitute the public keys and certificates between node and trusted server (TSR). Each and every client should request a certificate from a trusted server before entering into the network. At least one certificate has to set out from TSR, after securely authenticate the desired host. Trusted server provides a certificate (certfA) to node A as follows:

$$TSR \rightarrow A : \mathrm{certf}_A = \{ IPA_A,\ PUK_A,\ TS,\ ED,\ PUK_T \} \quad (1)$$

Certificate of Node A consists of IP address of node A (IPAA), public key (PUKA), timestamp (TS) at which time certificate was created, and expiry date (ED) at which time the certificate get expires. Each and every client should maintain fresh certificates with the certificate distribution center.

2.2 Secure Route Discovery

Secure route discovery process involves finding a safe and energy efficient shortest routing path using AODV based routing protocol. Step 1: Source node A desires to deliver a package to the destination node X using AODV protocol. Node A sends a Route Request (RDP) packet along with IP address of the destination (IPA_X), Nounce value of the user A(N_A), POWER Field (P_A) and Certificate of user A ($Certf_A$) and count field are used to authenticate the user. Here the content of the certificate and POWER filed values are encrypted using HECC encryption algorithm having public key of the user A (PUK_A) to reduce communication and computational complexity.

$$A \rightarrow B : \{ RDP,\ IPA_X,\ N_A,\ P_A,\ count,\ \mathrm{certf}_A,\ PUK_{A-} \} \quad (2)$$

Step 2: Once node B receives the RREQ message from Source node A. Node B decrypts the message using its private key with the help of HECC based decryption algorithm and checks whether it is a destination node, and also checks its expiry date for avoiding invalid user. It updates its own information such as POWER value, adding Certificate, checking the validity, and authenticate the user.

$$B \rightarrow C : \{RDP,\ IP_X,\ N_A,\ P_{(A+B)},\ count,\ certf_A,\ certf_B,\ PUK_{A-},\ PUK_{B-}\} \quad (3)$$

Step 3: After Node C receives Route Discover Packet (RDP) from Neighbor Node B, it validates the signature using HECC for both A and B. Node C eliminates B's certificate and signs the content of message A using HECC and append its own certificate. Then rebroadcast the RDP with POWER value and Certificate. Every node presents in the network repeat the same step until reach the destination node X.

$$C \rightarrow D : \{RDP,\ IP_X,\ N_A,\ P(A+B+C),\ count,\ certf_A,\ certf_c,\ PUK_{A-},\ PUK_{C-}\} \quad (4)$$

2.3 Secure Route Setup

Step 1: After receiving the RDP from the destination, it calculates the average power of each link and finds a path which is having a maximum average power with minimum hop count. RREP message will be rebroadcasted in reverse manner to the same source node on the same route. Once it arrives at the destination node X, RREP message is sent to the node D in addition to average power (PAV) as shown below.

$$X \rightarrow D : \{RREP,\ IPA_A,\ N_A,\ P_{AV},\ certf_X,\ PUK_X,\ count\} \quad (5)$$

Step 2: Node D forwards the RREP packet back to same node C from which they get the RDP message. The RREP message includes hop count, average power, IP address of Source node A, Certificate of Destination node x and nonce value. Each node delivers the RREP message along the same path to the source by signing the message using HECC algorithm. Prior to forwarding the RREP message to next node C, it appends its own certificate to authenticate each node in the network as shown below:

$$D \rightarrow C: \{RREP, IPA_A, N_A, PUK_X, certf_X, PUK_D, certf_D, P_{AV},\ count\} \quad (6)$$

Step 3: Once RREP message reaches node C, it checks the validity of D's signature from the received message using HECC based signature verification algorithm. After verifying the node, it detaches the signature and certificate. Then it signs the content of the message using HECC and appends its own certificates before replay to B. While returning the RREP packet, every node checks its nonce value and signature using HECC algorithm. This will avoid impersonation attack and a replay of X's message by malicious node.

$$C \rightarrow B: \{RREP, IPA_a, N_A, PUK_X, certf_X, PUK_C, certf_c,\ P_{AV}, count\} \qquad (7)$$

Step 4: Once the RREP message is received by the source node, HECC based signature verification algorithm verifies the destination signature and nonce value. It also compares the power level of the entire link and seeing the path which is having maximum power level and minimum hop count. Selected path is used to transfer the packet from source to destination. The above steps are repeated to obtain the path for all links in the network. Then the path for all the links are stored in back up table of the source node.

2.4 *Route Maintenance*

If a link failure happened it would not go for new route discovery process. Instead it will select a path from backup route table which is available in source node. If any node is failed due to power loss or movement of node it would send error message (ERR) to the nearby node. This RERR message is returned to source node on the same route followed by node B. Nonce Nc value gives assurance that RERR message is fresh.

$$C \rightarrow B : [RERR,\ IPA_A,\ IP_X,\ N_C]\ PUK_C,\ certf_C \qquad (8)$$

3 Results and Discussion

A proposed SEER protocol has been implemented using GloMoSim software tool simulation. Considered network area value is 1000 × 1000 m and number of nodes taken to implement the algorithm is 20 nodes. The constant data bit transmitted between sources to destination is 512 bytes. Figures 1, 2, 3, 4, 5 and 6 shows the observed result for 20 node network. Each data point is an average of 20 simulation runs with identical configuration but different randomly generated mobility pattern. As shown in Fig. 2 the PDR for SEERP is 95 % higher than existing algorithms. This implies that SEERP is highly effective in finding and maintaining routes for discovering of data packets, even with relatively high node mobility. Power field related routing metrics significantly reduces the power consumption in SEERP is shown in Fig. 3. This shows that SEERP is consuming less power than SRP which increases the stability of nodes and lifetime of the network. Figure 4 shows that remaining battery power in the nodes of the system as function of time. SEERP carries more packets because it was less operational oversees (RREP, RREQ,

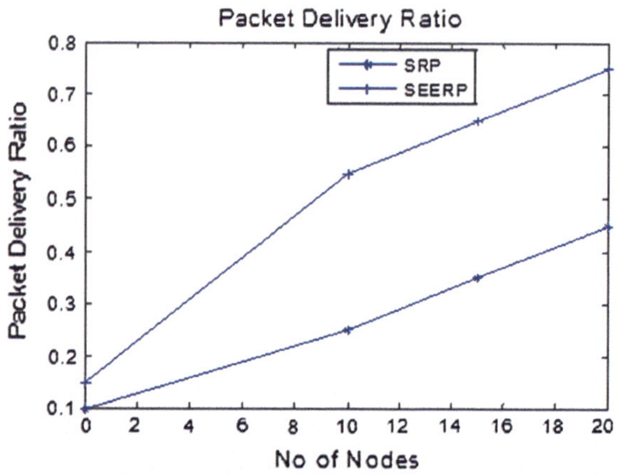

Fig. 1 Packet delivery ratio

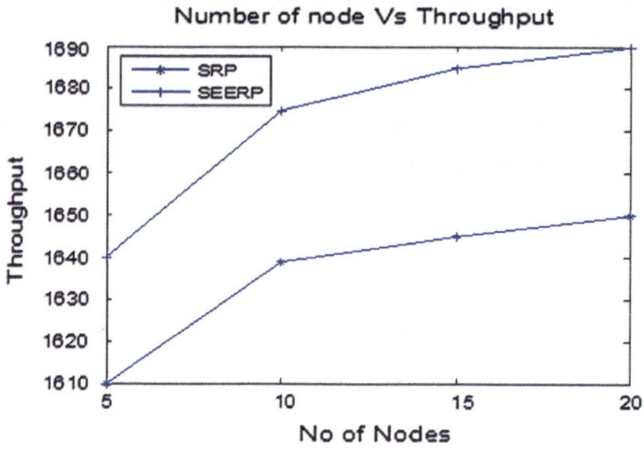

Fig. 2 Average path length versus node speed

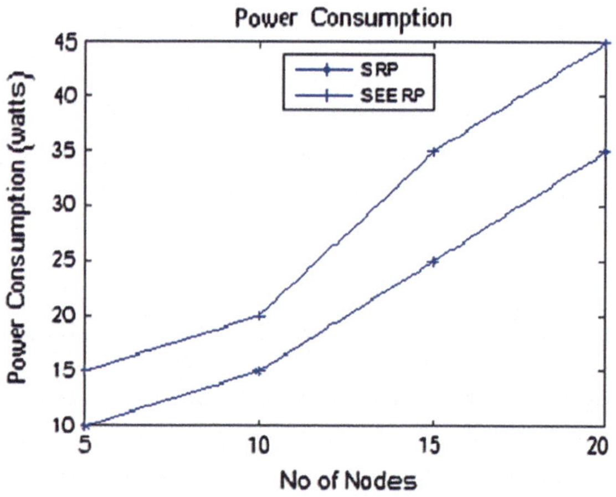

Fig. 3 Power consumption

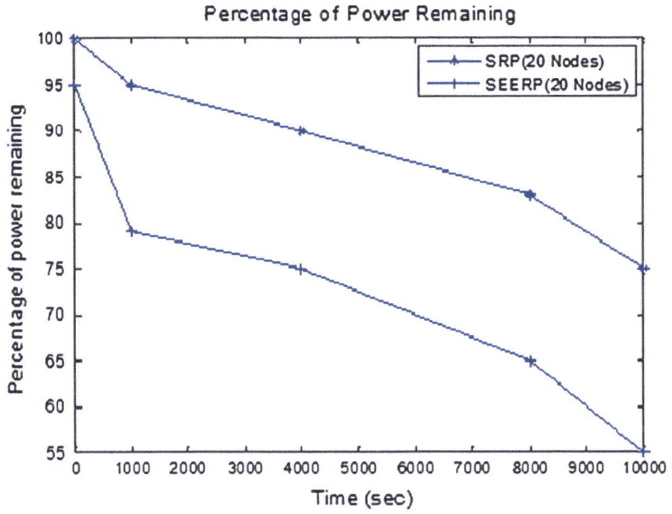

Fig. 4 Percentage of power consumption

RERR) than the SRP. It shows the tradeoff between powers saved by SEERP is less than the SRP. Figure 5 shows that the average end-to-end delay versus node speed where node speed increases. Average data packet latency gets decreased; this will improve the packet delivery ratio and throughput. Reduced transmitting power of the nodes results in reduction of the total transmit power, hence the network lifetime has to be increased. This can be observed from the Fig. 6.

Fig. 5 End to end delay

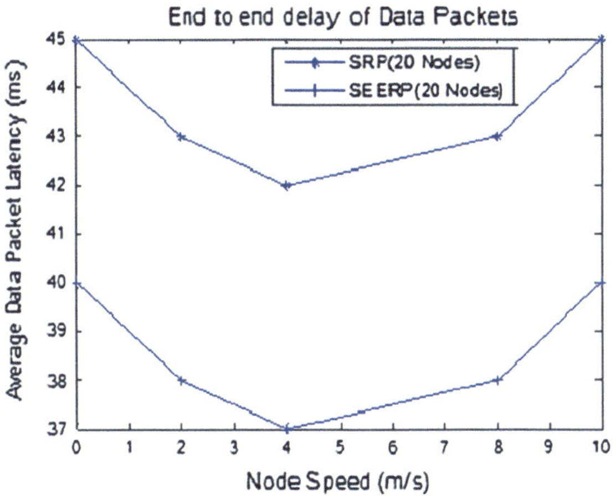

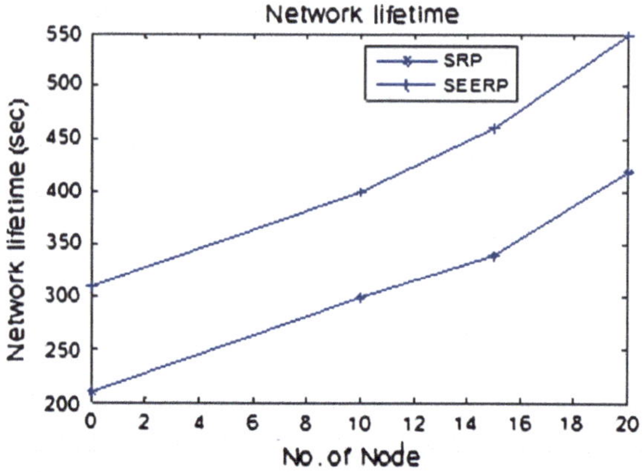

Fig. 6 Network lifetime

4 Conclusion

Finally, SEER protocol reduces the number of RREQ packet in AODV routing protocol, which consumes less power, processing time, end-to-end delay and high packet delivery ratio. The proposed broadcasting algorithm helps to increase the throughput by decrease the packet loss due to non-availability of node having enough battery power to retransmit the data packet to next node. It is also helpful to see an optimal path without the new route discovery process, in addition to this it routes packets from source to destination in a secure manner using Hyper Elliptic Curve Cryptography technique.

References

1. Vijaya I, Mishra PB, Dash AR, Rath AK (2011) Influence of routing protocols in performance of wireless mobile adhoc network. In: The IEEE proceedings of second international conference on emerging applications of information technology (EAIT), pp 340–344
2. Lai J-Y, Huang C-T (2011) Energy-adaptive dual-field processor for high-performance elliptic curve cryptographic applications. In: The IEEE transactions on very large scale integration (VLSI) systems 19(8)
3. Tutanescu I, Anton C, Ionescu L, Caragata D (2012) Elliptic curves cryptosystems approaches. IEEE Proc Inf Soc 357–362
4. Yingzhan K (2011) Extended fault analysis on elliptic curve cryptosystems against repeated doubling. In: The IEEE proceedings of international conferences of instrumentation, measurement, computer, communication and control, pp 545–548
5. Fan X, Wollinger T, Gong G (2007) Efficient explicit formulae for genus 3 hyperelliptic curve cryptosystems. IEEE J Proc Inf Secur 1(2):65–81
6. Wollinger T, Pelzl J, Paar C (2005) Cantor versus Harley: optimization and analysis of explicit formulae for hyperelliptic curve cryptosystems. IEEE Trans. Comput. 54(7):861–872
7. Fang Y, Wu Z (2012) A new parallel processor architecture for genus 2 hyperelliptic curve cryptosystems. In: The proceedings of IEEE computer society annual symposium on VLSI (ISVLSI), pp 177–182

8. Nizamuddin N, Ch SA, Amin N (2011) Signcryption schemes with forward secrecy based on hyperelliptic curve cryptosystem. In: The IEEE proceedings of high capacity optical networks and enabling technologies (HONET), pp 244–247
9. Jalil KA (2011) Securing routing table update in AODV routing protocol. In: The IEEE proceedings of IEEE conference on open systems (ICOS), pp 116–121
10. Vaidya B, Makeakis D (2011) Provisioning secure on-demand protocol in mobile ad hoc network. In: The IEEE proceedings of international conference on internet (AH-ICI), pp 1–5

An IoT Based Remote Monitoring of Landfill Sites Using Raspberry Pi2

K. Tharun Kumar Reddy, P. Ajay Kumar Reddy, P. Siva Nagendra Reddy and G.N. Kodanda Ramaiah

Abstract Now a day's global warming is increasing due to the emission of green house gases from landfill site and industries. The main green house gases are Carbon Dioxide, Methane and Carbon Monoxide etc [1]. The long term exposure to these gases causes lung and heart diseases to the people who are living near to the land fill sites and industries. It is also harmful to animals and plant life. In this paper we describe an IoT based remote monitoring system for measuring the concentration of green house gases emitting from landfills and industries [6]. This proposed system measure the concentration of Methane, Carbon Dioxide and Carbon Monoxide using semiconductor gas sensors and also measures humidity, temperature, pressure etc. The data from the sensors are processed by using Raspberry Pi2 and sensor values are uploaded to the internet using IoT technology. This system will send alert message to relevant authorities and peoples when concentrations of harm full gases reaches permissible levels using GSM Modem/internet SMS gateway. This system has given access to everyone who wish to know the concentration of gases emitted from landfill site or industrial area by sending a request message to the system.

Keywords Internet of things · Raspberry Pi2 · Land fill sites · GSM/GPS · Gas sensors

1 Introduction

The industries and landfill sites are becoming the sources for environmental degradation. The air quality in surrounding areas of landfill sites and industries is reducing at very fast rate due to human activities. These industries and landfill sites

K. Tharun Kumar Reddy (✉)
Vemu Institute of Technology, Pakala, A.P., India
e-mail: kethireddy.tharun@gmail.com

P. Ajay Kumar Reddy · P. Siva Nagendra Reddy · G.N. Kodanda Ramaiah
Department of ECE, Kuppam Engineering College, Kuppam, A.P., India

K.R. Attele et al. (eds.), *Emerging Trends in Electrical, Communications and Information Technologies*, Lecture Notes in Electrical Engineering 394,
DOI 10.1007/978-981-10-1540-3_23

Table 1 Typical composition of landfill gas

S.No.	Constituent gas	Range	Average
1	Methane (CH_4)	35–60 %	50 %
2	Carbon dioxide (CO_2)	35–55 %	45 %
3	Nitrogen (N_2)	0–20 %	5 %
4	Oxygen (O_2)	0–2.5 %	<1 %
5	Hydrogen sulfide (H_2S)	1–1700 ppm	21 ppm
6	Halides	NA	132 ppm
7	Water vapor (H_2O)	1–10 %	NA
8	Non Methane organic compounds (NMOCs)	237–14.249 ppm	2700 ppm

increasing water pollution by releasing heavy metals harmful chemicals into water and also increasing air pollution by releasing toxic gases into the atmosphere [1]. The landfill sites are the dumping sites which contain organic materials, toxic waste and recyclable material. The landfill sites releasing toxic gas into atmosphere. The main components of landfill gas are the green house gases, methane and carbon dioxide. The fact that methane is about 20 times more harm full to the environment than carbon dioxide [2]. The main constituents of landfill gas and their proportions are shown in Table 1.

2 Defining System Functionality

This proposed system employs different sensors, Raspberry Pi2, Wi-Fi Dongle, GSM/GPRS module. The block diagram of remote monitoring system is shown in Fig. 1.

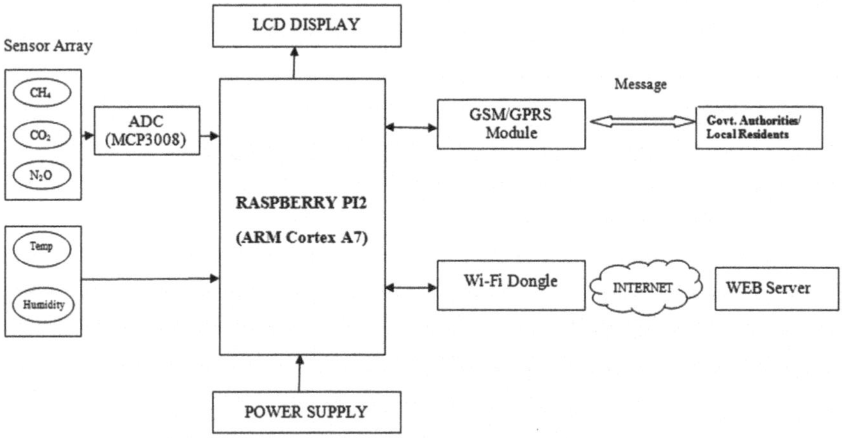

Fig. 1 Block diagram of remote environmental monitoring system

The proposed system will measure concentration of different gases like Carbon Monoxide, Carbon Dioxide and Methane etc. Besides that the system will also measure Humidity, Pressure and Temperature [3]. The Metal Oxide Semiconductor sensors MQ5, MQ135, MQ7 are used to measure Methane, Carbon Dioxide and Carbon Monoxide respectively. The system employs DHT22, BMP180 sensors which are used to measure Humidity, Temperature and Pressure [4]. The output of Gas sensor (MQ5, MQ135, MQ7) is in analog nature. The Raspberry Pi will accept only the digital signals. Hence the output of gas sensors is connected to the Analog to Digital Converter (MCP3008). The output of ADC is connected to GPIO pins of Raspberry Pi. The Humidity and Temperature sensors (DHT22), Pressure sensor (BMP180) are directly connected to GPIO pins of Raspberry Pi, because the output of DHT22 and BMP180 is in Digital form.

The Raspberry Pi will process the data from sensors and compare the values with safety levels. It will alert the Govt. Authorities and surrounding people by sending concentration of gases through Short Message Service when the concentration of gases exceeds the safety level [5]. Here the Raspberry Pi is connected wirelessly to the Internet using Wi-Fi Dongle. Now the Raspberry Pi will upload real time measured values to the Xively IoT platform. The uploaded data can access any one at anywhere and anytime by visiting the corresponding Xively web address.

The flow chart for proposed system is shown in Fig. 2.

3 Hardware Requirements

Raspberry PI2:

The Raspberry Pi2 is a small sized, single board portable computer designed for promoting the teaching of basic computer science in schools. The Raspberry Pi2 is a main processing device in remote environment monitoring system. The Raspberry Pi2 is based on Broadcom BCM2836 system on chip (SoC), which has quad core ARM Cortex A7, 900 MHz processor, 1 GB RAM and Video Core IV dual core GPU. The Raspberry Pi2 model B is shown in Fig. 3.

The other features are 4 USB ports, full HDMI port, Ethernet port, 40 pin GPIO connector, 15 pin camera interface and serial display interface. The Raspberry Pi operates with 5 V supply. The Raspberry Pi2 can run Linux–kernel based operating systems and also windows 10.

Sensors Array:

Methane Sensor (*MQ5*):

The MQ5 is a metal oxide semiconductor type gas sensor. The MQ5 gas sensor is used to detect methane (CH_4) due its high sensitivity to methane concentrations [8]. The Tin Oxide (SnO_2) is the sensitive layer in the design of MQ5 [9]. It has 6 pins. In this 4 pins are used to fetch the signals and remaining 2 pins are used for providing heating current. The other features of MQ5 sensor is fast response, simple drive circuit, stable and long life (Fig. 4).

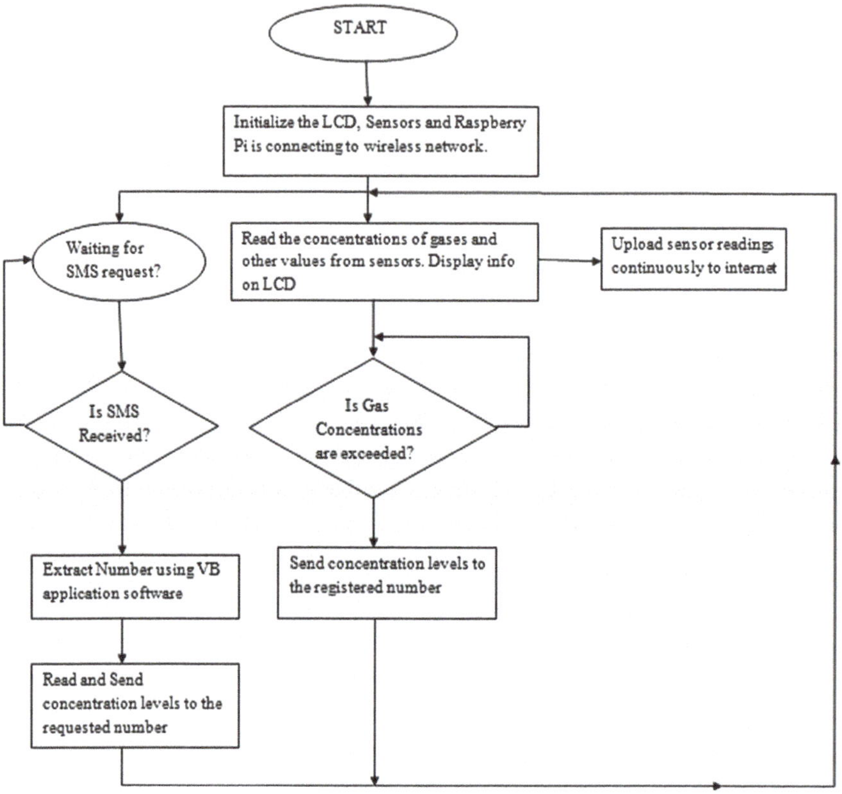

Fig. 2 Flow chart

Fig. 3 Raspberry Pi2 (model B)

Carbon Monoxide Sensor (*MQ7*):
The MQ7 gas sensor is used to detect carbon monoxide due to its high sensitivity to methane concentrations [10]. The MQ7 is designed by micro AL_2O_3 ceramic tube,

Fig. 4 MQ5 methane sensor

Fig. 5 MQ7 carbon monoxide sensor

measuring electrode, tin dioxide sensitive layer and heater. The MQ7 measuring range is from 20 to 2000 ppm (Fig. 5).

Carbon Dioxide Sensor (*MQ135*):
MQ135 is a metal oxide semiconductor type gas sensor which has high sensitivity to Carbon dioxide (CO_2) so here MQ135 is used to measure CO_2 Concentrations [11]. The MQ135 measuring range is from 0 to 1000 ppm (Fig. 6).

Pressure Sensor (*BMP180*):
The BMP180 is a digital barometric pressure sensor with high performance. It consumes 3u Amp power. The BMP180 is based on peizo resistive technology. The measuring pressure range is 300–1100 hpa and temperature range is −40 to 850 °C.

Fig. 6 MQ135 carbon dioxide sensor

It has seven pins and is available in LGA package. The BMP 180 sensor can communicate with processor through I2C bus. It will operate on voltages from 1.8 to 3.6 V (Fig. 7).

Humidity and Temperature Sensor (*DHT22*):
DHT22 is a capacitive type humidity sensor. It measures relative humidity and temperature also. It measures relative humidity and temperature also. The output is a digital calibrated signal. The DHT22 produces 40 bit data output, which includes humidity, temperature and checksum data (Fig. 8).

Analog to Digital Converter (MCP3008):
The MCP3008 is a successive approximation type 10 bit analog to digital converter. The MCP3008 has 8 input channels, on chip sample and hold circuit and SPI serial interface. The MCP3008 is offered in 16 pin SOIC and PDIP packages. The resolution of MCP3008 is 10 bit (Fig. 9).

Fig. 7 BMP 180 pressure sensor

Fig. 8 DHT 22 humidity and temperature sensor

Fig. 9 MCP3008 ADC

4 Implementation and Results

The Raspberry Pi2 is interfaced with the sensors array, GSM module and display unit. The Raspberry Pi2 will upload sensor values to the internet into Xively IoT platform by sending an HTTP request to the Xively server. The SMS alerts are sent to the respective authorities and anyone who wish to access the real time values can send SMS to the authorized number [12]. The sensors values are sent as a SMS to the corresponding number through internet SMS gateway using Visual Basic application.

The proposed system is practically verified at landfill site located in Kuppam. The measured real time values are can be easily be logged in Xively in the address https://personal.xively.com/feeds/1999131271. The graphs are plotted on a web page between measured values (Y-axis) and time period (X-axis). The results are shown in Figs. 10 and 11.

Fig. 10 Screen shots of uploaded gas concentrations at land fill site on a web page

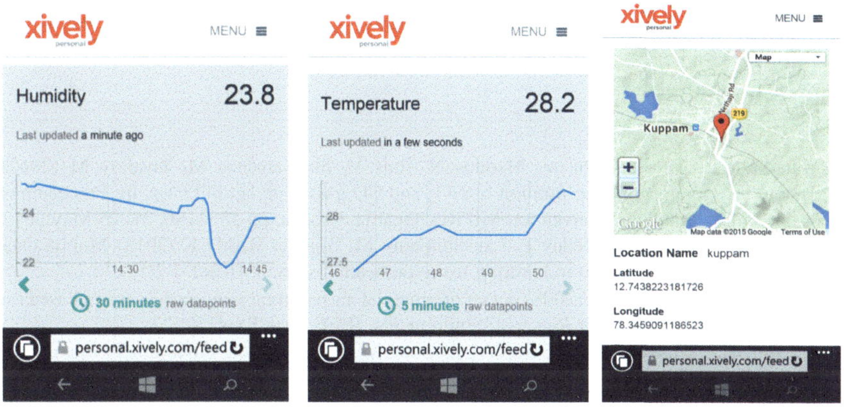

Fig. 11 Screen shots of uploaded humidity and temperature readings and location of a land fill site on a web page

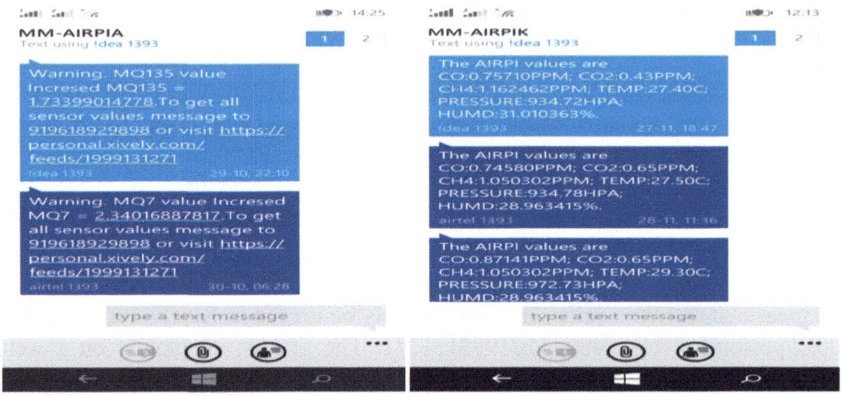

Fig. 12 Screen shot of SMS alert received on a mobile

The screen shot of warning alert received on a mobile when concentrations increased above safety level and message was received on a user request are shown in Fig. 12.

5 Conclusion and Future Scope

The proposed system is successfully designed and implemented using IoT technology and Raspberry Pi. The measured gases concentrations are uploaded to the internet for visualizing the data to the end users. The Raspberry Pi is proven high speed, low cost, smart and efficient platform for implementing the remote monitoring system. In future we can also capture and forward the landfill sites images to the Govt. Authorities through Social networks.

Acknowledgments This work has been carried out with the financial support of The Institution of Engineers (India), to which authors are very grateful.

References

1. Mihajlovic Z, Milosavljevic V, Maodus N, Rajs V, Slankamenac M, Zivanov M (2012) System for monitoring concentration of NO_2 and CO gasses on landfill sites. In: Proceedings of the 35th international convention MIPRO, Opatija, Croatia, pp 183–186, 21–25 May 2012
2. Manjula S, Ajay Kumar Reddy P, Lakshmipathy M, Bhaskar Reddy K (2013) Mobile data acquisition for air pollution monitoring using embedded system. IJERT J 2(11)
3. Kularatna N, Sudantha BH (2008) An environmental air pollution monitoring system based on the IEEE 1451 standard for low cost requirements. IEEE Sens J 8:415–422
4. Mihajlovic Z, Milosavljevic V, Rajs V, Živanov M (2013) Remote environment monitoring system for application in industry and landfill sites. In: 2nd mediterranean conference on embedded computing, MECO–2013, Budva, Montenegro

5. CPCB (Central Pollution Control Board) (2005) Management of municipal solid wastes. http://cpcb.nic.in/pcpdiv_plan4.htm. Accessed 5 July 2006
6. IIR (India Infrastructure Report) (2006) Urban infrastructure New Delhi. Oxford University Press, Oxford
7. IPCC (Intergovernmental Panel on Climate Change) (2006) IPCC Guidelines for National Greenhouse Gas Inventories
8. Kumar A, Singh IP, Sud SK (2011) Energy efficient and low-cost indoor environment monitoring system based on the IEEE 1451 [9]. Standard. Sens J IEEE 11:2598–2610
9. Carvalho V, Lopes JG, Ramos HG, Alegria FC (2009) City-wide mobile air quality measurement system. In: IEEE SENSORS conference, Christchurch, pp 546–551, 25–28 Oct 2009
10. Beirne S, Kiernan BM, Fay C, Foley C, Corcoran B, Smeaton AF, Diamond D (2010) Autonomous greenhouse gas measurement system for analysis of gas migration on landfill sites. In: Sensors applications symposium (SAS), Limerick, Ireland, pp 143–148, 23–25 Feb 2010
11. Sekgoele K, Chowdhury SP, Chowdhury S (2011) Technical and economic assessment of power generation from landfill gas in South Africa. Power and energy society general meeting, San Diego, USA, pp 1–8
12. Mihajlovic Z, Milosavljevic V, Rajs V, Zivanov M (2012) Application of GPRS modules in data acquisition and control of devices for air quality monitoring. Telecommunications forum (TELFOR), Belgrade, Serbia, pp 1020–1023, 20–22 Nov 2012

Design of Common Source Amplifier Using Amorphous Silicon TFT

G. Srikanth, B.S. Kariyappa and B.V. Uma

Abstract Thin Film Transistors (TFTs) are now being used for a variety of applications. Their success in displays, ease of fabrication and low cost production has attracted the attention of researchers and academics all over the world. They have found applications in memory, sensors, flexible electronics etc. In this work an Amorphous Silicon Based TFT has been designed and simulated which in turn is used in the design of Common Source amplifier. It has been found that though the gain achieved by TFTs is low compared to Metal Oxide Semiconductor Field Effect Transistor (MOSFETs), they can still be used in the design of circuits where low cost and less fabrication time is a major criteria. The design and simulation of the device is carried out using Silvaco ATLAS, and that of the circuit is carried out using Silvaco gateway.

1 Introduction

TFTs are analogous to MOSFETs in operation in that they too are three terminal devices with the voltage applied at the Gate, Source and Drain terminal controlling the flow of current between the source and drain region. TFTs have a wide variety of structures. The bottom gate top contact structure are the most studied ones. They have a significant advantage over MOSFETs when it comes to the cost and time involved in fabrication. Besides, the number of steps required to fabricate TFTs is far less in number and less complicated compared to that of MOSFETs. They also suffer from a few disadvantages in that the mobility of the semiconductor layer is low, and hence their ON current is less compared to that of MOSFETs. The major

G. Srikanth (✉) · B.S. Kariyappa · B.V. Uma
R V College of Engineering, Bangaluru 59, India
e-mail: srikanthg2007@rediffmail.com

B.S. Kariyappa
e-mail: kariyappabs@rvce.edu.in

B.V. Uma
e-mail: umabv@rvce.edu.in

K.R. Attele et al. (eds.), *Emerging Trends in Electrical, Communications and Information Technologies*, Lecture Notes in Electrical Engineering 394,
DOI 10.1007/978-981-10-1540-3_24

application of TFTs so far is in displays [1]. It also has a significant contribution in the study of sensors, and memories.

TFT differs from the MOSFET in that the conducting channel is induced in the accumulation regime rather than through the formation of an inversion layer. The first hydrogenated amorphous silicon (a-Si:H) TFTs were actually designed to measure the mobility of the material, which was at that time difficult to access by other techniques. It was only later that the technological importance of the device was recognized in applications in which large area is required and where single crystalline silicon can no longer be used.

Hydrogenated amorphous silicon has evolved into a ubiquitous materials system in large-area electronics for many commercial applications ranging from liquid-crystal displays to medical imaging. Amorphous silicon (also a-silicon or a-Si) is a form of silicon that has no crystalline structure. a-Si is critical to producing TFTs because it is without any crystal structure that allows TFTs to be vapour-deposited onto large substrates. So, amorphous silicon is ideal to use as thin film transistors. Polysilicon (P-Si) is a material consisting of small silicon crystals. P-Si is an attempt to overcome the poor performance of Amorphous Silicon TFT by introducing a low level of crystalline structure into the semiconductor material. A drawback to polysilicon TFT is that it is manufactured using laser annealing, an expensive but necessary processing step.

The large scale manufacturing of a-Si:H TFTs forms the basis of the active matrix flat panel display industry. Poly-Si TFTs facilitate the integration of electronic circuits into portable active matrix liquid crystal displays, and are increasingly used in active matrix organic light emitting diode (AMOLED) displays for smart phones. The recently developed AOS TFTs are seen as an alternative option to poly-Si and a-Si:H for AMOLED TV and large AMLCD TV applications, respectively.

In this paper a TFT has been designed using amorphous silicon as the semiconductor layer and silicon nitride, and silicon dioxide as the dielectric layers. This device is then used to design a common source amplifier. The design and simulation of the TFT is carried out in Silvaco ATLAS and that of the common source amplifier is carried out in Silvaco Gateway. The gain of the device is considerably less when compared to that of the MOSFETs for the same voltages. However, the ease of fabrication and low cost production of TFTs can be considered a significant advantage.

The rest of the paper is organized as follows. The second section describes the design of amorphous silicon based TFT and common source amplifier. In the third section the simulation results of the TFT and the amplifier are discussed. The fourth section gives a conclusion of the observations obtained from the simulation results.

2 Design

The transistor designed has a bottom-gate top-contact structure. It uses amorphous silicon as the semiconductor layer. Two layers of dielectrics, silicon nitride and silicon dioxide are used. Silicon nitride has high strength over a wide temperature range, good thermal shock resistance and good chemical resistance. Silicon dioxide is easily deposited on various materials and grown thermally on silicon wafers and is resistant to various materials during the etching of other materials. Both the materials have considerably high dielectric strength which makes them good insulators. The Gate, Source and Drain electrodes are chosen to be aluminium (Al). The structure of the TFT designed is as shown in the Fig. 1. The thickness of each layer is as given in Table 1.

The passivation layer is added to prevent the oxidation of the amorphous silicon layer. Two dielectrics silicon dioxide and silicon nitride are used between gate and

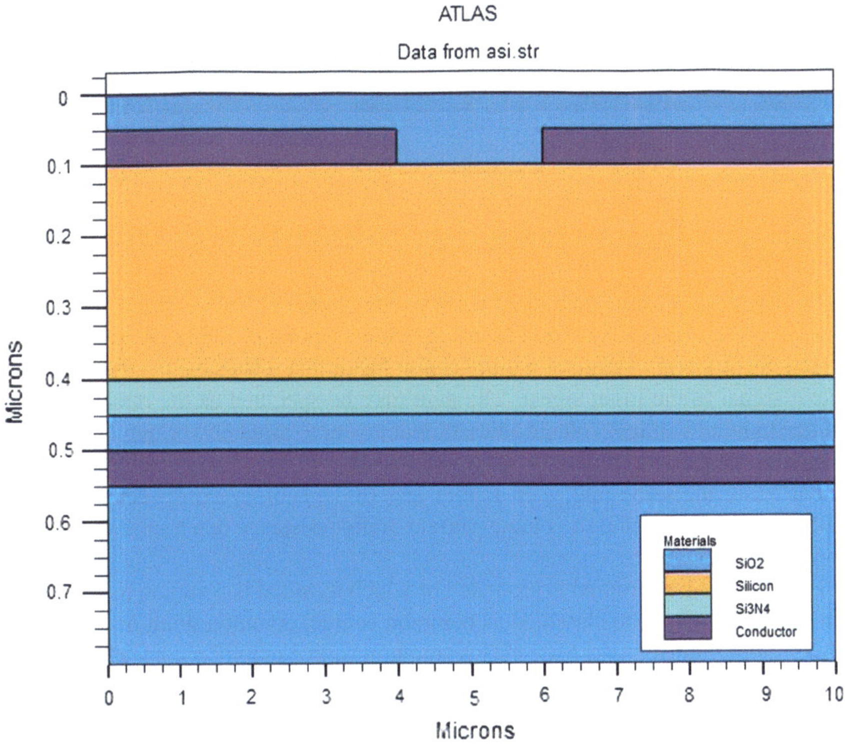

Fig. 1 Structure of the TFT

Table 1 Thickness of each layer

Layer	Thickness (um)
Substrate	0.25
Gate/Source/Drain	0.05
Silicon dioxide	0.05
Silicon nitride	0.05
Amorphous silicon	0.3

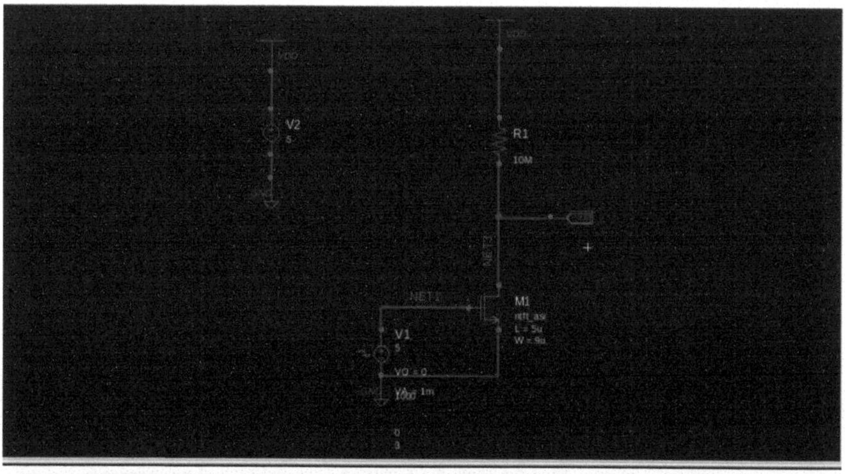

Fig. 2 Schematic of the common source amplifier

the amorphous silicon layer. The TFT shown in Fig. 1 is used to design a common source amplifier the schematic of which is shown in Fig. 2.

An amplifier is an electronic circuit that increases the power or strength of the signal. Amplifiers are described according to their input and output properties. They exhibit the property of gain, or multiplication factor that relates the magnitude of the output signal to the input signal. The gain may be specified as the ratio of output voltage to input voltage (voltage gain), output power to input power (power gain), or some combination of current, voltage, and power. In many cases, with input and output being the same unit, gain is unit less (though often expressed in decibels (dB)). An amplifier can be a voltage, current, trans-resistance or a trans-conductance amplifier [2].

Based on which terminal is grounded or which terminal is common to both input and output, they can be classified as common source, common drain or a common gate amplifier. In case of a MOSFET a common source amplifier converts the input voltage into current and then drives this current to the load to get the output voltage. The basic principle of operation of a TFT based common source amplifier is similar to that of a MOSFET. However, due to the lower mobility of amorphous silicon layer, we can expect this value to be comparatively less in case of TFT. The schematic of the common source amplifier that was designed is as shown in the Fig. 2. A resistor of value 10 M is used as the load. This high value of the load is required because of the very low mobility of the charge carriers of the device. The dc voltage of 5 V is applied to the circuit and a sinusoidal voltage of 1 mV is applied at the input. The output is taken at the drain of the device.

As can be seen from the figure the source of the device is grounded and is common to both the input and output network of the circuit. A lot of analysis has been done of common source amplifiers based of Bipolar Junction Transistors (BJTs) and MOSFETs. They are usually used for small signal amplification and in two stage operational amplifiers where the first stage which is usually a differential amplifier gives a high gain and the second stage which is the common source amplifier gives a high voltage swing. In typical MOSFET amplifiers, the value of the gain for a common source amplifier is usually around 10–30. However, for TFTs it is considerably less.

For the amplifier that is designed the gain of the amplifier is measured. The gain is defined as the ratio of the output voltage to the input voltage. The major focus in the design was to increase the gain as much as possible.

3 Results and Discussion

The simulation of the device is carried out in the Silvaco ATLAS tool. ATLAS is a physically-based two and three dimensional device simulator. It predicts the electrical behaviour of specified semiconductor structures and provides insight into the internal physical mechanisms associated with device operation. It can be used standalone or as a core tool in SILVACO's VIRTUAL WAFER FAB simulation environment. In the sequence of predicting the impact of process variables on

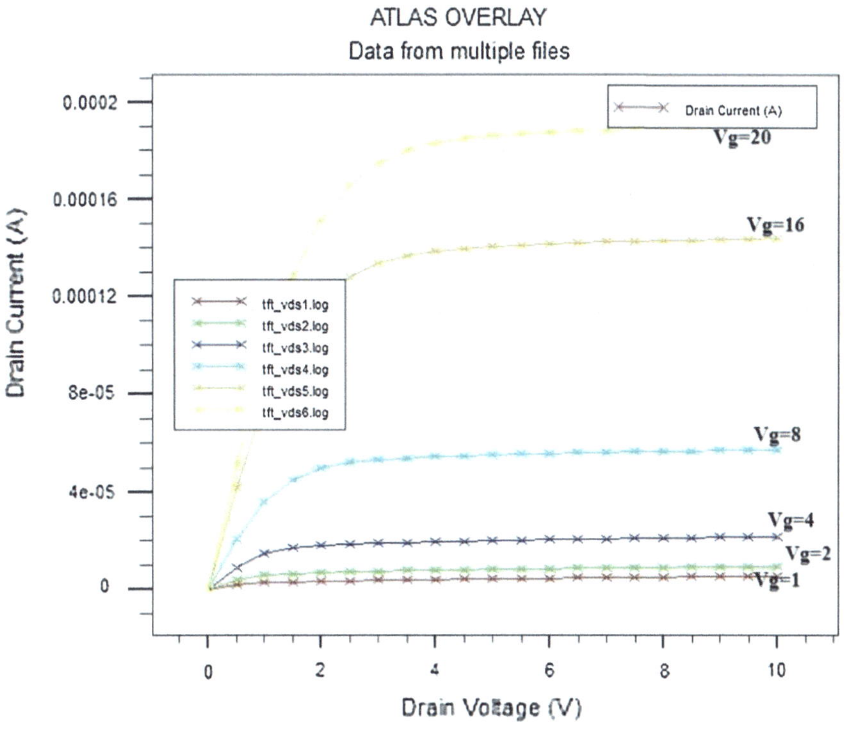

Fig. 3 IV characteristics of the TFT

circuit performance, device simulation fits between process simulation and SPICE model extraction [3].

The I-V characteristic of the device is plotted and is as shown in the Fig. 3. It can be seen that the device characteristics is almost identical to that of a MOSFET except that the on current is low as a result of low mobility of the carriers in the amorphous silicon layer.

The I-V characteristics clearly shows the distinct regions of operations of the transistor. When the gate voltage is very low, the output current is almost zero. This region is similar to the cut-off region in the MOSFET. Similarly for a small V_d the drain current raises linearly, just as in the case of a MOSFET for the linear region of operation. And for a significantly large V_d the drain current saturates resembling the saturation region of the MOSFET I-V characteristics.

The TFT is then then used for the design of the schematic. The design and simulation of the schematic is carried out in the Silvaco Gateway. Gateway is a schematic capture tool and is the entry point for the IC design flow. Gateway consists of a GUI for schematic entry and is integrated with the following

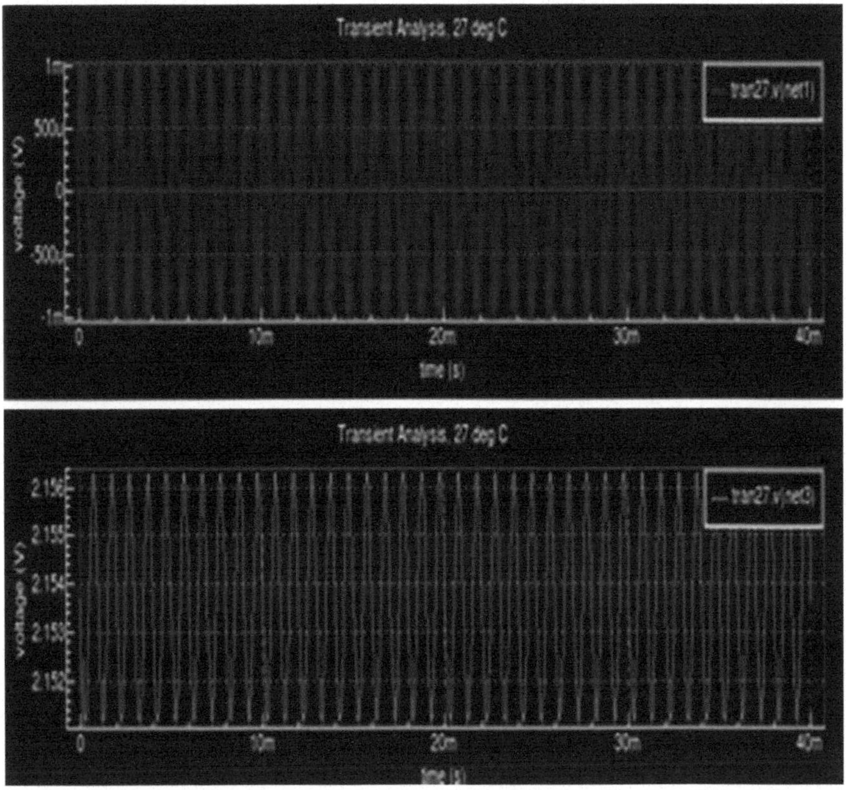

Fig. 4 Transient plot of the common source amplifier

simulators: Silvaco SmartSpice (Analog), Silvaco SmartSpice RF (RF), Silvaco Silos (Verilog), Silvaco Atlas (TCAD MixedMode), Synopsys HSPICE (Analog).

Figure 4 shows the transient plot of the common source amplifier for a sinusoidal input of amplitude 1 mV.

The input is shown in the blue colour and the output is as shown in the red colour. The gain of the device is calculated by measuring the peak to peak voltage of the input and output and then taking the ratio of output voltage to the input voltage. The value is found to be 2.7.

The circuit effectively acts as an amplifier. However, its gain is not as high as that of an amplifier when designed with a MOSFET device. The gain can be increased by increasing the output resistance. It should be noted that the resistance needed at the output is very high, in the mega-ohm range. This is because the trans-conductance of the device is very much less compared to that of a MOSFET.

4 Conclusion

TFTs have been successfully used in display applications. They have also found application in sensors, wearable and flexible electronics. From the simulation we can observe that the I-V characteristics are very much similar to that of MOSFETs which are currently being used for the design of VLSI circuits. The results for the simulation of the common source amplifier prove that TFTs can be used for the design of analog circuits. The gain for the given circuit is calculated to be 2.7, for a MOSFET this value can be easily in the range of 10–30. Though the performance of the TFTs in the design of analog circuits is considerably low when compared to a MOSFET, the low cost and low fabrication of TFTs may in turn evaluate to a significant advantage. With the use of organic and polymer materials for the conduction layers, they can be good candidates to flexible and wearable electronics. If the mobility of the TFTs can be increased and its threshold voltage lowered, they can easily be used in the design of analog circuits.

References

1. Brody TP (1984) The thin film transistor-a late flowering bloom. IEEE Trans Electron Devices 31–11:1614–1628
2. Razavi B (2001) Design of Analog CMOS Integrated Circuits
3. ALTAS User Manual (2006) Device Simulation Software
4. Gilles H (2004) OTFTs: from theory to real devices. 19(7):1946–1962. (2002) The IEEE website. [Online]. http://www.ieee.org/
5. Stadlober B, Zirkl M, Beutl M, Leising G (2005) High-mobility pentacene organic field-effect transistors with a high-dielectric-constant fluorinated polymer film gate dielectric. Appl Phys Lett **86**(24):242902-1–242902-3
6. Sze SM (2007) Physics of semiconductor devices. Wiley, New York, USA

7. Voigt MM, Guite A, Chung DY, Khan RU, Campbell AJ, Bradley DD, Meng F, Steinke JH, Tierney S, McCulloch I, Penxten H, Lutsen L, Douheret O, Manca J, Brokmann U, Sonnichsen K, Hulsenberg D, Bock W, Barron C, Blanckaert N, Springer S, Grupp J, Mosley A (2009) Polymer field-effect transistors fabricated by the sequential gravure printing of polythiophene, two insulator layers, and a metal ink gate. Adv Funct Mater 19:1–8
8. Wu Y, Li Y, Gardner S, Ong BS (2005) Indolo [3,2-b] carbazolebased thin-film transistors with high mobility and stability. J Am Chem Soc **127**(2):614–618

Design of Ultra Low Power Asynchronous Domino Logic Pipeline Using Critical Data Path

K. Nirmala, P. Prasanth Babu, K. Prasanth and D. Maruthi Kumar

Abstract This paper presents Design of ultra low-power asynchronous domino logic pipeline method, which targets to introduce design of latch-free pipe-line targeting to latch-free pipeline. To construct data paths, both dual rail and single rail domino gates are used. Dual-rail domino gates are mainly used to construct critical data paths. Hence the handshake signals are reduced greatly, using critical data path. This pipeline offers low power consumption and high throughput. A 16 × 16 array style multiplier is used for evaluating the proposed pipeline method. Asynchronous static pipeline method is compared with the proposed pipeline method, it saves up to 83.0 and 16.4 % of power.

1 Introduction

During the last decade, the research on asynchronous technology has been continued. VLSI systems tend to be more complex because of scaling in continued CMOS technology [1]. The issues which come under physical design, known as high power consumption and large area are become complex problems. Even though the technology scaling provides high integration possibilities, some difficulties occur at physical level. Asynchronous design does not use global handshake signals, it uses local handshake signals. Hence it is considered as a better solution for avoiding the problems which relates to global clock.

K. Nirmala · P. Prasanth Babu (✉) · K. Prasanth · D. Maruthi Kumar
Electronics and Communication Engineering, Srinivasa Ramanujan Institute of Technology, Ananthapuramu 515701, A.P., India
e-mail: prasanthbabu.padigala@gmail.com

K. Nirmala
e-mail: nirmala.ns4@gmail.com

K. Prasanth
e-mail: prashanth.k443@gmail.com

D. Maruthi Kumar
e-mail: maruthikumar2015@gmail.com

K.R. Attele et al. (eds.), *Emerging Trends in Electrical, Communications and Information Technologies*, Lecture Notes in Electrical Engineering 394,
DOI 10.1007/978-981-10-1540-3_25

The attractive features are given below:

(1) Low power consumption.
(2) High speed of operation.
(3) Provides no skew and no distribution problems in clock.
(4) Small area occupation.

However asynchronous design, has the choice of handshake protocols that affects the circuit implementation like area, speed, robustness, and power. Asynchronous circuits use two most popular protocols, one is four phase bundled-data protocol and other is four phase dual-rail protocol. The four phase bundled-data protocol design is similar to the design of synchronous circuits. Local clock pulses are generated using handshake signals and they use delay matching to identify valid signal. Handshake circuits uses timing assumptions, hence it leads to most efficient circuits. However, in four-phase dual-rail protocol design, the encoded data is combined with handshake signal. The encoded data is detected by handshake circuits so that they lead to allow correct operation in the presence of data path delays. This feature allows advanced VLSI systems to provide data path delay variations. This paper provides a method for designing Asynchronous domino logic pipeline, which targets to increase the circuit efficiency which is used for wide range of applications, and achieves an area-efficient and ultra low-power asynchronous domino logic pipeline. Asynchronous domino logic pipeline can entirely eliminate explicit storage elements between stages by avoiding the latching functionality of implicit domino logic gates.

2 Background

Asynchronous domino logic pipeline uses PS0 style of implementation and it is an important foundation for most proposed styles [2, 3].

(1) Structure of PS0: Figure 1 represents the block diagram of PS0. In PS0, it consists of one function block and one completion detector. Each function block is designed using dual-rail domino logic and each completion detector generates a local handshake signal. Using the pipeline the handshake signals can control the flow of data and it will be transferred to pre-charge or evaluation port of the previous pipeline (Fig. 2).
(2) Four-phase dual rail protocol: PS0 is designed using the four-phase dual-rail protocol. Figure 3 represents an example for data transfer using the four-phase dual-rail protocol, and Table I represents the code table for encoding using four-phase dual-rail. The wires (*w_t*, *w_f*) are used to encode a request signal into the data signal. The data bit 0 is encoded as (0, 1), bit 1 is encoded as (1, 0) and spacer is encoded as (0, 0); (1, 1) is not used. A valid data can be obtained using a receiver by observing the two wires. The dual rail encoded data path is also known as the delay in sensitive data path, because the sender and the receiver can communicate without bothering the delays between the logic blocks [4, 5].

Figure 2 represents an example of the dual-rail domino AND gate and completion detector of 2-bit. Two input AND gate is used to generate a bit done signal by observing the outputs of dual rail domino gate, hence it behaves as a 1-bit completion detector [6]. To implement a 2-bit completion detector, C-element is used to combine the bit done signals. When the total bit done signals from entire data paths, it forms a full completion detector as represented in Fig. 1.

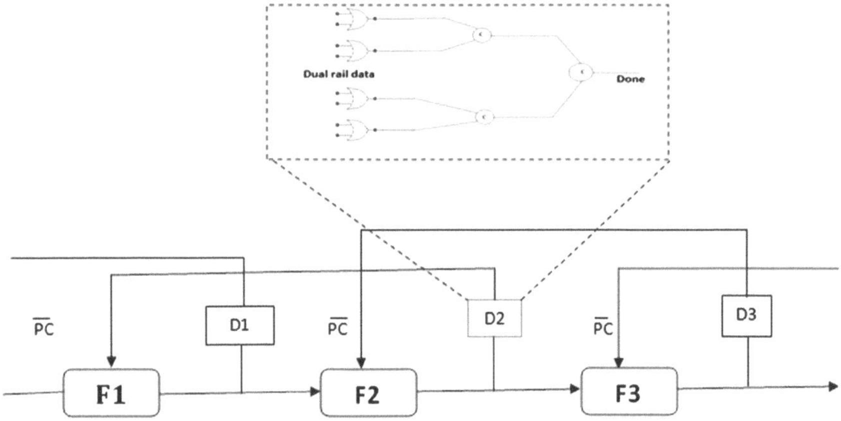

Fig. 1 Block diagram of PS0

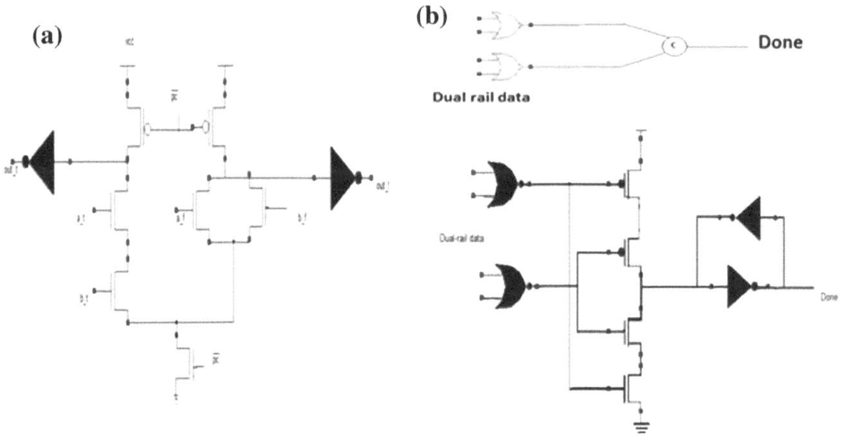

Fig. 2 **a** Dual-rail domino AND gate, **b** 2-bit completion detector

Fig. 3 Data transfer based on four-phase dual-rail protocol

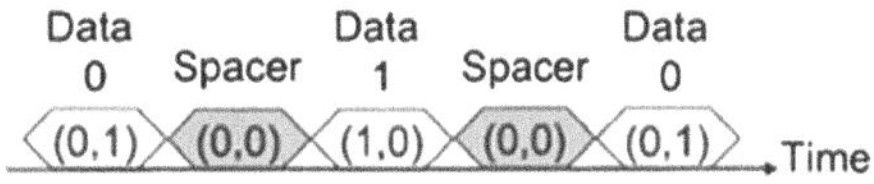

Table 1 Four-phase dual-rail encoding code table

	Code word (W_1, W_1)
Data 0	(0, 1)
Data 1	(1, 0)
Spacer	(0, 0)
Not used	(1, 1)

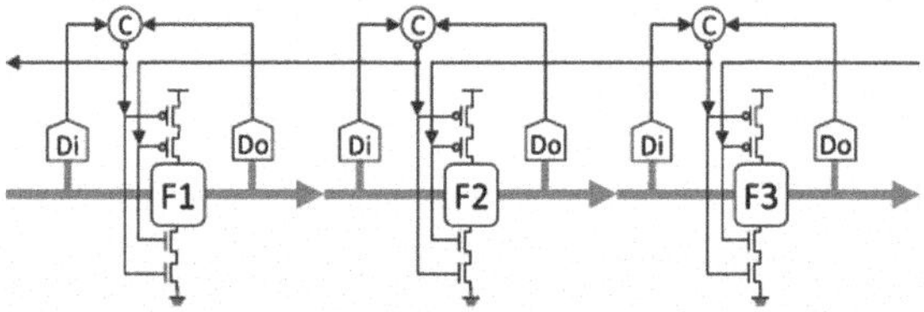

Fig. 4 Block diagram of PCHB

Pre-charge Half-Buffer Pipeline: Figure 4 represents the block diagram of Pre-charge Half-Buffer pipeline (PCHB). PCHB uses two completion detectors: one on the input side is Di and one on the output side is Do. In PCHB stage and PS0 pipeline, the complete cycle of events are same. If the detector detects valid input bits, the PCHB stage starts evaluation. This design eliminates the skew across individual bits in the stage: one at the input side (Di) and other at the output side (Do). The only difference between PCHB pipeline and PS0 is that, PCHB stage verifies its input bits. The verification bits at input side are done using the input completion detector (Di), hence PCHB stage start evaluation when all input bits are valid.

LP2/2: LP2/2 pipeline style uses both dual-rail protocol design and bundled-data protocol design. It is a high throughput pipeline style. Figure 5a represents the

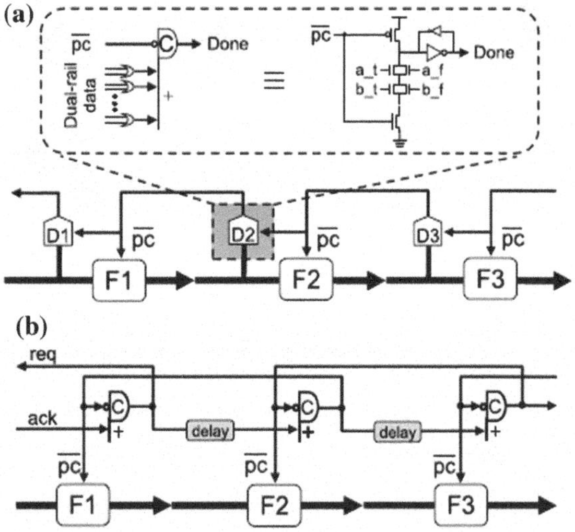

Fig. 5 Block diagrams of LP2/2. **a** LP2/2 based on dual rail protocol, **b** LP2/2-SR

diagram for LP2/2 using dual-rail protocol. LP2/2 minimizes the sequential of handshake events hence; it improves the throughput of PS0. However, the overhead problems are not solved in handshake control logic [7, 8].

Figure 5b represents the diagram for LP2/2 using bundled-data protocol (LP2/2-SR). A single extra bundling signal is used in LP2/2-SR so it avoids the detection overhead problems. The bundling signal acts as completion signal, it matches the delay in function blocks.

3 Design of Asynchronous Pipeline Based on Critical Data Path

A. Overview

Figure 6 represents a method for asynchronous method. Special dual-rail logic is used to construct the pipeline. The encoded signal and data signal are transferred using critical data paths. Single-rail logic is used to construct the non-critical data paths, which transfers data signal. A static NOR gate detects the dual-rail [9].

Critical data path generates a total done signal for each pipeline stage. Pre-charge ports of the previous stages are connected to the NOR gate outputs.

The total done signal in APCDP is generated by detecting the critical data path. This type of design has two advantages. First, a single NOR gate is used to modify the completion detector, and the data path width will not increase the detection overhead. Second, single-rail logic in non-critical data paths reduces the overhead problem in function block logic. Hence the overhead problems in control logic and function logic are reduced using APCDP. It improves the throughput and power consumption.

The important factor in function blocks is to find a stable critical data path. Traditional gates can be used to construct critical data path, but it becomes difficult. In traditional logic gate, the gate delay data dependence problem will occur. The intuitive way of constructing stable critical data path is by adding the delay elements. But this method requires complex timing analysis and it cause large overhead of delay elements. The efficient solution for constructing critical data paths is

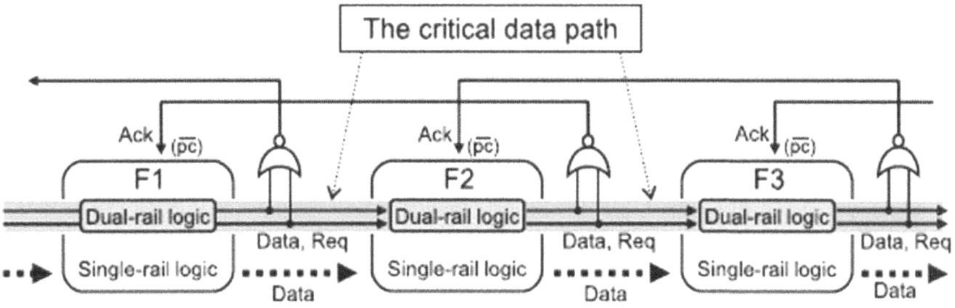

Fig. 6 Block diagram of APCDP

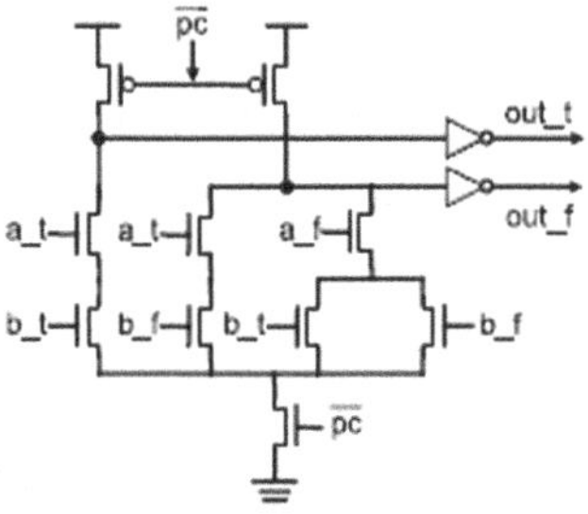

Truth table of dual-rail AND logic

$\overline{pc}$	a_t	a_f	b_t	b_f	out_t	out_f
0	–	–	–	–	0	0
1	0	1	0	1	0	1
1	0	1	1	0	0	1
1	1	0	0	1	0	1
1	1	0	1	0	1	0

Fig. 7 Synchronizing AND gate and the truth table of dual rail AND gate

Synchronizing Logic Gates (SLGs). When valid data arrives, then only SLGs starts evaluation, it solves the gate delay data dependence problem. Hence, the design provides less area and low power consumption.

B. Logic Gates

(1) Synchronizing Logic Gates: SLGs with dual-rail domino gates avoids the gate-delay data dependence problem [10, 11]. Figure 7 represents the synchronizing AND gate and the truth table of dual rail AND logic. Table 2 shows that there are four transistor paths: (1) [*a_t*, *b_t*]; (2) [*a_t*, *b_f*]; (3) [*a_f*, *b_t*]; and [*a_f*, *b_f*]. For every input condition, two paths will get active. Hence, the delays in that path will be considered. However when logic gate increases the delay becomes more, it effects the outputs.

(2) Synchronizing Logic Gates with a Latch Function:

Because of the delays in SLGs, SLGLs are extended. The latch function for Synchronizing AND gate and the latch states are shown in Fig. 8. An SLGL has an enable port (en_*t*, en_*f*). The opaque and transparent state of the SLGL is controlled by enable port. The SLGLs cannot start evaluation without the presence of the enable signal. Using this feature SLGLs are used for constructing critical data paths [12].

C. Structure of APCDP

Figure 9 shows the structure of APCDP. The constructed critical data path is represented using solid arrow. The non-critical data paths are represented with dotted arrow. The critical data paths are constructed using dual-rail data path and the non-critical data paths are constructed using single rail data paths. A total done signal is generated using 1-bit completion detector, which is constructed with NOR

Table 2 States of pull down transistor paths on different data patterns

Data patterns (a_t, a_f, b_t, b_f)	Pull-down transistor paths						
	Figure 2a conventional			Figure 6 synchronizing AND gate			
	[a_t, b_t]	[a_f]	[b_f]	[a_t, b_t]	[a_t, b_f]	[a_f, b_t]	[a_f, b_f]
(0, 1, 0, 1)	Off	**On**	**On**	Off	Off	Off	**On**
(0, 1, 1, 0)	Off	**On**	Off	Off	Off	**On**	Off
(1, 0, 0, 1)	Off	Off	**On**	Off	**On**	Off	Off
(1, 0, 1, 0)	**On**	Off	Off	**On**	Off	Off	Off

gate. The completion detector is available at each pipeline stage. The connection between single-rail domino gate and dual-rail domino gate is established using encoding converter.

Each pipeline stage's SLG are linked together for selecting Lin gate in each pipeline stage. It is best to select the Lin gate that is originally linked to the Lin gate in the following pipeline stage. SLGs are naturally linked after converting the Lin gates to SLGs.

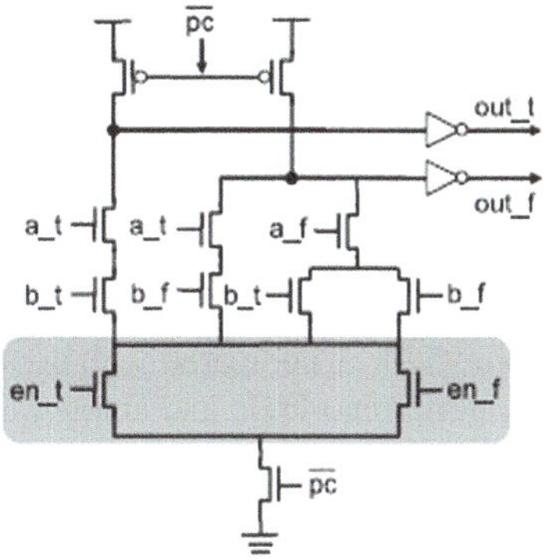

The states of latch

$\overline{pc}$	en_t	en_f	States
0	–	–	Opaque
1	0	0	Opaque
1	0	1	Transparent
1	1	0	Transparent
1	1	1	No used

Fig. 8 Synchronizing AND gate with LATCH function and the truth table of latch states

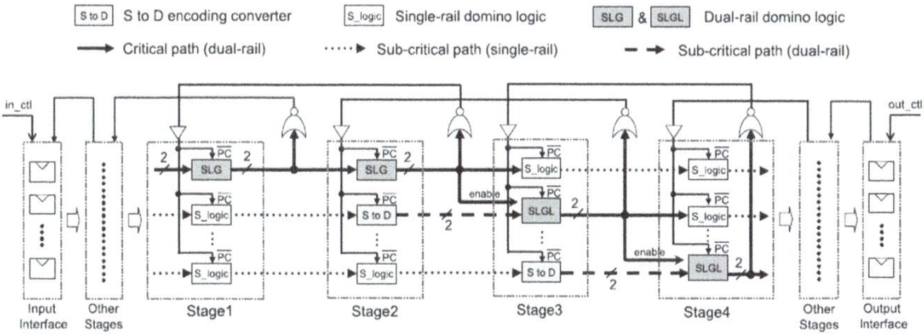

Fig. 9 Structure of APCDP

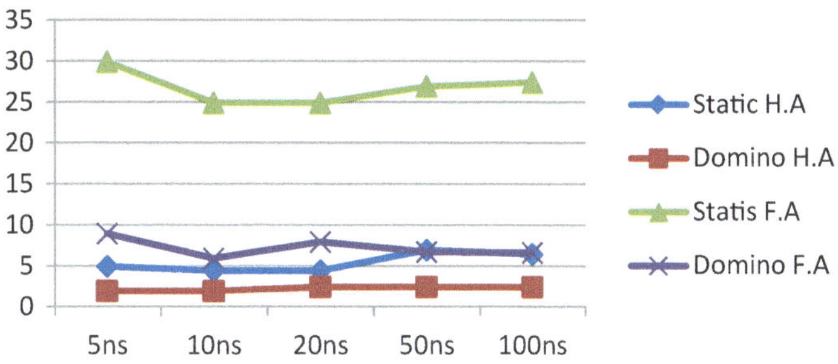

Fig. 10 Power consumption results for H.A and F.A

4 Evaluation

A 16 × 16 array style multiplier is designed using APCDP. Conventional dual-rail asynchronous pipelines require large area and are not selected because the large function blocks cannot be designed (such as the 16 × 16 array style multiplier).

Results:

Tables 3 and 4 shows the evaluation results of the 16 × 16 array style multiplier. The performances of throughput are evaluated without considering design margins, which are ideal results.

The evaluation results show that APCDP using domino gates has less power consumption, and the smallest transistor count. Figure 10 represents the power consumption performances at different time intervals. The results prove that APCDP is the most power efficient design.

(1) Transistor Count: Table 3 shows that APCDP, respectively, reduces the transistor count are reduced to 67.32 % from 95.98 % compared with domino gates and static gates. The combination of dual-rail domino logic and single-rail domino logic are used in APCDP. The non-critical data paths are designed using single-rail domino logic that saves a lot of the transistor count. The SLGs and SLGLs which are used in APCDP require more transistors, but they have small impact on the transistor count, since they are in small quantity.

The total FET width also calculated in addition to the transistor count, hence the total area can be calculated. Table 4 shows that APCDP reduces the total FET width to 49.4 % from 85.68 % compared with domino logic and static logic respectively.

Table 3 Evaluation of 16 × 16 array multiplier (power)

Logic gate	Domino gate	Static gate
Transistor counts	7420	10584
H.A counts	252 (14 × 18)	280 (14 × 20)
F.A counts	7168 (224 × 32)	10304 (224 × 46)
H.A power consumption	0.63 (mw)	4.06 (mw)
F.A power consumption	1.83 (mw)	7.85 (mw)
Total power consumption	2.46 (mw)	11.19 (mw)

Table 4 Evaluation results (area) of 16 × 16 array multiplier

Logic gate	Domino gate (in mm)	Static gate (in mm)
Area for 1 H.A	2.64	3.6
Area for 1 F.A	2.04	4.68
Area for 14 H.A	36.96	50.4
Area for 22 F.A	456.96	1048.32
Total area	493.92	1098.72

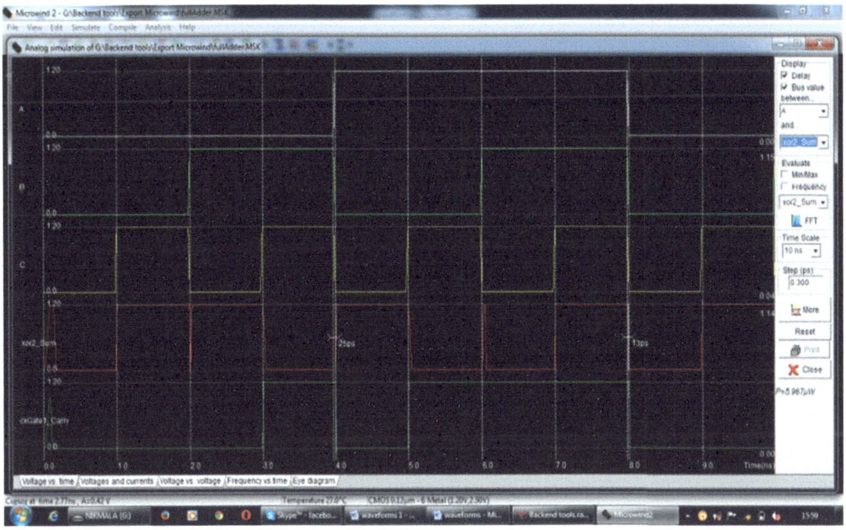

Fig. 11 Simulation results for full adder using static gate

(2) Power: The power consumption of VLSI circuits relates to the number of transistors. The results show that APCDP using domino gates consumes much less energy than APCDP using static gates. Figure 11 shows the power consumption of full-adder using domino gates. Figure 12 shows the power consumption of full-adder using static gates. As a result, the power consumption of domino gates is less compared with the static gates.

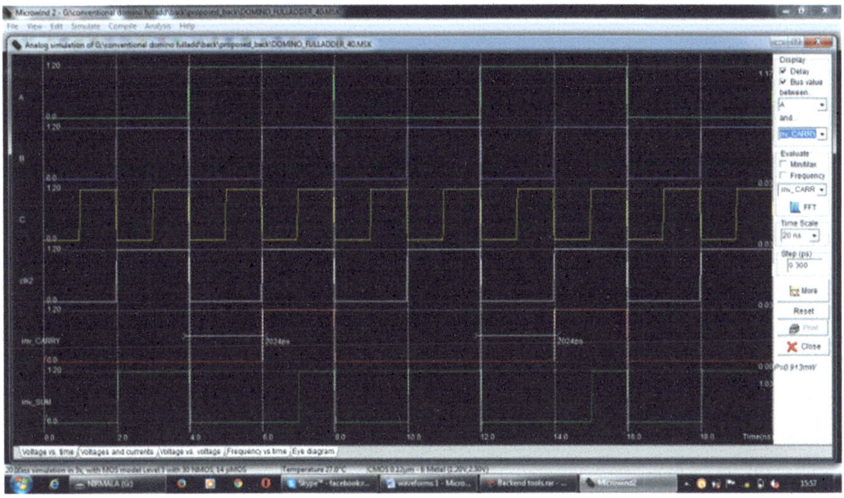

Fig. 12 Simulation results for full adder using domino gates

C. Further Discussion

The APCDP is designed based on schematic simulation. To design large function modules, APCDP is not applicable. When APCDP is used to design large function modules, design automation becomes important issue.

Another issue is the design automation for layout. This optimization is not suitable for APCDP, because the constructed critical data path robustness is reduced. In specific P&R method, it increases the delay on the constructed path, hence it avoids the robustness.

One more issue is timing verification. For verifying domino circuits there are no EDA support. In APCDP, the timing analysis problem is avoided using known critical data path and the dual rail handshake.

5 Simulation Results

See Figs. 11 and 12.

6 Conclusion

This paper introduced a method to design ultra-low power asynchronous domino logic pipeline. The design is realized using domino gates and using static gates. The overhead of handshake control logic as well as function block logic are greatly reduced. The APCDP increases the pipeline throughput, also decreases the power consumption and area. The evaluation results show that the APCDP design using Domino gates has better performance than APCDP design using static gates.

References

1. Calhoun BH, Cao Y, Li X, Mai K, Pileggi LT, Rutenbar RA (2008) Digital circuit design challenges and opportunities in the era of nanoscale CMOS. Proc IEEE 96(2), 343–365
2. Sparsø J, Furber S (2001) Principles of asynchronous circuit design: a systems perspective. Kluwer, Boston, MA, USA
3. Krstic M, Grass E, Gurkaynak FK, Vivet P (2007) Globally asynchronous, locally synchronous circuits: overview and outlook. IEEE Des Test Comput 24(5):430–441
4. Martin AJ, Nystrom M (2006) Asynchronous techniques for systemon-chip design. Proc IEEE 94(6):1089–1120
5. Teifel J, Manohar R An asynchronous dataflow FPGA architecture. IEEE Trans Comput 53 (11):1376–1392
6. Williams TE (1991) Self-timed rings and their application to division. Ph.D. dissertation, Department of Computer Science, Stanford University, Stanford, CA, USA
7. Lines AM (1998) Pipelined asynchronous circuits. Department of Computer Science, California Institute of Technology, Pasadena, CA, USA, Technical Report

8. Singh M, Nowick SM (2007) The design of high-performance dynamic asynchronous pipelines: lookahead style. IEEE Trans. Very Large Scale Integr (VLSI) Syst. 15(11):1256–1269
9. Nowick SM, Singh M (2011) High performance asynchronous pipelines an overview. IEEE Des Test Comput 28(5):8–22
10. Low HS, Shang D, Xia F, Yakovlev A Variation tolerant AFPGA architecture. Hariyama M, Ishihara S, Kameyama M
11. Li L, Wang W, Choi K, Park S, Chung M-K (2010) SeSCG: selective sequential clock gating for ultra-low-power multimedia mobile processor. In: Proceedings of IEEE International Conference on EIT, May 2010, pp 1–6
12. Singh M, Tierno JA, Rylyakov A, Rylov S, Nowick SM (2010) An adaptively pipelined mixed synchronous-asynchronous digital FIR filter chip operating at 1.3 gigahertz. IEEE Trans. Very Large Scale Integr. (VLSI) Syst. 18(7):1043–1056

Contrast Based Color Plane Selection for Binarization of Historical Document Images

M.E. Paramasivam and R.S. Sabeenian

Abstract This paper primarily focuses on establishing that the document image processing and natural image processing domains are mutually dependent, which many researchers have not experimented. For a given color image, the contrast-per-pixel (CPP) for each color channel is computed and the channel that exhibits highest CPP value is binarized. To evaluate the proposed method, the color image is also converted to a grayscale image using a weighted color-to-grayscale conversion and then binarized. Otsu Algorithm is preferred for binarization. Images from DIBCO and H-DIBCO datasets were used for evaluating the proposed algorithm. The resultant binary images were appraised based on precision metrics which shows that the highest CPP exhibits better performance. Experimentally, the extracted color channels performed marginally better than the weighted color-to-grayscale converted image, which clearly indicates that image binarization depends on natural image processing.

Keywords Document image processing · Natural image processing · Otsu binarization · Contrast measure · Binarization metrics · Precision

1 Introduction

Historical Document Image processing is one of the most challenging areas of image processing till date. The reasons behind this challenge can be listed as follows: severe degradation of the document, unstructured character positioning, variations in the same document, variation of appearance and bleed-through in some document with same and different color inks [1, 2].

M.E. Paramasivam (✉) · R.S. Sabeenian
Department of Electronics and Communication Engineering,
Sona College of Technology, Sona Nagar, T.P.T. Road, Salem 636005, India
e-mail: sivam.sct@gmail.com

R.S. Sabeenian
e-mail: sabeenian@gmail.com

K.R. Attele et al. (eds.), *Emerging Trends in Electrical, Communications and Information Technologies*, Lecture Notes in Electrical Engineering 394,
DOI 10.1007/978-981-10-1540-3_26

Since 2009, the DIBCO and H-DIBCO [3] contests have encouraged a number of researchers to contribute in the domain. As a first step in document image processing, the image is binarized to get a clear classification of text and background. This would enable quicker segmentation of text from the images. We have considered Otsu [4] method for binarization, due to its capability that almost after two decades, [5] have analyzed properties for the algorithm and indicated it to be the most excellent global binarization method.

Though, the problem is primarily to split the image into two broad sets, one indicating the text and the other the background; it is significant to note that images vary in the aspect of color, contrast, brightness, etc. Hence, it is worth that the image is analyzed for its variation in the above parameters based on Natural Image Processing and then subjected to the best performing binarization algorithm.

2 Algebra of Image Processing

Let us consider, a value set φ_w, which shall represents all the possible values in a given image data structure, where w represents the maximum possible weight of the value when represented in binary. When $w = 8$, then the value set is represented as $\varphi_8 = \{v : v \in \mathbb{N}\, and\, 0 \leq v \leq 255\}$. The cardinal number of the set can be defined as $n(\varphi_8) = 2^w = 2^8 = 255$. On a minimal case, if $w = 1$ then, $n(\varphi_w) = 2$ and $\varphi_2 = \{v : v \in \mathbb{B}\, and\, 0 \leq v \leq 1\}$.

Let τ be the point set representing the spatial domain and let $n(\tau) = 2$. Thus the elements in the point set can be represented as $\{p, q\}$. In case of a color image the point set is $\tau = \{\{x, y\}, z : x, y, z \in \mathbb{N}\, with\, x > 0, y > 0\, and\, 1 \leq z \leq 3\}$, where $\{x, y\}$ represents a single plane dimension and z extends of the $\{x, y\}$ plane to the third dimension. The value $z = 1, 2\, and\, 3$ represents the R, G and B planes respectively in a RGB Color-space. The data structure representation of a color image $\mathbb{C}$in the form $A^B = \{f : f\ is\ a\ function\ from\ B\ to\ A\}$ can be given as

$$\mathbb{C} = (A)^B = \{(A, B) : \mathrm{A} \in \tau \text{ and } B \in \varphi_w\} \tag{1}$$

Then an element (pixel) in the image can now be represented as (A, B), where the A represents the location of the element and B represents the corresponding value.

2.1 *Color-to-Grayscale Conversion*

With the enormous growth of digital representation, color images are nowadays very common, but for that sake, gray images are not ignored. Grayscale images are widely used for the economic reasons of printing and also to reduce the algorithmic complexity. Color-to-grayscale, under the domain of '*Natural Image processing*', is an open issue of dimension reduction.

In terms of image algebra [6], the goal of color-to-grayscale conversion is to decrease the value of z to 1, thereby reducing the point set representation as $\tau = \{(\{x, y\}, z^*) : x, y, z \in \mathbb{N}\, with\, x > 0,\ y > 0\, and\, z^* = 1\}$ where $z^* = \sum_{i=1}^{3} f(z_i)$ in case of a weighted grayscale conversion (WGC) method. It is mandatory that the neither the values of set φ_w, nor the element $\{x, y\}$ of set τ are concerned during this process of dimension reduction.

It also required that during this process of dimension reduction, it is necessary to preserve meaningful virtual experience of the original three dimensions into a single dimension. However, the resultant of the process is that, the obtained reduced single dimension image loses vital information such as contrast, sharpness, shadow and structure and in most cases not suited for the next level of image processing.

A number of researchers have done a lot of work on weighing the planes for obtaining a single plane grayscale image [7], while very few of them have looked on each plane of $\mathbb{C}$ separately. The red channel as a separate grayscale image can be represented as $\mathbb{G}_{\mathcal{R}} = (A)^B = \{(A, B) : \mathrm{A} \in \tau \text{ and } B \in \varphi_8\}$, where $\tau_{\mathcal{R}} = \{(\{x, y\}, z) : x, y, z \in \mathbb{N}\ with\ x > 0,\ y > 0\ and\ z = 1\}$. Similarly, the other channels as grayscale image may be represented as $\mathbb{G}_{\mathbb{G}}$ and $\mathbb{G}_{\mathbb{B}}$ when $z = 2$ and $z = 3$. Alternatively, any two color channels are nullified to visualize only one channel as a grayscale image.

2.2 Contrast-Per-Pixel (CPP)

The human perception of evaluating a given image can be termed as Human Vision System (HVS). We conducted a number of experiments to identify the equivalent computational parameter of Human Vision System and found that Contrast-Per-Pixel (CPP) of the image can be a basic parameter. For a grayscale image $\mathbb{G}$, the contrast of each pixel in the image may be defined as $\epsilon \mathrm{A}^{\mathrm{B}}$, where,

$$\Phi = \sum_{i=1}^{x} \sum_{j=1}^{y} [(\mathbb{G}\{i, j\} \oplus s) / 8] \tag{2}$$

and the template ‘s’ in a Moore Neighborhood is $s\ (\tau\{x, y\}) = \begin{cases} 1, & if\ \tau = \{x, y\} \\ -1, & otherwise \end{cases}$

The average of all these variations over the image gives Contrast-Per-Pixel (CPP).

$$\mathrm{CPP} = \frac{1}{x * y} \sum_{i=1}^{x} \sum_{j=1}^{y} [\Phi] \tag{3}$$

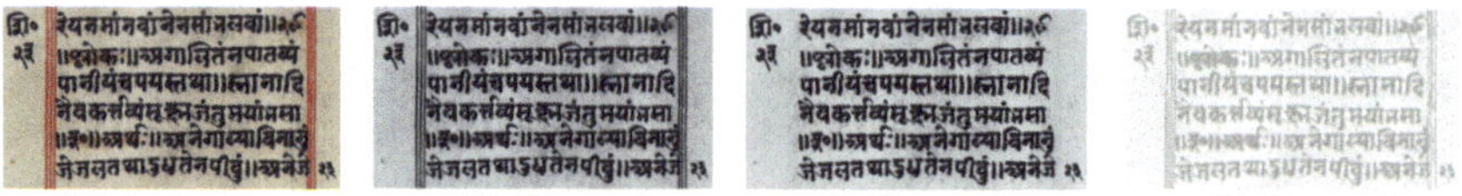

Original color image | R-Channel CPP : 0.9389 | G-Channel CPP: 0.8556 | B-Channel CPP : 0.7200

Fig. 1 Color image and it corresponding R, G and B channels with CPP values

Higher the value of the CPP, the HVS is more satisfactory. Figure 1 shows an example image taken from the historical text '*Shikshapatri*'.[1] The CPP values for each channel indicate that the R—Channel has the highest CPP value and correspondingly, the HVS also indicates that this grayscale image is more appealing when compared to the other grayscale images.

3 Algebra of Image Binarization

Binarization of a grayscale image is primarily carried out for easy segmentation. Consider a grayscale image $\mathbb{G} = (A)^B = \{(A, B)\colon \mathrm{A} \in \tau \text{ and } B \in \varphi_8\}$. The process of binarization of a grayscale image $\mathbb{G}$ involves reducing w from 8 to 1. During this process, the values of the element $\{x, y\}$ in the set τ are not affected. We shall primarily focus on Otsu Global Binarization method.

The main aim of Global binarization involves classifying the value set φ_w into C different classes distinguishable by $(C - 1)$ thresholds. If there are only two groups, the thresholding algorithms focus on determining a single value of T in such a way that φ_w can be dichotomized into $\varphi_{w0} = \{v\colon v \in \mathbb{N}\, and\, 1 \le v \le T\} \subseteq \varphi_8$ and $\varphi_{w1} = \{v\colon v \in \mathbb{N}\ and\ T + 1 \le v \le 255\} \subseteq \varphi_8$ representing the text and background class respectively. The binarized image is represented as

$$\mathbb{L} = \{(A, B)\colon \mathrm{A} \in \tau \text{ and } B \in \varphi_2\} = \begin{cases} 1, \{(A, B)\colon \mathrm{A} \in \tau \text{ and } B \in \varphi_8\} > T \\ 0, \{(A, B)\colon \mathrm{A} \in \tau \text{ and } B \in \varphi_8\} < T \end{cases} \quad (4)$$

where T represents the threshold range.

Otsu proposed a method based on Discriminant Analysis of Statistics [4], which states that the ratio between the total variance (σ_T^2) of the image with that of the with-in class variance (σ_w^2) between the two sets is expressed as $\kappa = \frac{\sigma_T^2}{\sigma_w^2}$.

The maximum value of κ for all possible gray level is the threshold value (i.e.) $T^* = \arg \min\limits_{t \in [1, L]} (\sigma_w^2)$, when σ_T^2 is equal to unity.

[1]Courtesy: http://www.shikshapatri.org.uk/.

4 Proposed Method

The pseudo code of the proposed method is given below.

```
BEGIN

  GIVEN Color image ℂ

  EXTRACT color channels 𝔾_R, 𝔾_G and 𝔾_B
  COMPUTE CPP values for color channels
  IF (CPP_R = CPP_G = CPP_B) THEN

    SET grayscale image 𝔾 to 𝔾_R || 𝔾_G || 𝔾_B
  ELSE
    SET CPP[i] with CPP_R, CPP_G and CPP_B as elements
    INITIALIZE max to 0 and maxIndex to 0
    FOR i = 1 to 3
      IF (CPP[i] > max)
          maxIndex = i;
          max = arr[i];
      ENDIF
    ENDFOR
    CASE maxIndex OF

        1: SET 𝔾 = 𝔾_R

        2: SET 𝔾 = 𝔾_G

        3: SET 𝔾 = 𝔾_B
    ENDCASE
  ENDIF

  CALL Otsu with 𝔾, RETURNING Binarized Image 𝕃
END
```

5 Results and Discussion

The algorithm was tested on the DIBCO and H-DIBCO datasets. The grayscale image with highest CPP exhibited a higher value of Precision after the binarization process, when compared to the weighted grayscale conversion (WGC) method. We used GIMP [7] method $(\mathbb{G} = \mathbb{G}_{\mathcal{R}} * 0.3 + \mathbb{G}_{\mathbb{G}} * 0.59 + \mathbb{G}_{\mathbb{B}} * 0.11)$ of weighted conversion for evaluation.

Fig. 2 **a** Color image. Otsu Algorithm applied on **b** WGC Grayscale Image [CPP = 0.41547, Precision = 0.64295], **c** R—Channel Grayscale Image [CPP = 0.47394, Precision = 0.74324]

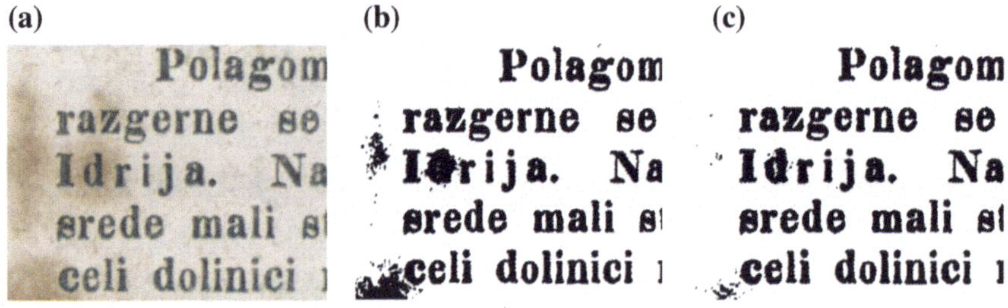

Fig. 3 **a** Color image. Otsu Algorithm applied on **b** WGC Grayscale Image [CPP = 0.23473, Precision = 0.34087], **c** R—Channel Grayscale Image [CPP = 0.25060, Precision = 0.35388]

Since, the objective of the process is goal specific (i.e.) best binarization; we have considered precision metrics for evaluation of the binarized image. A few of the sample results are shown in Figs. 2 and 3. Our objective was to determine if the process of weighted grayscale conversion would influence the binarization process and we can definitely say that it does influence.

6 Conclusion

The Table 1 gives the values of CPP of the three color channels as grayscale images along with the image obtained by weighted grayscale conversion and the corresponding Precision values of the binarized images. The values in bold indicate the best ones as perceived by HVS and it clearly indicates that the grayscale image with highest CPP has exhibited the best binarization. The values in italics and underline indicate no change in CPP and precision in all the four cases, which is barely 25 % of the dataset. We may now conclude, it is necessary to work initially on the natural image processing and then carry out the necessary binarization for better results, rather than using the 'rgb2gray' command of MATLAB for converting a color image to grayscale image.

Table 1 CPP and precision values on DIBCO 2013 dataset

DIBCO 2013								
Image name	R-Channel		G-Channel		B-Channel		WGC	
	CPP	Precision	CPP	Precision	CPP	Precision	CPP	Precision
HW01	**0.17679**	**0.98032**	0.17018	0.97633	0.15557	0.97763	0.17039	0.97692
HW02	**0.36994**	**0.96180**	0.34691	0.94298	0.30553	0.89188	0.34910	0.94402
HW03	**0.36591**	**0.96349**	0.34904	0.94906	0.32406	0.89670	0.35109	0.95429
HW04	*0.18353*	*0.95578*	*0.18353*	*0.95578*	*0.18353*	*0.95578*	*0.18353*	*0.95578*
HW05	*0.22412*	*0.40803*	*0.22412*	*0.40803*	*0.22412*	*0.40803*	*0.22412*	*0.40803*
HW06	*0.21700*	*0.90916*	*0.21700*	*0.90916*	*0.21700*	*0.90916*	*0.21700*	*0.90916*
HW07	*0.27676*	*0.99133*	*0.27676*	*0.99133*	*0.27676*	*0.99133*	*0.27676*	*0.99133*
HW08	**0.21238**	**0.93785**	0.19783	0.93116	0.15998	0.91220	0.93655	0.19786
PR01	**0.23294**	**0.89614**	0.20847	0.90501	0.17809	0.90432	0.2132	0.90190
PR02	**0.32935**	**0.95954**	0.30956	0.95708	0.28969	0.92292	0.31320	0.95633
PR03	0.27954	0.98555	**0.30580**	**0.99757**	0.25863	0.33921	0.29256	0.99624
PR04	**0.25060**	**0.35388**	0.23137	0.33909	0.20959	0.31804	0.23473	0.34087
PR05	**0.47319**	**0.79373**	0.44708	0.79646	0.42546	0.76710	0.45202	0.79437
PR06	**0.29739**	**0.61402**	0.27465	0.63097	0.23598	0.59545	0.27712	0.62713
PR07	**0.63900**	**0.97076**	0.61334	0.96917	0.55254	0.95677	0.61390	0.96962
PR08	**0.47394**	**0.74324**	0.41452	0.63306	0.29285	0.49196	0.41547	0.64295

References

1. Govindaraju V et al (2005) Text extraction from gray scale historical document images using adaptive. Local connectivity map. In: Proceedings of 8th IEEE international conference on document analysis and recognition. IEEE Press, NY, pp 794–798
2. Baird H (1994) Background structure in document images. In: Document image analysis. World Scientific, Singapore, pp 17–34
3. Pratikakis I et al (2014) ICFHR2014 competition on handwritten document image binarization (H-DIBCO 2014). In: 14th international conference on frontiers in handwriting recognition (ICFHR). IEEE Press, New York, pp 803–313
4. Otsu N (1979) A threshold selection method from gray-scale histogram. IEEE Trans Syst Man Cybernet 9:62–66
5. Xu X et al (2011) Characteristic analysis of Otsu threshold and its application. Pattern Recogn Lett 32:956–961
6. Ritter GX, Wilson JN (1996) Handbook of computer vision algorithms in image analysis. CRC Press (1996)
7. Kanan C, Cottrell GW (2012) Color-to-grayscale: does the method matter in image recognition? PLoS One 7–1:1–7

Signature Wavelet Identification of Sounds of Musical Instruments Using RLS Algorithm

Raghavendra Sharma and V. Prem Pyara

Abstract Filter bank theory is used to identify the approximation and detail coefficients of the wavelet filter that are used to identify the scaling and wavelet function of the wavelet. If the filter bank coefficients of the sounds of musical instruments are calculated then it is possible to identify the signature wavelet of the sound signal, which can be used to reconstruct the original signal with negligible error. The filter bank coefficients can be identified with adaptive algorithms viz. LMS, NLMS and RLS. Among the three algorithms, RLS algorithm perform better in all regard and the algorithm converges very fast i.e. number of iterations are less. Hence an algorithm based on RLS algorithm is developed to find out scaling and wavelet functions of the sounds of musical instruments with better accuracy and speed of convergence.

1 Introduction

Of late, a lot of work has been done to identify wavelets matched to the signal to provide best representation of the signal, mostly for deterministic signals [1, 2]. The problem of identification of wavelet matched to the signal has been handled in tho ways: (1) starting with a set of wavelets, one identifies the best match to the signal under consideration; (2) by identifying a wavelet which is matched to the signal using adaptive algorithms. A wavelet can be reconstructed from its approximation and detail coefficients [3, 4]. If approximation coefficients are obtained, then the detail coefficients can be calculated using approximation coefficients [5]. Most of the wavelet applications are dealt with the approximation and detail coefficients and are represented as quadrature mirror filters (QMF) [6], having mirror image spectra.

R. Sharma (✉)
Department of Electronics and Communication, Anand Engineering College, Agra 282007, India
e-mail: raghsharma2000@yahoo.com

V.P. Pyara
Department of Electrical Engineering, Dayalbagh Educational Institute, Agra 282010, India

K.R. Attele et al. (eds.), *Emerging Trends in Electrical, Communications and Information Technologies*, Lecture Notes in Electrical Engineering 394,
DOI 10.1007/978-981-10-1540-3_27

The detail coefficients have been obtained with the help of various optimization techniques viz. mean square error adaptive filter algorithms LMS and NLMS [7]. An adaptive algorithm alters filter parameters iteratively to minimize the error according to some prescribed criterion. But the values of filter coefficients obtained by the two aforementioned algorithms (LMS & NLMS) are not precise and take more time for computation [8]. One of the challenges associated with wavelet synthesis is to compute wavelet coefficients accurately in shortest time.

The paper is organized as follows; In Sect. 2, adaptive algorithms are used to identify the filter bank coefficients of the standard wavelet, and an algorithm based on RLS algorithm is developed in Sect. 3, which identifies the signature wavelet of various musical instrument sounds. In Sect. 4, we have discussed the results obtained through the algorithm discussed in Sect. 3 and scaling and wavelet functions are plotted for various instruments sounds. Finally, concluding remarks are given in Sect. 5.

2 Identification of Filterbank Coefficients of Standard Wavelets

In the algorithms developed in [9, 10], h(-n) have been evaluated using LMS algorithm to reconstruct the wavelet. Here, the desired signal is down sampled and then up sampled before applying to the filter. However, h(-n) thus obtained vary significantly with step size as shown in Table 1. This problem of sensitivity to step size is overcome using NLMS algorithm as shown in Table 2. The number of iterations and the accuracy of the result can be improved with RLS algorithm as shown in the Table 3. The filter coefficients h(-n) are used to find out the scaling and wavelet function of the wavelet using adaptive filter. In adaptive filter the stream of impulses is given as input to the filter, up-sampling and then passing the estimated output iteratively to the filter helps in calculating the scaling function. The iteration method shown in Fig. 1 is implemented on coif2 and db3 wavelets.

Table 1 Filter coefficients obtained by LMS algorithm of wavelet db3

Actual h(n)	Step size	Filter coefficients obtained by LMS algorithm
[0.0498 −0.1208 −0.1909 0.6504 1.1411 0.4705]	0.01	[0.0476, −0.1180, −0.1903, 0.6409, 1.1440, 0.4654]
	0.05	[0.0188, −0.1201, −0.2223, 0.6320, 1.1483, 0.5070]
	0.10	[−0.3817, −0.1855, −0.1970, 0.5478, 0.6875, 0.0337]
	0.15	[−0.6961, 1.2182, −0.03197, −1.6593, 1.9380, 0.0424]
	0.30	[−1.013, 48.713, 2.9040, −1.6613, −1.1017, 1.0041] $\times 10^{200}$

Fig. 1 Flow chart for scaling function

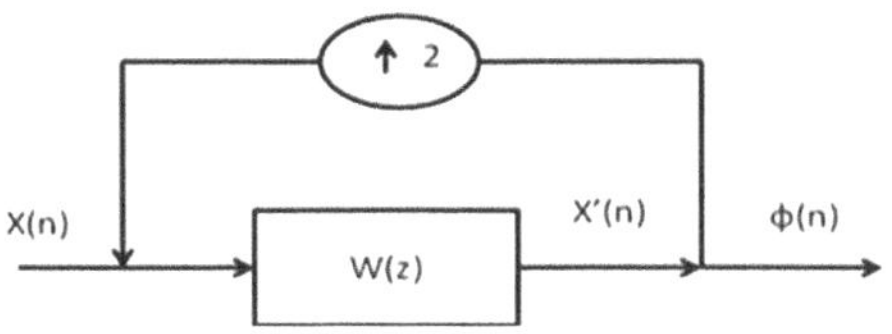

2.1 Filterbank Coefficients Using LMS Algorithm

LMS algorithm is implemented for determining the filter bank coefficients of wavelet db3 and coif2. The selection of step size is very crucial in case of LMS algorithm [11, 12]. For wavelet db3, if we choose a step size more than 0.1662, then, the filter coefficients will never converge, and the reconstruction of the wavelet is not possible due to absurd values as shown in the Table 1.

2.2 Filterbank Coefficients Using NLMS Algorithm

The same experiment is repeated for determination of the filter bank coefficients of db3 with NLMS algorithm which are not sensitive to step size as shown in the Table 2.

2.3 Filterbank Coefficients Using RLS Algorithm

The same experiment is further carried out for constructing the wavelets db3 and coif2 with RLS algorithm and the reconstruction is successful, with minimum convergence time among the three algorithm and the filter bank coefficients are also much closer to the desired values as shown in the Table 3.

Table 2 Filter coefficients obtained by NLMS algorithm of wavelet db3

Actual h(n)	Step Size	Filter coefficients obtained by NLMS algorithm
[0.0498 −0.1208 −0.1909 0.6504 1.1411 0.4705]	0.01	[0.0458, −0.1239, −0.1936, 0.6161,1.0973, 0.4435]
	0.05	[0.0497, −0.1273, −0.1982, 0.6493, 1.1343, 0.4703]
	0.10	[0.0452, −0.1334, −0.1967, 0.6598, 1.1490, 0.4661]
	0.15	[0.0488, −0.1061, −0.1686, 0.6523, 1.1342, 0.4834]
	0.30	[0.0332, −0.1227, −0.1426, 0.6560, 1.1793, 0.4834]

Table 3 Filter coefficients obtained by RLS algorithm of wavelet db3

Actual h(n)	Step Size	Filter coefficients obtained by RLS algorithm
[0.0498 −0.1208 −0.1909 0.6504 1.1411 0.4705]	0.01	[0.0497, −0.1246, −0.1876, 0.6486, 1.1416, 0.4704]
	0.05	[0.0463, −0.1249, −0.1958, 0.6451, 1.1401, 0.4689]
	0.10	[0.0470, −0.1244, −0.1917, 0.6478, 1.1363, 0.4690]
	0.15	[0.0497, −0.1230, −0.1910, 0.6488, 1.1442, 0.4709]
	0.30	[0.0461, −0.1226, −0.1988, 0.6445, 1.1374, 0.4682]

Table 4 Iterations used by adaptive algorithms

Algorithm	No of iterations
LMS	4N
NLMS	2N
RLS	N

The number of iteration used for obtaining the desired values of the filter coefficients of all types of the wavelets with minimum error is shown in the Table 4.

Where N is the number of filter bank coefficients. This variation in iteration time is very severe in case of filter with larger number of coefficients. The mean square error obtained by RLS algorithm is minimal and it converges much faster than the other two algorithms.

3 Filter Bank Coefficients of Sounds of Musical Instruments Using RLS Algorithm

It is seen in the previous section that among the three adaptive algorithms (LMS, NLMS and RLS), RLS algorithm perform better if used for identification of the filter bank coefficients of the standard wavelets because the filter bank coefficients are closer to the standard values and the algorithm converges very fast i.e. number of iterations are less [13–15]. Hence an algorithm based on RLS algorithm is the best algorithm to find the filterbank coefficients of sound of musical instruments and an algorithm is developed to find out scaling and wavelet functions of the sounds of musical instruments with better accuracy and speed of convergence.

3.1 RLS Based Algorithm for Identifying Scaling and Wavelet Function of Sounds of Musical Instruments

An algorithm is developed in this section which uses RLS algorithm for identification of filter bank coefficients of the sound produced by musical instruments. These filter bank coefficients are used for the generation of scaling and wavelet function of the sounds. The steps for the algorithm are as follows:

1. Take the signal in the wav format.
2. Choose the sampling rate and the part of the signal is taken for analysis depending upon the sampling rate.
3. The order of the filter is decided, larger the order of the filter the shape of the wavelet function will be smooth.
4. Call RLS algorithm.
5. Check the error after each iteration, if it is within the specified limits then stop the convergence process; if not then go back to the step 3.
6. If the error criterion is satisfied then record the filter coefficients.
7. Use these filter coefficients to draw the scaling and wavelet function.

4 Results and Discussions

The algorithm developed in the previous section is applied to the sound samples of the various musical instruments. The sound signal is first sampled at 44.1 K samples and the part of signal taken for the each category of the instrument sound is different. For wind instrument and string instrument the size of the signal is 10 ms and for percussion and keyboard instruments the size is 100 ms. The order of the filter is taken as 32 because as the order of the filter is more the shape of the obtained wavelet function is smooth. The average value of the filter bank coefficients h(n) for the various instrument sound signals is shown in the Table 5.

The average values of h(n) obtained by the previous algorithm are used to draw the scaling and wavelet functions of the different musical instrument sounds. Scaling and wavelet functions are presented with amplitude on Y axis and compact support of 3 on X axis. Scaling and wavelet functions obtained through RLS algorithm are different from the standard wavelets and if we use these scaling and wavelet function to synthesize the sound of musical instruments then synthesized sound better resemble the original signal. The scaling and wavelet functions of the musical instruments are plotted in the Figs. 2, 3, 4 and 5.

Table 5 Average value of the coefficients h(n) for different instrument sounds

Instrument sound	Average h(n)
Flute	[0.0713 0.0375 0.0289 0.0213 0.0106 0.0021 −0.0259 −0.0038 −0.0297 −0.0368 −0.0282 −0.0371 −0.0430−0.0272 −0.0438 −0.0514 −0.0645 −0.0543 −0.0576 −0.0623 −0.0595 −0.0495 −0.0629 −0.0522 −0.0493 −0.0488 −0.0397 −0.0517 −0.0383 −0.0151 −0.0064 −0.0135]
Sitar	[0.0113 −0.0021 −0.0064 −0.0066 −0.0062 −0.0017 −0.0076 −0.0104 −0.0131 −0.0173 −0.0236 −0.0299 −0.0283 −0.0240−0.0195 −0.0192 −0.0177 −0.0158 −0.0166 −0.0154 −0.0173 −0.0203 −0.0180−0.0088 −0.0011 −0.0003 0.0037 0.0025 0.0042 0.0051 0.0026 0.0059]
Dafli	[0.0302 0.0311 0.0244 0.0234 0.0203 0.0209 0.0243 0.0255 0.0256 0.0289 0.0291 0.0295 0.0284 0.0319 0.0288 0.0263 0.0236 0.0302 0.0236 0.0220 0.0258 0.0189 0.0164 0.0129 0.0156 0.0141 0.0150 0.0145 0.0143 0.0173 0.0163 0.0143]
Harmonium	[0.0193 0.0163 0.0062 0.0041 −0.0009 −0.0069 −0.0035 −0.0075 −0.0158 −0.0166 −0.0134 −0.0156 −0.0176 −0.0204 −0.0170 −0.0087 −0.0087 −0.0089 −0.0131 −0.0086 −0.0103 −0.0068 −0.0041 −0.0005 −0.0089 −0.0107 −0.0084 −0.0037 0.0016 −0.0013 −0.0043 −0.0056]

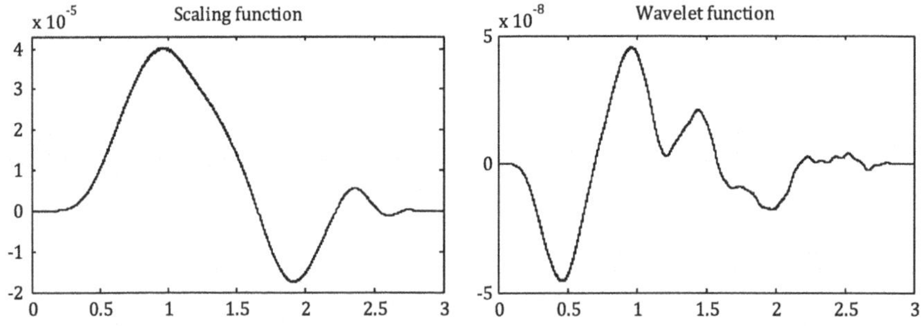

Fig. 2 Scaling and wavelet function of flute sound

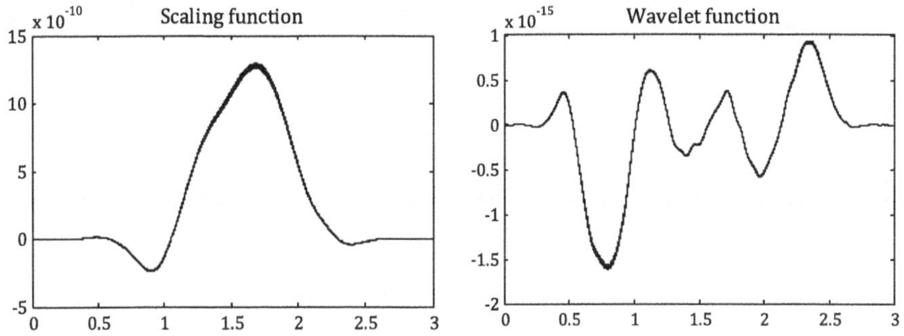

Fig. 3 Scaling and wavelet function of sitar sound

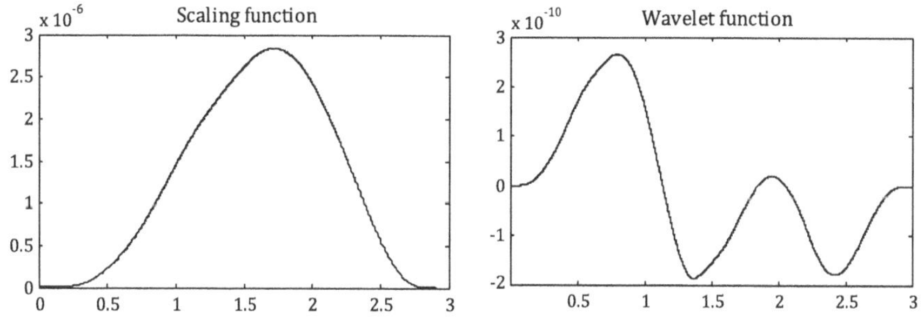

Fig. 4 Scaling and wavelet function of dafli sound

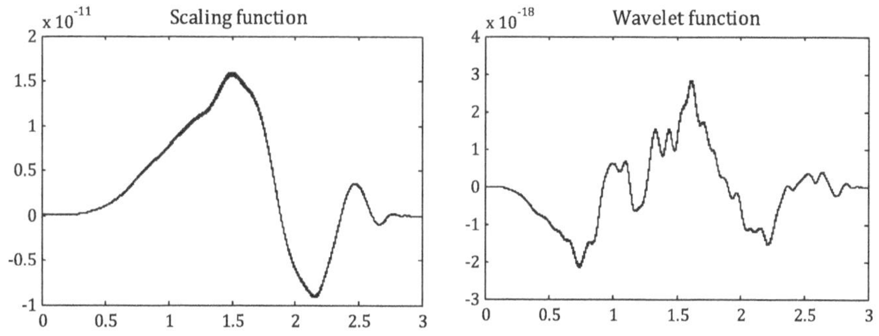

Fig. 5 Scaling and wavelet function of harmonium sound

5 Conclusion

In this paper, the identification of filter bank coefficients of sounds of musical instruments is calculated with the help of RLS based algorithm and the scaling and wavelet functions are plotted. The three adaptive algorithms can be used to identify the filter bank coefficients of the standard wavelets, but RLS algorithm gives filter bank coefficients closer to the actual values and the convergence is faster as compared to other two algorithms. Hence RLS based algorithm is used to identify the filter bank coefficients of the sounds of musical instruments. The identified filter bank coefficients are used to draw the scaling and wavelet functions of the sound of wind instruments (flute), string instruments (sitar), percussion instruments (dafli) and keyboard instruments (harmonium). Hence it is concluded that the signature wavelet of musical instrument sound is identified with the help of RLS based algorithm with better accuracy and less iteration time.

References

1. Ramchandran K, Vetterli M, Herley C (1996) Wavelets sub band coding and best bases. Proc IEEE 84(4):541–560
2. Vetterli M, Harley C (1992) Wavelets and filter banks theory and design. IEEE Trans Signal Process 40(9)
3. Chapa JO, Rao RM (2000) Algorithms for designing wavelets to match a specified signal. IEEE Trans Signal Process 48(12):3395–3406
4. Sinith MS, Nair MN, Nair NP, Parvathy S (2010) Identification of wavelets and filter bank coefficients in musical instruments. In: IEEE conference on audio language and image processing, pp 727–731
5. Tewfik A, Sinha D, Jorgensen P (1992) On the optimal choice of a wavelet for signal representation. IEEE Trans Inf Theory 38(2):747–766
6. Gianpolo E (1993) Pitch synchronous wavelet representation of speech and music signals. IEEE Trans Signal Process 41(12):3313–3330
7. Soria E, Calpe J, Chambers J, Martinez M, Camps G, Guerrero JDM (2008) A novel approach to introducing adaptive filters based on the LMS algorithm and its variants. IEEE Trans 47:127–133
8. Markus R (1993) The behavior of LMS and NLMS algorithms in the presence of spherically invariant process. IEEE Trans Signal Process 41(3):1149–1160
9. Razzaq MS, Khan NM (2010) Performance comparison of adaptive beam forming algorithms for smart antenna systems. World Appl Sci J 11(7):775–785
10. Chen T, Vaidyanathan PP (1994) Vector space framework for unification of one and multidimensional filter bank theory. IEEE Trans Signal Process 42(8):2006–2021
11. Douglas SC, Markus R (2000) Convergence issue in the LMS adaptive filter. CRC Press LLC
12. Sharma R (2015) Denoising, coding and identification of sounds of some musical instruments, PhD Thesis
13. Sharma R, Pyara VP (2012) Denoising of sounds of musical instruments by RLS adaptive algorithm. Int J Adv Res Comput Sci 3(5):1–4
14. Sharma R, Pyara VP (2012) Comparative study of adaptive algorithms for identification of filter bank coefficients of wavelets. In: Proceedings, international journal of computer applications® (IJCA), pp 21–25
15. Sharma R, Pyara VP (2012) A comparative analysis of mean square error adaptive filter algorithms for generation of modified scaling and wavelet function. Int J Eng Sci Technol 4 (4):1402–1407

Study on Influence of Hip Trajectory on the Balance of a Biped Robot

Ravi Kumar Mandava and Pandu R. Vundavilli

Abstract Balancing of a biped robot plays an important role, as it has to walk on two feet while moving from one place to another to execute the task assigned to it. The present research paper is mainly focusing on the study of influence of hip trajectory on the balance of an 18-DOF biped robot in single support phase (SSP) while walking on the flat floor. Two different types of trajectories, namely straight line and particle swarm-based cubic polynomial are considered for the hip joint. The gaits for the lower and upper limbs of the robot are generated by using the concept of inverse kinematics after considering the said options for the hip trajectory. The balance of the robot is determined by calculating the Dynamic Balance Margin (DBM) of the generated gait. Further, the two developed approaches are tested for their capability to generate dynamically balanced gaits in computer simulations.

1 Introduction

Humanoid robots are more anthropomorphic robotic systems which try to mimic the capabilities of a human being and perform the challenging jobs. The objective of present research is not only to design suitable trajectories for generating gaits on a particular terrain, but also to see whether the gaits generated are dynamically balanced or not. Further, DBM is used to measure the dynamic balance of the two legged vehicle. For calculating the dynamic balance of the biped robot, different researchers proposed different methods. Vukobratovic et al. [1] proposed the concept of zero moment point (ZMP) for calculating the amount of balance of the two legged robot. Gowswami [2] developed the concept of foot rotation indicator (FRI),

R.K. Mandava (✉) · P.R. Vundavilli (✉)
School of Mechanical Sciences,
IIT Bhubaneswar, Bhubaneswar 751013, Odisha, India
e-mail: rm19@iitbbs.ac.in

P.R. Vundavilli
e-mail: pandu@iitbbs.ac.in

K.R. Attele et al. (eds.), *Emerging Trends in Electrical, Communications and Information Technologies*, Lecture Notes in Electrical Engineering 394,
DOI 10.1007/978-981-10-1540-3_28

which was used to generate the dynamically balanced gaits for a single support phase walking cycle. It is important to note that the balance is one of the important concerns during the design of the gaits of any biped robot. For getting stability of the biped robot, most of the researchers designed various trajectories for the hip and foot joint of the robots to follow. Huang et al. [3] planned a scheme for the development of plane walking configurations for a two legged robot. This scheme made it possible to develop a hip motion that is smooth and extremely stable without first determining the anticipated ZMP trajectory. Later on, Chow and Jacobson [4] studied the optimal biped locomotion of the hip, and suggested that the hip trajectory be synthesized prior to joint angle profiles. Then, Hwang et al. [5] developed gait for a biped robot that climb up a sloping surface. In that study, the slope of the surface changes while the biped robot walks. The walking trajectory is modified during double support phase for transitional motions. In addition to the above works, Udai [6] designed the optimum hip trajectory for a biped robot during single support using genetic algorithm (GA). The optimization is performed with the gradual increments of time. This methodology can be extended for generating the real time trajectory. Moreover, Pandu and Pratihar developed GA-NN and GA-FL gait planners for a dynamically balanced biped robot for negotiating sloping surfaces [7] and crossing a ditch surface [8]. They used fuzzy logic, analytical and neural network based approaches to solve the gait generation problems. The generated gait was tested for its balance by calculating the position of zero moment point (ZMP). Further, Rodrigues et al. [9] used steady state GA for a five-link biped robot and also found required torques at each joint of the robot to obtain a preferred trajectory for robot trunk center of mass. Moreover, Kim [10] developed a systematic methodology for online joint trajectory generation of a human-like biped walking. The authors had used PSO for the generation of full-body model of the humanoid robot. Recently, Haun and Anh [11] developed novel stable walking for humanoid robot using PSO.

In the present paper an attempt is made to study the influence of hip joint on the balance of the biped robot. The gait of the biped robot on the flat surface has been generated by using the concept of inverse kinematics along with swing foot and hip trajectories. The hip joint is allowed to follow two different trajectories, namely straight line trajectory developed based on analytical method and cubic polynomial trajectory obtained after PSO-based optimization.

2 Mathematical Formulation of the Problem

In the present work, an attempt is to made to study the influence of hip trajectory on the balance of a 18-DOF biped robot in sagittal plane during single support phase (SSP) only. The robot considered in the present study (ref. to Fig. 1), consists of two arms with 3-DOF in each and two legs with 6-DOF in each leg. The gait generation is only possible when the wrist, foot and hip joint are allowed to follow a definite trajectory. Both the wrist and swing foot are assumed to follow cubic

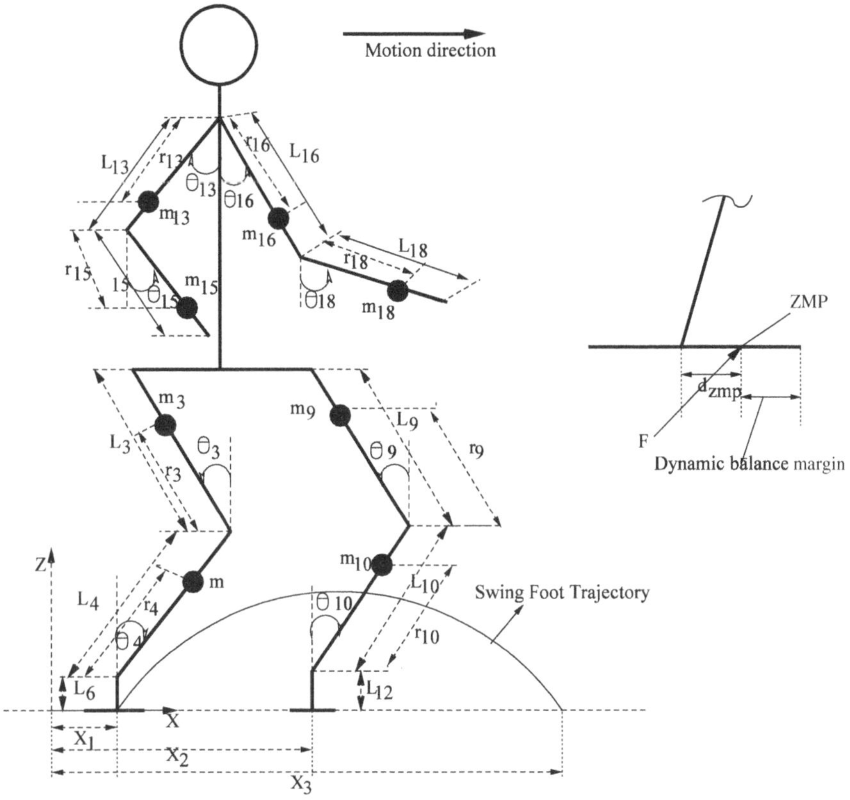

Fig. 1 Schematic diagram showing the structure of the biped robot

polynomial trajectories and the hip joint is considered to follow two different trajectories, namely a straight line trajectory and PSO optimized cubic polynomial trajectory. The following cubic polynomial trajectory and boundary conditions are used for both swing leg and wrist of the hand.

$$z = c_0 + c_1 x + c_2 x^2 + c_3 x^3 \quad (1)$$

Foot trajectory:

- at $x = x_1$, $z = 0$;
- at $x = x_1 + (x_2 - x_1)/2$, $z = f_s/6$;
- at $x = x_2 + (x_3 - x_2)/2$, $z = f_s/6$;
- at $x = x_3$, $z = 0$.

Wrist trajectory:

- at $x = x_{11}$, $z = h$;
- at $x = (x_{11} + x_{12})/2$, $z = h + f_w/8$;
- at $x = ((x_{13} - x_{12})/2) + x_{12}$, $z = h + f_w/8$;
- at $x = x_{13}$, $z = h$.

where the factors c_0, c_1, c_2, and c_3 represent the coefficients of cubic polynomial, z indicates the swing foot and wrist height. Further, the coefficients of the cubic polynomial are to be determined by using the boundary conditions given below.

The closed form of inverse kinematics based solution is used to generate gaits of the biped robot. The angles θ_3 and θ_4 that appear in that Sagittal plane are derived after utilized the concept of inverse kinematics.

$$\theta_4 = sin^{-1}\left(\frac{h_1 L_3 \sin\psi + l_1(L_4 + L_3\cos\psi)}{(L_4 + L_3\cos\psi)^2 + (L_3\sin\psi)^2}\right) \tag{2}$$

where $h_1 = L_4\cos\theta_4 + L_3\cos\theta_3$, $l_1 = L_4\sin\theta_4 + L_3\sin\theta_4$, $\psi = \theta_4 - \theta_3 = \text{arcos}((h_1^2 + l_1^2 - L_4^2 - L_3^2)/2\,L_4L_3)$. Thus, θ_4 can be calculated from the equation $\theta_3 = \theta_4 - \psi$. Further, no joint angle variation is considered for the swing foot. Similarly, the angles θ_9 and θ_{10} are also calculated. The position of the ZMP from the ankle joint of the ground leg in x-direction can be determined using the following equations.

$$x_{ZMP} = \frac{\sum_{i=1}^{n}(I_i\dot{\omega}_i + m_i x_i(g - \ddot{z}_i) - m_i\ddot{x}_i z_i)}{\sum_{i=1}^{n}(m_i(\ddot{z}_i - g)} \tag{3}$$

where I_i and $\dot{\omega}_I$ denotes the moment of inertia (kg-m^2) and angular acceleration (rad/s^2) of the link i, m_i indicates the mass of the link (kg), $\ddot{x}_i$ and $\ddot{z}_i$ denote the acceleration in x- and z-directions, respectively, and g is the acceleration due to gravity (m/s^2). Further, the dynamic balance margin in the direction of motion can be determined by using the expression given below.

$$x_{DBM} = \left(\frac{f_s}{2} - |x_{ZMP}|\right) \tag{4}$$

where f_s and x_{zmp} length of foot support and zero moment point in x-direction, respectively.

3 Optimization of Hip Trajectory by Using PSO Algorithm

In the present study, particle swarm optimization (PSO), which is an optimization method that imitates the behavior of biological population is used to generate the optimal cubic polynomial trajectory for the hip joint of the biped robot while walking on the flat floor. It is a global search and optimization algorithm first introduced by Eberhart and Colleagues in 1995 [12]. This algorithm works based on the behavior of swarm, such as fish schooling and bird flocking. The working principle of PSO algorithm is shown in Fig. 2.

In present study, four parameters of the biped walking, such as two coefficients of cubic polynomial, height and length of the hip joint form the reference position are considered as the position of swarm. It is important to note that the other two boundary conditions required to solve the coefficients of the cubic polynomial are taken from the beginning and end of the trajectory motion. As there are four parameters, the total number of swarm in each population is coming out to be equal to four.

The particles in each iteration are updated to get the *Pbest* and *Gbest*. For initial position Pbest and Gbest are different. The modified velocity and position of each

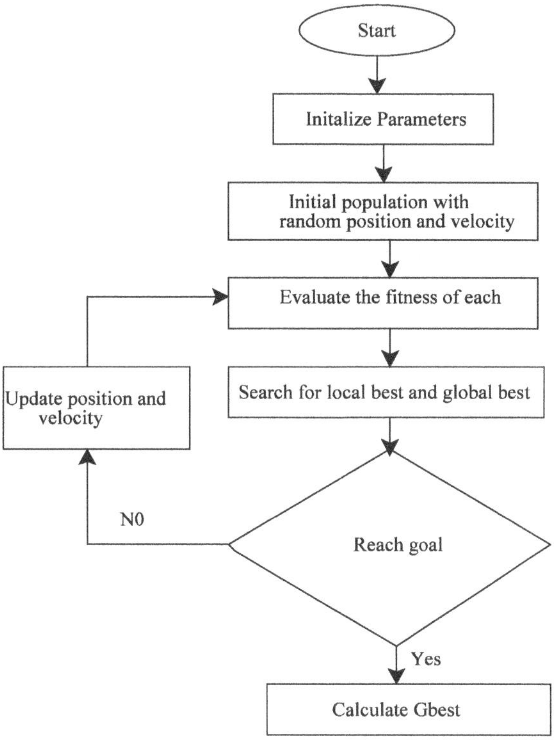

Fig. 2 Flow chart of PSO algorithm

particle can be calculated by using current velocity and position given in the subsequent equations.

$$V_{i,D}^{t+1} = W * V_{i,D}^{t} + C_1 * R_1 * \left(P_{i,D}^{t} - X_{i,D}^{t}\right) + C_2 * R_2 * \left(G_{i,D}^{t} - X_{i,D}^{t}\right)$$

$$X_{i,D}^{t+1} = X_{i,D}^{t} + V_{i,D}^{t+1} \tag{5}$$

where i varies = 1,2, …, N, W indicates inertia weight, $P_{i,D}^{t}$ and $G_{i,D}^{t}$ represents P*best* and G*best* respectively. Moreover $V_{i,D}^{t}$ and $X_{i,D}^{t}$ indicates velocity and position of the particle j at iteration t, respectively. Further C_1 and C_2 indicates cognitive and social parameter and R_1 and R_2 represents random number in the range (0–1) respectively.

4 Results and Discussions

Once the algorithms for different hip trajectories (that is, straight line and PSO-based cubic polynomial) are developed, the performance of the algorithms in terms of generating the dynamically balanced gaits for the 18-DOF biped robot is tested in simulations. To find the optimal parameters, such as two coefficients of cubic polynomial, height and length of hip joint of the biped robot, the parameters of PSO, namely swarm size and number of generations are kept equal to 40 and 60, respectively. The optimal values of two coefficients of cubic polynomial, height and length of the hip joint of the biped robot are seen to be equal to 1.9876, 3.9821, 146.59, 0.0245 mm, respectively. The straight line and cubic polynomial trajectories obtained for the hip joint are shown in Fig. 3.

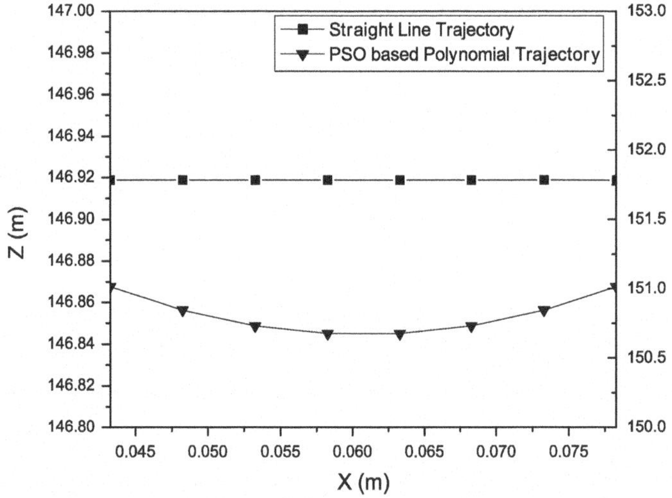

Fig. 3 Hip trajectories generated during the walk of biped robot on flat surface

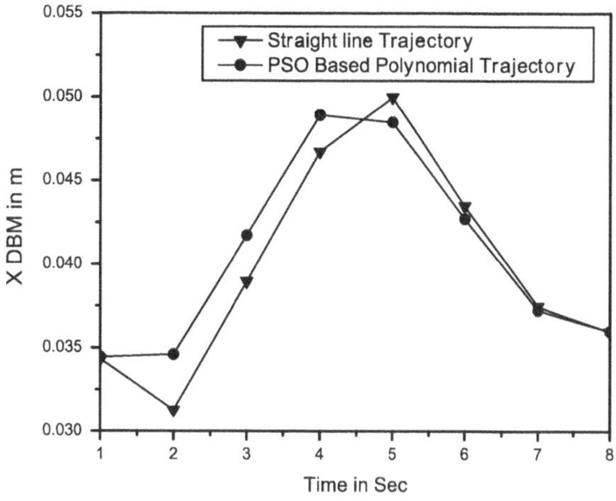

Fig. 4 Schematic diagram showing the variation of DBM

From the graph, it can be clearly seen that the hip trajectory obtained using PSO algorithm is cubic polynomial in nature. Further, Fig. 4 shows the variation of DBM for both the cases. The average dynamic balance margin for the hip joint with straight line and PSO-based cubic polynomial are seen to be equal to 0.0405 and 0.0397 m, respectively. From these results, it can be observed that the gait generated using straight hip trajectory is more dynamically balanced than the cubic polynomial trajectory. This may be due to the reason that the cubic polynomial obtained is having a negative slope and resulted in lower hip height when compared with the straight line hip trajectory. This observation is falling in-line with the observation of human beings, when we try to walk with keeping the height of the hip joint at a lower level than what it could be.

5 Conclusions

In the present study, an attempt is made to study the influence of hip trajectory on the balance of the biped robot. Two trajectories, namely straight line and PSO-based cubic polynomial are developed for the hip joint of the biped robot. Further, the balance of the generated gait is tested with the help of DBM. It has been observed that straight line hip trajectory is found to be more balanced when compared with the cubic polynomial trajectory.

Acknowledgments The authors thank IIT Bhubaneswar, India for sponsoring this project under seed money grant.

References

1. Vukobratovic M, Frank AA, Juricic D (1970) On the stability of biped locomotion. IEEE Trans Biomed Eng BME 17(1):25–36
2. Gowsami A (1999) Foot rotation indicator point: a new agit planning tool to evaluate postural stability of biped robots. In: Proceedings of IEEE international. conference of robotics and automation, Detroit, Michigan, May 1999, pp 47–52
3. Huang Q, Yokoi K, Kajita S, Kaneko K, Arai H, Koyachi N, Tanie K (2001) Planning walking patterns for a biped robot. IEEE Trans Robot Autom 17(3):280–289
4. Chow CK, Jacobson DH (1971) Studies of locomotion via optimal programming. Math Biosci 10:239–306
5. Hwang SW, Yeon JS, Park JH (2013) Trajectory generation method for biped robots to climb up an inclined surface. In: IEEE international conference on robotics, Oct-2013, pp 1–5
6. Udai AD (2008) Optimum hip trajectory generation of a biped robot during single support phase using genetic algorithm. In: IEEE international conference on emerging trends in engineering technology, pp 739–744
7. Vundavilli PR, Pratihar DK (2008) Soft computing-based gait planners for a dynamically balanced biped robot negotiating sloping surfaces. Appl Soft Comput 9:191–208. Elsevier
8. Vundavilli PR, Pratihar DK (2010) Dynamically balanced optimal gaits of a ditch-crossing biped robot, vol 58, no. 4, pp 349–361
9. Rodrigues L, Prado M, Travares P, da Silva K, Rosa A (1996) Simulation and control of biped locomotion—GA optimization. In: Proceedings of IEEE international conference on evolutionary computation, pp 390–395
10. Kim JW (2014) Online joint trajectory generation of human-like biped walking. Int J Adv Robot Syst INTECH
11. Haun TT, Anh HPH (2015) Novel stable walking for humanoid robot using particle swarm optimization algorithm. In: International conference on artificial intelligence and industrial engineering, pp 322–325
12. Kennedy J, Eberhart R (1995) Particle swarm optimization. In: Proceedings of IEEE international conference on neural networks, Perth, 27 November–1 December 1995, pp 1942–1948. doi:10.1109/ICNN.1995.488968

A Novel Technique for Edge Detection Using Gabor Transform and K-Means with FCM Algorithms

D. Maruthi Kumar, K. Prashanth, Praneel Kumar Peruru and P. Charishma Kumar Reddy

Abstract In this paper, a technique has been proposed for detecting edge in medical images throughput from computed tomography and magnetic resonance imaging devices. The proposed technique is Gabor wavelet transform along with two clustering methods i.e. Fuzzy c-means with k-means which is used to adorn the edge information while suppressing noise.

Keywords Image processing · Histogram · Gabor filter · Fuzzy-means · K-means · Edge segmentation

1 Introduction

The edge detection is a mathematical technique and is fundamental tool in image processing for the detection of edges in the given image at which changes in the very sharp brightness or at discontinuities. Actually, edge detection is one of stage in image processing, computer vision and machine vision especially in the fields of attribute detection and attribute extraction. Magnetic Resonance Imaging and computed

D. Maruthi Kumar (✉) · K. Prashanth · P.K. Peruru
Electronics and Communication Engineering,
Srinivasa Ramanujan Institute of Technology,
Ananthapuramu 515701, A.P., India
e-mail: maruthikumar2015@gmail.com

K. Prashanth
e-mail: prashanth.k443@gmail.com

P.K. Peruru
e-mail: praneel.cse@srit.ac.in

P. Charishma Kumar Reddy
Computer Science and Engineering,
Srinivasa Ramanujan Institute of Technology,
Ananthapuramu 517001, A.P., India
e-mail: mail2charishma@gmail.com

K.R. Attele et al. (eds.), *Emerging Trends in Electrical, Communications and Information Technologies*, Lecture Notes in Electrical Engineering 394,
DOI 10.1007/978-981-10-1540-3_29

tomography are equipments used presently as diagnostic tools in the field of medical. These diagnostic tools provide information about biological tissues and anatomical structures [1–3]. The determination of tissue premises plays an important role in the medical images, to identify the deviant in anatomical structures and tissues, which help clinicians for treatment purpose in the defect areas such as surgery or radiotherapy. But it is very time consuming process for experts in radiology to process manually. So for this, purpose only edge detection techniques has been proposed i.e. Gabor wavelet transform. There are classical edge detection methods like sobel gradient detector [4], prewitt edge detector, laplacian. In all these methods having one con's i.e. determining edge is not done perfectly due to presence of noise. So, this problem can be solved by using Gabor wavelet transform [5] which has filter banks to suppress noise and detect the edge. information and the clustering techniques to convert grayscale image into binary image. The denoising filter like median filter or bilateral filter which reduces the noise while preserving the edge information and also less time consuming process which can be used perfectly for real time applications. The first step is segmentation of an given image into the different regions to identify edge regions. Thus, next step is to find out edge points from grayscale image. The most familiar method of determining the points according to threshold value which is one of the difficult process because there is not necessary to segment for few images so here threshold value needed to be adaptive in nature. To overcome this some denoising method has adopted like adaptive median filter to denoise the image. The edge determine by human easily has be modeled as filter bank. The filter bank consists of Gabor functions like orientation and frequencies. So, the output function can be accepted as human visual system by using Gabor wavelet transform which has demonstrated good performance in texture representation and discrimination. The adopted technique i.e. Gabor wavelet transform and clustering techniques like fuzzy c-means and k-means [6, 7], has overcome the arises issues with good pro's.

2 Proposed Algorithm

2.1 Gabor Wavelet Transforms

In the detection of an edges in image processing, a linear filter such as Gabor filter is used. Here filter bank which is similar to human visual system consists of frequency and orientation representations of Gabor filters, especially in texture analysis and discrimination [1–3]. It can also be represented as a function of Gaussian kernel alleviator by plane wave of sinusoidal. In portrayal, there is a vital amount of computer vision applications using Gabor functions, such as texture segmentation, image analysis and discriminations.

A Gabor filters with varied orientation and frequency can be needful for extirpate useful attribute from an given image. Gabor wavelets proclaim the image directional features by adjusting the frequency properties [8, 9]. The main theme of frequency adjusting for denoising the medical images during this process preserving of edge information is most vital [13].

It has the multiplication-convolution property in which convolution of harmonic function Fourier transform and Gaussian function Fourier transform which proclaim the Fourier transform of Gabor's filter. The filter has orthogonal directions in terms of real and imaginary components. The distinguishing of 2D Gabor wavelet are appropriate to derive the directional features, and waveform used to preserve the edge pels while reducing he noise, that has been present in medical images (Fig. 1).

2D Gabor wavelet representation as

$$G(x, y, \theta, u, \sigma) = \frac{1}{2\pi\sigma^2} \exp\left\{ -\frac{x^2 + y^2}{2\sigma^2} \right\} \exp\{2\pi j(ux \cos\theta + uy \sin\theta)\}, \quad (1)$$

where μ is the sinusoidal wave frequency
θ is sinusoidal wave orientation
σ is the standard deviation in the x and y direction and $j = \sqrt{-1}$.

The output response of Gabor filtering is given in the 2D convolution of input image I(x, y) and G(x, y) for particular μ, θ and σ.

2.2 K-Means Clustering

Clustering is the method of fragmenting set of valid data points according to the similarities of their features into small number of groups. Generally data points are fragmented in clusters, the main aim is to assign a set of data points to the cluster. In this K-means cluster [6], to find positions of cluster which minimizes the distance from cluster to the set of data points. The K-means clustering uses of the Euclidean distance.

A set of observations $(x_1, x_2, \ldots, x_n)$ partition into k sets $c = \{c_1, c_2, \ldots, c_i\}$ to reduce the sum of squares distance within cluster.

$$C_{\mathrm{MSE}} = \sum_{i=1}^{k} \sum_{x_j \in c_i} \|x_j - c_i\|^2, \quad (2)$$

Here C_{MSE} is the mean square error value of examined pel to assign cluster. It can be predict that x is present in cluster if $\|x_j - c_i\|$ is minimum of all the k distances. K-means algorithm can be overviewed as: Initializations of centre location (c_1, c_2, c_3, c_k), placing x_i to its nearest cluster centre c_k. Placing the membership of each pixel to the k clusters, where centroid is nearest to that pixel. And finally c_i to be the centre of mass of all points for all k cluster centres.

2.3 Fuzzy C-Means Clustering

In this method, grouping of pels according to similarities of their features, because image can be represented by its features such as histogram properties [9]. A set of data points into set of c fuzzy clusters according to some desired criterion. The minimum value can be calculated with help of cost function which nothing but the distance between cluster centers and pels. The FCM based segmentation preserve more information than other segmentation methods like crisp and hard segmentation. Suppose if we can't consider the spatial information of image then FCM has drawback due to sensitive of noise. So in this we have to consider the spatial information and also repositioned the cluster center after calculating the C-means [10–12]. Each pel assigned to particular group by using membership function after calculating cost function i.e. minimum value using Euclidean distance between examined pel and centre to be assigned. The membership can be defined as probability of a pel belonging to cluster, where probability depends on distance of pel to cluster centre as shown in Eq. (4)

$$E = \sum_{j=1}^{C} \sum_{i=1}^{C} \mu_{ij}^{k} \left\| p_j - c_i \right\|, \tag{3}$$

$$c_i = \frac{\sum_{j=1}^{N} \mu_{ij}^{k} \mu_{ij}^{k} x_j}{\sum_{j=1}^{N} \mu_{ij}^{k}},$$
$$\mu_{ij} = \left(\sum_{m=1}^{C} \left(\frac{\left\| x_j - c_i \right\|}{\left\| x_j - c_m \right\|} \right)^{2/(k-1)} \right)^{-1}. \tag{4}$$

Where = cost of an examined pel to assign cluster

μ_{ij} and c_i = membership of a pixel to cluster and cluster centre
||.|| = absolute value operator
K = adjust the fuzziness as a constant.

So, the proposed algorithm been tested on several images from medical equipments like CT and MRI. The proposed algorithm processed in three steps:

1. Gabor wavelet transformation to encounter directional edge information.
2. A clustering algorithm to convert grayscale image into binary image which even though consists unrelated pels.
3. A morphological operation to remove unrelated pels in binary images.

3 Parameters for Image Assessment in Edge Detection Method

Now the thing is to obtain the quality parameter so as to estimate the integrity of edge detection. Several methods have been proposed in the literature. To evaluate the performance, two parameters can be used:

1. Mis Classification Rate (MCR)
2. Pratt's Figure Of Merit (FOM)

these parameter just only measure the similarity between human manually edge detected and edge detected by algorithm.

$$\text{MCR} = \frac{\sum |B_A \cap B_D| + \sum |F_A \cap F_D|}{\sum (B_A + F_D)} \times 100\,\%, \tag{5}$$

F_A and F_D = Foreground pels of actual and detected image

B_A and B_D = Background pels of actual and detected image

The less value of MCR denotes good detection.

$$\text{FOM} = \frac{1}{\max(N_t, N_d)} \sum_{i=1}^{N_d} \frac{1}{1 + \alpha L(i)^2}, \tag{6}$$

N = number of edge pels

d, t = detected edge, accurate edge
L(i) = distance between *i*th accurate edge pel and detected edge pel
α = scaling factor which always equal to 1/9
Case i: If edge detected, then FOM = 1
Case ii: If edge not detected, then FOM value increases
Otherwise, it may decrease to zero.

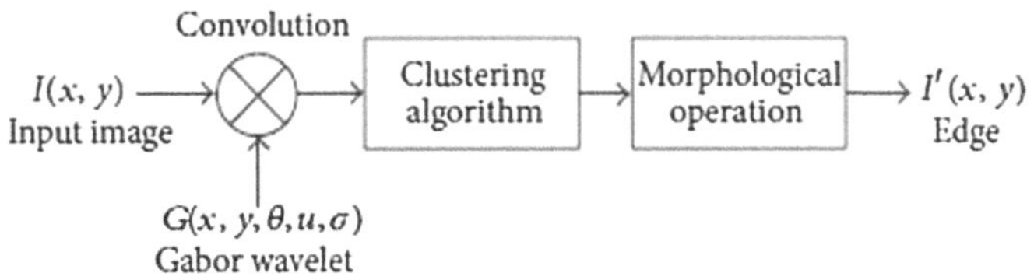

Fig. 1 Proposed method for edge detection

4 Experimental Results

The edges of the CT images are identified by using GWT. Figure 2 shows the edges using K-means and FCM clustering after GWT. Some irrelevant pixels in the binary images are present of which can be removed by morphological operations. The Fig. 2e is the resultant image of the GWT which has total edge related data. Figure 2 represents the binary images from K-means and FCM methods and their respective skeletons for the brain images.

Table 1 represents the MCR and FOM results and comparison. These values gives that the performance of the proposed method is better than the other methods. The technique Canny edge detection is very sensitive for noisy images.

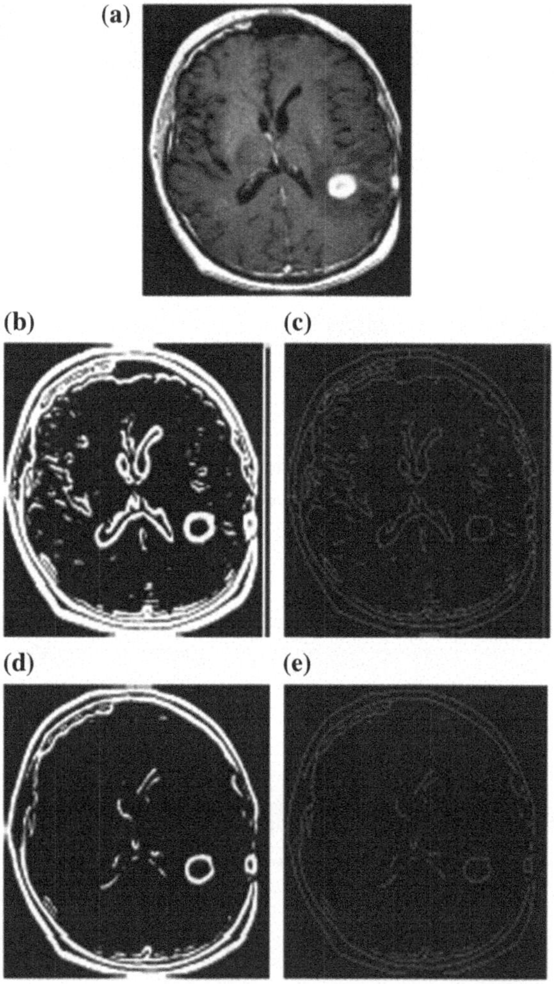

Fig. 2 Edges using *k*-means and FCM clustering after GWT; **a** original, **b** *k*-means clustering, **c** skeleton of (**b**), **d** FCM clustering, and **e** skeleton of (**d**)

Table 1 MCR and FOM results for the manually ground truth edge

MCR results					FOM results				
K means	FCM	Prewitt	Canny	Sobel	K means	FCM	Prewitt	Canny	Sobel
5.69	4.87	6.18	9.305	6.27	0.72	0.788	0.71	0.509	0.7089

Table 2 MCR and FOM results for the detected image

PSNR	MCR results					FOM results				
	K means	FCM	Prewitt	Canny	Sobel	K means	FCM	Prewitt	Canny	Sobel
35.03	0.94	0.93	0.88	0.793	0.85	0.973	0.974	0.973	0.9753	0.971
30.04	0.917	0.905	0.901	16.85	0.853	0.971	0.97	0.97301	0.1903	0.9719
25.03	0.49	0.96	0.9105	18.05	0.935	0.9701	0.9702	0.9615	0.0766	0.963
23.5	0.23	1.32	2.97	21.26	2.98	0.95	0.899	0.4789	0.065	0.467

Table 2 shows the proposed algorithm can detect the more precisely edges than the classical method though the noise is present. The K-means clustering algorithm is less sensitive than FCM algorithm.

5 Conclusion

This work presents the edge detection techniques using Gabor Wavelet Transform along with k-means and fuzzy c-means has great performance even during noisy conditions.

References

1. Ma Z, Tavares JMRS, Jorge RN, Mascarenhas T (2010) A review of algorithms for medical image segmentation and their applications to the female pelvic cavity. Comput. Methods Biomech. Biomed. Eng. 13(2):235–246
2. Jorge R-R, Eduardo B-C (2007) Medical image segmentation, volume representation and registration using spheres in the geometric algebra framework. Pattern Recogn 40(1):171–188
3. Abbas Q, Celebi ME, Gar´cıa IF (2013) Breastmass segmentation using region-based and edge-based methods in a 4-stage multiscale system. Biomed Signal Process Control 8 (2):204–214
4. Zhang J-Y, Yan C, Huang X-X (2009) Edge detection of images based on improved sobel operator and genetic algorithms. In: Proceedings of the international conference on image analysis and signal processing (IASP '09), pp 31–35, April 2009
5. Shen L, Bai L (2006) A review on Gabor wavelets for face recognition. Pattern Anal. Appl. 9 (2–3):273–292. [38] ˇZalik KR (2008) An efficient k-means clustering algorithm. Pattern Recogn Lett 29(9):1385–1391

6. Herrera PJ, Pajares G, Guijarro M (2011) A segmentation method using Otsu and fuzzy k-Means for stereovision matching in hemispherical images from forest environments. Appl Soft Comput J 11(8):4738–4747
7. Ji Z, Xia Y, Chen Q, Sun Q, Xia D, Feng DD (2012) Fuzzy c-means clustering with weighted image patch for image segmentation. Appl Soft Comput J 12(6):1659–1667
8. Lu X, Sun Y, Yuan Y (2011) Optimization for limited angle tomography in medical image processing. Pattern Recogn 44(10–11):2427–2435
9. Zhang Z, Ma S, Liu H, Gong Y (2009) An edge detection approach based on directional wavelet transform. Comput Math Appl 57(8):1265–1271
10. Chuang K-S, Tzeng H-L, Chen S, Wu J, Chen T-J (2006) Fuzzy c-means clustering with spatial information for image segmentation. Comput Med Imaging Graph 30(1):9–15
11. Balafar MA, Ramli AR, Saripan MI, Mashohor S (2008) Medical image segmentation using fuzzy C-mean (FCM), bayesian method and user interaction. In: Proceedings of the international conference on wavelet analysis and pattern recognition (ICWAPR '08), pp 68–73, August 2008
12. Cai W, Chen S, Zhang D (2007) Fast and robust fuzzy c-means clustering algorithms incorporating local information for image segmentation. Pattern Recogn 40(3):825–838
13. Jiang W, Lam K-M, Shen T-Z (2009) Efficient edge detection using simplified Gabor wavelets. IEEE Trans Syst Man Cybern B: Cybern 39(4):1036–1047

Part III
Trends in Renewable Energy and Control Systems, Power Systems, Power Electronics, Power Quality and FACTS

Prims Aided Floyd Warshall Algorithm for Shortest Path Identification in Microgrid

O.V. Gnana Swathika and S. Hemamalini

Abstract Reconfiguration of microgrid results in dynamic variation in topology of the network. Due to this, the conventional protection scheme is no longer applicable in microgrid. Hence the microgrid network demands an adaptive protection scheme, which is a key challenge to protection engineers. It is critical for the protection engineers to know the current topology of the microgrid before a suitable relay coordination technique is deployed on it. This paper proposes a Prims aided Floyd Warshall algorithm, where the Prims algorithm identifies the current topology of the network at any instant of time. In the event of fault occurrence, the output of Prims algorithm is a list of active nodes: utility grid, loads, Distributed Generators (DG) and critical circuit breakers in the network. This list aids the Floyd Warshall algorithm in identifying the shortest path from the node closer to the fault to the utility grid (in grid connected mode) or point of common coupling (in islanded mode). The algorithm ensures minimum portion of network disconnection for fault isolation. The proposed algorithm is tested and validated on an IEEE 21-bus microgrid network in islanded mode and an IEEE 69-bus distribution network with DG included at certain nodes in grid connected mode.

Keywords Microgrid protection · Prims · Floyd Warshall · Grid connected mode · Islanded mode

---

O.V. Gnana Swathika (✉) · S. Hemamalini
School of Electrical Engineering, VIT University, Chennai, India
e-mail: gnanaswathika.ov@vit.ac.in

S. Hemamalini
e-mail: hemamalini.s@vit.ac.in

K.R. Attele et al. (eds.), *Emerging Trends in Electrical, Communications and Information Technologies*, Lecture Notes in Electrical Engineering 394,
DOI 10.1007/978-981-10-1540-3_30

1 Introduction

Distributed generators (DG's) penetrated microgrid is the solution for the ever growing power demand. This causes the distribution system to be no longer radial in nature, since there is bidirectional power flow in microgrid [1]. Few issues faced by protection engineers due to DGs in microgrid include: false tripping in generators and feeders, blind spot and variation in fault current magnitudes depending on the mode of operation of microgrid namely grid connected or islanded mode. The dynamic nature of microgrid is accelerated due to the reconfiguration of microgrid. Thus an adaptive protection scheme is in demand, wherein the relay settings are altered based on the current topology and the mode of operation of microgrid. The protection of DG penetrated distribution network is realized using Multilayer Perceptrons (MLPs) neural networks [2]. The negative sequence component is used as a measuring quantity to identify a range of asymmetrical faults in the microgrid [3]. Global Positioning System (GPS) is used for adaptive protection of microgrid [4]. Fault current based adaptive protection scheme is derived in [5]. Adaptive monitoring of mode of operation of microgrid and issuing trip signals to circuit breaker are implemented in [6, 7]. Relay hierarchy detection based on fault current coefficients and adaptively varying suitable relay settings is done in [8, 9]. The adaptive techniques and algorithms discussed above are subjected to the assumption that the topology of the network is already known. This paper proposes Prims aided Floyd Warshall algorithm to isolate the fault in the microgrid by disconnecting a minimum portion of the network and without intervening the supply to the healthy portion of network. Prims algorithm identifies the active nodes of the microgrid network after each reconfiguration. With the prior knowledge of active nodes of network, the Floyd Warshall algorithm identifies the shortest path that exists from the active node closest to the fault to the utility grid. The proposed algorithm adaptively varies the relay settings based on current topology and achieves appropriate relay coordination in the microgrid.

2 Shortest Path Identification Problem

Reconfiguration in microgrid maybe triggered due to inclusion of new load or DG, fault occurrence in the network, islanded operation and maintenance requirements. The network topology of the microgrid changes due to reconfiguration. Adaptive variation of relay settings is necessary to obtain suitable relay coordination scheme in the network.

The primary objective is the minimization of distance from the faulted point to the nearest operating source, with minimum portion of load center disconnection. This can be formulated as a minimization problem:

Table 1 Time complexity using prims algorithm

Minimum edge weight data structure	Time complexity
Adjacency matrix for dense or sparse graph	$O(V^2)$
Binary heap and adjacency lists for sparse graph	O(V log V)
Binary heap and adjacency lists for dense graph ($E = V^2$)	$O(V^2 \log V)$
Fibonacci heap and adjacency list	O(E + V log V)

$$d = \min(P) \tag{1}$$

where,

d distance from the faulted point to nearest operating source
n total number of nodes of the network
p path through the graph between each pair of vertices

subjected to the constraint that the shortest path identified from the network using the proposed algorithm should be a radial network. The proposed Prims aided Floyd Warshall algorithm, is a rigorous algorithm with mathematical proof in which the optimization is based on the conditions and assumptions that are applied. If the run-time is 'V' and infinite, then the time complexity of Prims algorithm is as shown in Table 1. Dijkstra, the Double-Sweep and the Dantzig algorithms maybe used to identify the shortest path problem. But they are heuristic algorithms that extract optimal solution based on certain non-mathematical rules.

3 Methodology

In this paper, the proposed Prims aided Floyd Warshall algorithm is used to identify the minimum spanning tree from the point of fault to the nearest operating source in the microgrid network. Floyd-Warshall algorithm is a graph theory algorithm that identifies the shortest path in a weighted graph with positive or negative edge weights. It assists in determining transitive closure of a relation R. The Floyd–Warshall algorithm derives all possible paths through the graph between each pair of vertices. It compares the paths that exist between each pair of vertices until an optimal shortest path between two vertices is achieved.

4 Prims Aided Floyd Warshall Algorithm

Prims algorithm is a greedy algorithm that identifies a minimum spanning tree for a connected weighted undirected graph. It generates a subset of edges that creates a tree, where every node is included. It assures that net weight of all the edges in the

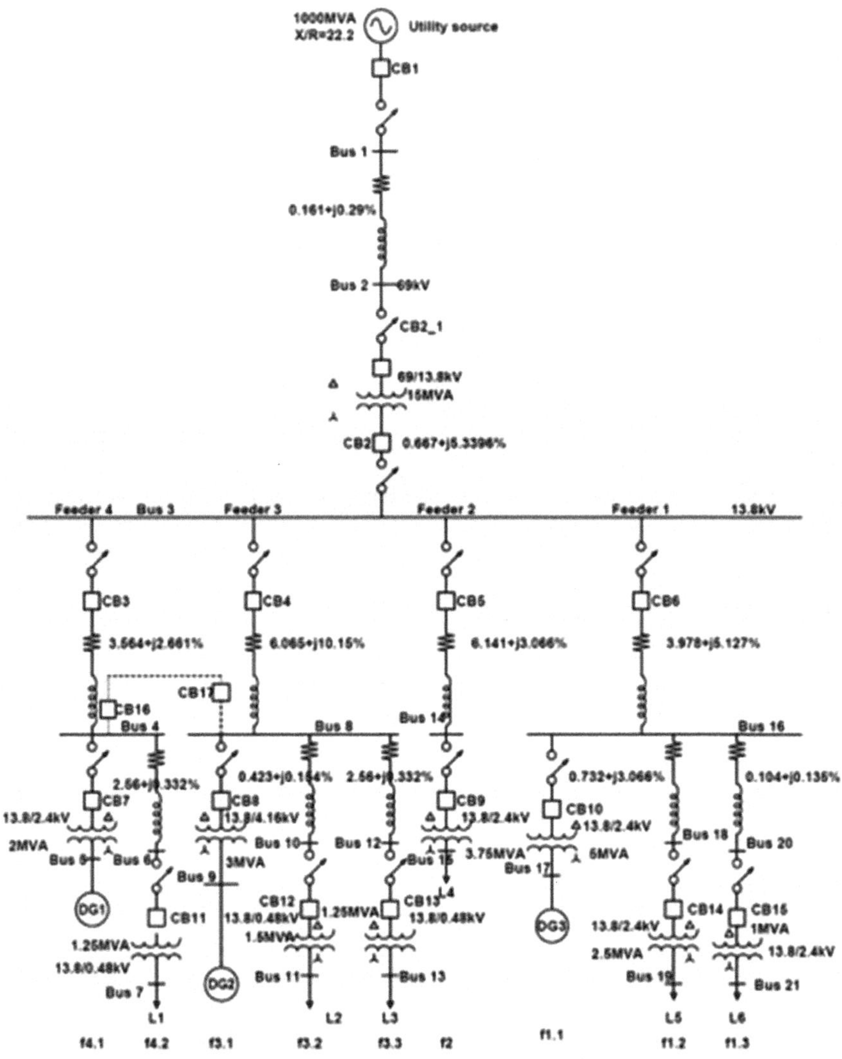

Fig. 1 IEEE 21-bus microgrid network

tree is minimal. The size of the tree is continuously increased, by appending one edge at a time, starting with a tree consisting of a single node, until it spans all nodes. This algorithm when used in conjunction with Floyd Warshall algorithm identifies the shortest route to clear the fault in current topology of microgrid. Consider the 21-bus microgrid network shown in Fig. 1, whose specification is given in Appendix. Depending on whether the microgrid is operating in grid connected or islanded mode, utility grid (UG) or point of common coupling (PCC) is considered as the base node. The path and distance parameters from any

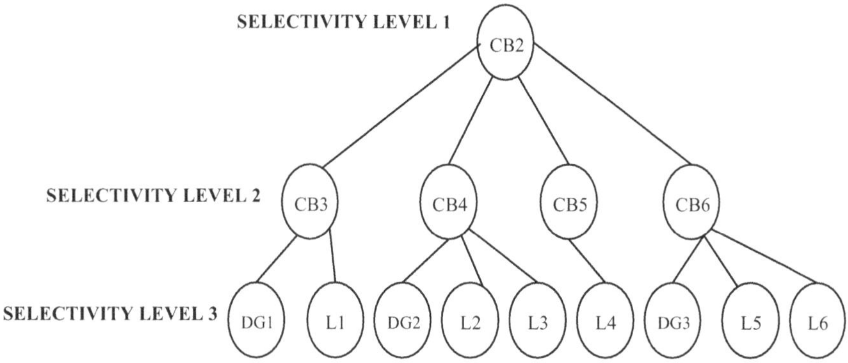

Fig. 2 Graph representation of IEEE 21-bus microgrid in islanded mode

node of consideration are derived with reference to the base node. Let the test system currently operate in islanded mode. The graph representation of the islanded network is as shown in Fig. 2. The active nodes of the network comprises of critical circuit breakers, DG's and loads of the test system and are 14 nodes in total. They are: CB2, CB3, CB4, CB5, CB6, DG1, L1, DG2, L2, L3, L4, DG3, L5 and L6. The weight of each edge is assumed to be '1'. The cost adjacency matrix of the graph acts as the input to the Prims algorithm. Based on the active nodes, the adjacency matrix dimension is 14 × 14. Once active nodes are identified, the Floyd Warshall algorithm is employed to identify the shortest path from faulted point to the PCC (CB2). The shortest path from various faulted points to the utility grid of the 21-bus microgrid network is indicated in Table 2.

Table 2 Shortest path identification using prims aided Floyd Warshall algorithm

Node closer to faulted point	Distance	Path
CB3	1	CB3-CB2
CB4	1	CB4-CB2
CB5	1	CB5-CB2
CB6	1	CB6-CB2
DG1	2	DG1-CB3-CB2
L1	2	L1-CB3-CB2
DG2	2	DG2-CB4-CB2
L2	2	L2-CB4-CB2
L3	2	L3-CB4-CB2
L4	2	L4-CB5-CB2
DG3	2	DG3-CB6-CB2
L5	2	L5-CB6-CB2
L6	2	L6-CB6-CB2

FLOYD WARSHALL ALGORITHM:

Consider a network M (S, T) with ‘a’ and ‘b’ as two random nodes in the network network has:

- node set S = {1, 2, . . . , n}
- arc set T = {(a, b) : a, b ∈ V, a != b} where |S| = n
- minimum of one cycle in the network

The square matrices P_c and Q_c for c = 0. . . n are calculated, holding the shortest path weights and the shortest routes between ‘a' and ‘b’, respectively.

Step 1. The network comprises of ‘n’ nodes and ‘c’ stage number. Two square n × n matrices, P_c and Q_c are formed.

Step 2. For c = 0 calculate P_0 and Q_0:

$P_0 = [p_{ab}]$, where

$p_{ab}=$

p_{ab} when a direct path connects node a and b
∞ when no direct path connects node a and b
0 when a and b are the same nodes

$Q_0 = [q_{ab}]$, where

$q_{ab} =$

b when a direct path connects a and b nodes
− when no direct path connects a and b node
− when a and b are the same nodes

Step 3. For j = 1. . . n, P_c and the Q_c matrices are computed as indicated. P_j and the Q_j matrices are now derived on the basis of the entities of the last matrices computed, i.e. the P_{j-1} and the Q_{j-1} matrices:

$P_c = [p_{ab}]$ where

$p_{ab} =$

p_{ab} when a = b = c
min (p_{ab}, $p_{ac} + p_{cb}$) otherwise

$Q_c = [q_{ik}]$ where

$q_{ik} =$

b when a = b, a = c, b = c
b when $p_{ik} \le p_{ac} + p_{cb}$
c when $p_{ik} > p_{ac} + p_{cb}$

Step 4. If P_n and Q_n are not arrived, then Step 3 is performed again; else the algorithm terminates.

5 Simulation Results

A 69-bus IEEE standard distribution network shown in Fig. 3 is considered in grid connected mode for analysis. DG’s maybe introduced at any node in the network. The topology of the network changes due to its dynamic behavior. For fault at any location, the shortest path for fault clearance with minimum load center disconnection is identified and validated using the proposed algorithm.

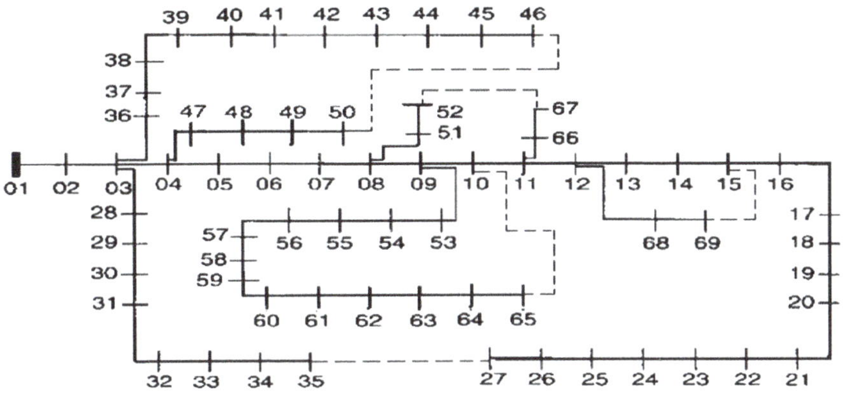

Fig. 3 IEEE 69-bus distribution network

Assume a fault is triggered in feeder closer to bus 35. The paths that exist between the faulted point to utility grid at node 1 are:

35-34-33-32-31-30-29-28-03-02-01;
35-27-26-25-24-23-22-21-20-19-18-17-16-15-14-13-12-11-10-09-08-07-06-05-04-03-02-01;
35-27-26-25-24-23-22-21-20-19-18-17-16-15-69-68-12-11-10-09-08-07-06-05-04-03-02-01;
35-27-26-25-24-23-22-21-20-19-18-17-16-15-14-13-12-11-10-65-64-63—62-61-60-59-58-57-56-55-54-53-09-08-07-06-05-04-03-02-01;
35-27-26-25-24-23-22-21-20-19-18-17-16-15-69-68-12-11-10-09-08-07-06-05-04-47-48-49-50-46-45-44-43-42-41-40-39-38-37-36-03-02-01;

The shortest path from bus 35 to the utility grid is identified using Floyd Warshall algorithm as shown in Fig. 4. There are 11 nodes involved in this shortest path. Hence the proposed algorithm ensures that the smallest portion of network is isolated when a fault occurs in the microgrid, unlike the conventional protection schemes which may cause discontinuity in supply for more consumers. The novel algorithm aids in adaptively setting the relay coordination of the network.

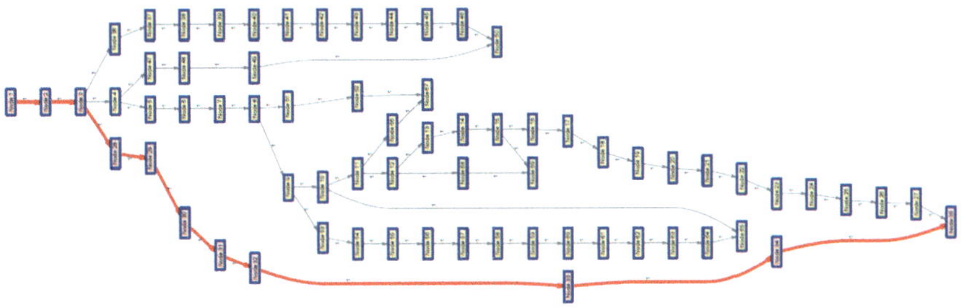

Fig. 4 Shortest path for fault clearance in an IEEE 69-bus distribution network

6 Conclusion

Prior knowledge of topology of microgrid after reconfiguration is a key parameter for protection engineers to apply suitable relay coordination techniques to the network. The proposed non-heuristic Prims aided Floyd Warshall algorithm is employed on the reconfigured microgrid to identify the shortest path to clear the fault in the network so that there is only minimum faulted portion disconnection in the network. This novel algorithm is tested and validated successfully on IEEE 21-bus microgrid in islanded mode and IEEE 69-bus distribution network in grid connected mode. Thus the algorithm maybe conveniently extended to any dynamic microgrid network.

Appendix

Base voltage: 69 kV; Base MVA: 1000 MVA; Frequency: 50 Hz; X/R ratio: 22.2; Power factor: 0.999;
Feederl: f1.1 13.8 kV/2.4 kV; f1.2 13.8 kV/2.4 kV; f1.3 13.8 kV/0.48 kV;
Feeder2: f2 13.8 kV/2.4 kV;
Feeder3: f3.1 13.8 kV/0.48 kV; f3.2 13.8 kV/0.48 kV; f3.3 13.8 kV/0.48 kV;
Feeder4: f4.1 13.8 kV/2.4 kV; f4.2 13.8 kV/0.48 kV;
DG1: 2.4 kV, 2 MVA; DG2: 4.16 kV, 3 MVA; DG3: 2.4 kV, 2.5 MVA;
Loads: L4 0.48 kV, 1.25 MVA; L3-2 0.48 kV, 1.5 MVA; L3-1 0.48 kV, 1.25 MVA;
L2: 2.4 kV, 3.75 MVA; L1-2: 2.4 kV, 2.5 MVA; L1-1: 0.48 kV, 1 MVA;
Transformer configuration: Δ-Υ (ideal mode).

References

1. Calderaro V, Milanovic JV, Kayikci M, Piccolo A (2009) The impact of distributed synchronous generators on quality of electricity supply and transient stability of real distribution network. Electr Power Energy Syst 79:134–143
2. Javadian SAM, Haghifam MR, Bathaee SMT, Fotuhi Firoozabad M (2013) Adaptive centralized protection scheme for distribution systems with DG using risk analysis for protective devices placement. Electr Power Energy Syst 44:337–345
3. Lin X, Zhang R, Tong N, Li X, Li M, Yang D (2015) Regional protection scheme designed for low-voltage micro-grids. Electr Power Energy Syst 64:526–535
4. Brahma SM, Girgis AA (2004) Development of adaptive protection scheme for distribution systems with high penetration of distributed generation. IEEE Trans Power Deliv 19:56–63
5. Han Y, Hu X, Zhang D (2010) Study of adaptive fault current algorithm for microgrid dominated by inverter based distributed generators. In: 2nd IEEE international symposium on power electronics for distributed generation systems, pp 852–854

6. Akhondi H, Saifali M. A new adaptive method for distribution system protection considering distributed generation units with simulated annealing method. http://cdn.intechopen.com/pdfs-wm/36831.pdf
7. Mahat Pukar, Chen Shen, Bak-Jensen Birgitte (2011) Claus Leth Bak: a simple adaptive overcurrent protection of distribution systems with distributed generation. IEEE Trans Smart Grid 2(3):428–437
8. Ustun TS, Ozansoy C, Zayegh A (2013) Fault current coefficient and time delay assignment for microgrid protection system with central protection unit. IEEE Trans Power Syst 28:598–606
9. Ustun TS, Ozansoy C, Zayegh A (2011) Implementation of Dijkstra's algorithm in a dynamic microgrid for relay hierarchy detection. In: Proceedings of IEEE smart grid communication, pp 481–486

A Comparative Study of Decoupler Design Techniques for TITO Control Processes

R. Hanuma Naik, D.V. Ashok Kumar and K.S.R. Anjaneyulu

Abstract In industrial control, the performance of multiple input and multiple output processes is constrained by an interaction in the plant dynamics. The interaction between the variables is abridged to the minimum degree by opting appropriate control configuration and designing the controller for each loop autonomously. Conversely, when interaction is modest, the controller often used with decoupler elements. This paper presents a comparative study of decoupler design for minimization of interaction among the controllers. The decoupler is designed and incorporated in control loop amid controller and process. This decomposes the MIMO system into multiple single loops and each loop controller is designed by using any existing SISO tuning method. This approach is straightforward, simple to know and implemented in practice. A case study is incorporated to validate the effectiveness of projected approach.

Keywords Control configuration · Independent loop · Interaction · Decoupler · TITO process · PID control · SIMC tuning method

1 Introduction

Recently, there is a great attention on Model Predictive Control (MPC) for controlling the multivariable processes in industry. This gives an optimized performance of system in terms of robustness when there is an interaction in the plant dynamics than decentralized control. However, it is handling of constraints [1], where as decentralized control may open the loop when input saturates. Hence, there is a need

R.H. Naik (✉) · D.V. Ashok Kumar
RGM College of Engineering and Technology, Nandyal, A.P., India
e-mail: rhnaik.1717@gmail.com

K.S.R. Anjaneyulu
JNTUA College of Engineering, Anantapuram, A.P., India

K.R. Attele et al. (eds.), *Emerging Trends in Electrical, Communications and Information Technologies*, Lecture Notes in Electrical Engineering 394,
DOI 10.1007/978-981-10-1540-3_31

to develop an advanced control scheme to bridge the gap with a decentralized control, which needs to regularly do numerical optimization. For these kinds of systems decoupling techniques gives better stability margins and diminish the consequence of input saturations.

The assortment of a decoupler design technique is a comparatively intricate job because each method has their merits and demerits. The common methods of decoupling are simplified, ideal, partial and inverted. Among these the simplified decoupling is more popular because of its simplicity in design. Ideal decoupler is infrequently used since it requires the inverse of model, which is not possible in most of square MIMO processes and greatly assist the design of controller parameters. Inverted decoupler is also not often used but presents benefit of equally the ideal and simplified decoupling approaches.

Several researchers have previously judge against ideal, simplified and inverted decoupler method. Luyben [2], Weischedel and McAvoy [3] have evaluated ideal and simplified decoupler methods and they accomplished that later decoupling scheme is further robust than all other decoupling. Waller [4], also discussed the stability of both decoupling schemes in detail with the help of an experiment using the same tuning method.

Arkun et al. [5] evaluated decoupling schemes and accomplished that ideal decoupling preserve be not as much of robust as simplified decoupling. Shinskey [6] detailed both simplified and inverted decoupling structures. Inverted decoupling has been used by many researchers and they commented that, it handles greatly the saturations in input and highly sensitive to model errors. Recently, Wang et al. [7], Wade [8], Tavakoli et al. [9], Nordfeldt et al. [10], Shen et al. [11–13], Maghade et al. [14], Rajapandin [15], Lee et al. [16], Garrido et al. [17] and Cai et al. [18] included various decoupling methods in their projected work.

This paper compares the three decoupling methods under the same strength of interaction exists in the process dynamics. The objective of this paper is: (1) pairing of input-output variables by the integration of RGA-RNGA-NI rules [19], (2) selection and design of decoupling scheme, (3) determination of equivalent transfer function of decoupled process, (4) design of individual loop controller using SIMC tuning technique [20].

2 Pairing of Loops

The prerequisite in design of decentralized control scheme is selection of loop pairing. Here it uses the concept of RGA-RNGA-NI for choosing the pair of manipulated variable (input)-controlled variable (output). The detailed discussion is given below.

2.1 Relative Gain Array (RGA)

The RGA, for fully cross coupled $n \times n$ process is,

$$\Lambda = G(0) \otimes G^{-T}(0) = \begin{bmatrix} \Lambda_{11} & \Lambda_{12} & \ldots & \Lambda_{1n} \\ \Lambda_{21} & \Lambda_{22} & \ldots & \Lambda_{2n} \\ \ldots & \ldots & \ldots & \\ \Lambda_{n1} & \Lambda_{n2} & \ldots & \Lambda_{nn} \end{bmatrix} \quad (1)$$

Here, $G(0)$ is fully cross coupled process transfer function matrix at $s = 0$ and $G^{-T}(0)$ is an inverse of its transpose.

2.2 Relative Normalized Gain Array (RNGA)

The normalized gain $(K_{N,ij})$ of a process transfer function element $g_{ij}(s)$ is [19],

$$K_{N,ij} = \frac{k_{ij}}{\sigma_{ij}} = \frac{k_{ij}}{\tau_{ij} + \theta_{ij}} \quad (2)$$

and RNGA (ϕ) using Eq. (2) can be determined as,

$$\phi = K_N \otimes K_N^{-T} \quad (3)$$

The loop pairing is done from RNGA elements whose value is closest to unity and positive. The pairing which directs unstable is keep away by Niederlinski's Index.

2.3 Niederlinski's Index (NI)

The Niederlinski's index (NI) for the intricate control configuration is designated by $N(G)$ and,

$$N(G) = \frac{|G(0)|}{\pi\, g_{ij}} \quad i, j = 1, 2, 3, \ldots n \quad (4)$$

where $|G(0)|$ indicate the determinant of matrix $G(0)$ and $\pi\, g_{ij}$ stand for product of diagonal elements of $G(0)$ for a completely cross coupled control system. For good stability of system, NI is should be larger than zero. Therefore better pairing can be chosen by using RGA-RNGA-NI rules [19, 21].

3 Equivalent Transfer Function (ETF) Model

To determine ETF, first we suppose to determine the Relative average residence time (γ_{ij}) i.e., the ratio of average residence time of loop $y_i - u_j$, when all other loops are closed to all other loops are open,

$$\gamma_{ij} = \frac{\hat{\sigma}_{ij}}{\sigma_{ij}} = \frac{\phi_{ij}}{\Lambda_{ij}} \tag{5}$$

In the form of array, it is known as relative average residence time array (RARTA):

$$\Gamma = \begin{bmatrix} \gamma_{11} & \gamma_{12} \\ \gamma_{21} & \gamma_{22} \end{bmatrix} = \begin{bmatrix} \phi_{11} & \phi_{12} \\ \phi_{21} & \phi_{22} \end{bmatrix} \otimes \begin{bmatrix} \frac{1}{\Lambda_{11}} & \frac{1}{\Lambda_{12}} \\ \frac{1}{\Lambda_{21}} & \frac{1}{\Lambda_{22}} \end{bmatrix} \tag{6}$$

where '$\otimes$' is element by element product. From Eq. (5), it can be rewrite as,

$$\hat{\sigma}_{ij} = \gamma_{ij}\sigma_{ij} = \gamma_{ij}\tau_{ij} + \gamma_{ij}\theta_{ij} = \hat{\tau}_{ij} + \theta_{ij} \tag{7}$$

Therefore, The ETF of a loop when all other loops closed is,

$$\hat{g}_{ij}(s) = \hat{k}_{ij}\frac{1}{\hat{\tau}_{ij}s + 1}e^{-\theta_{ij}s} = \frac{k_{ij}}{\Lambda_{ij}}\frac{1}{\gamma_{ij}\tau_{ij}s + 1}e^{-\gamma_{ij}\theta_{ij}s} \tag{8}$$

where θ_{ij} denotes time delay of approximated model (ETF). The PI controller parameters are determined based on the maximum weights of $\hat{g}_{ij}(s)$ and $g_{ij}(s)$, Where $g_{ij}(s)$ is actual open loop process transfer function which is in the form of FOPDT.

4 Decoupler Design

This section describes the design of decoupler of three types in detail. Figure 1 shows the decoupled decentralized control for two variable (TITO) processes.

$$\text{Here, } y_1 = g_{11}v_1 + g_{12}v_2 \text{ and } y_2 = g_{21}v_1 + g_{22}v_2 \tag{9}$$

where 'g_{ij}' represents process transfer function, 'y_i' is process output vector and 'v_j' is process input vector. The input to the process from controller after adding decoupler elements in the loops are,

$$v_1 = d_{11}u_1 + d_{12}u_2 \text{ and } v_2 = d_{21}u_1 + d_{22}u_2 \tag{10}$$

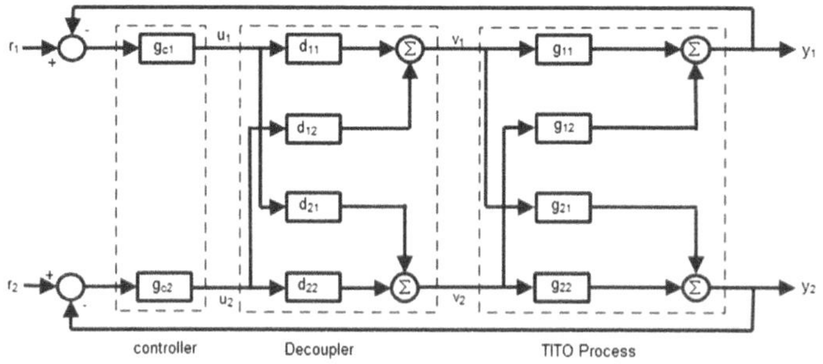

Fig. 1 Decoupled decentralized control system for 2 × 2 processes

4.1 Ideal Decoupler

With ideal decoupling, the purpose is to absolutely decouple the off-diagonal elements of process transfer matrix. Hence the apparent process becomes,

$$y_1 = g_{11}u_1 \text{ and } y_2 = g_{22}u_2 \tag{11}$$

From the Eqs. (9)–(11), decoupler elements d_{ij} can be obtained as,

$$d_{11} = \frac{g_{11}g_{12}}{g_{11}g_{22} - g_{12}g_{21}},\ d_{12} = \frac{-g_{12}g_{22}}{g_{11}g_{22} - g_{12}g_{21}},\ d_{21} = \frac{-g_{11}g_{21}}{g_{11}g_{22} - g_{12}g_{21}},\ d_{22} = \frac{g_{11}g_{12}}{g_{11}g_{22} - g_{12}g_{21}} \tag{12}$$

The ideal decoupler elements are quite difficult to determine. However, the apparent processes model be converted into diagonal by means of only the diagonal elements of $G_p(s)$ as on the diagonal.

4.2 Simplified Decoupler

As per the authors [22], the simplified decoupler is written as,

$$d = G_p^{-1}((G_p^{-1})_{diag})^{-1} \tag{13}$$

The simplified decoupler elements can be obtained from ideal decoupler by equating the main diagonal elements of decoupler to unity i.e., $d_{11} = d_{22} = 1$, thus,

$$d_{12} = \frac{-g_{12}}{g_{11}},\ d_{21} = \frac{-g_{21}}{g_{22}} \tag{14}$$

With this decoupler the goal is fulfilled, but it changes the apparent process. Therefore the controller needs to retune after installation of decoupler. The effected apparent processes becomes,

$$y_1 = \frac{g_{11}g_{22} - g_{12}g_{21}}{g_{22}} v_1 \quad \text{and} \quad y_2 = \frac{g_{11}g_{22} - g_{12}g_{21}}{g_{11}} v_2 \tag{15}$$

The simplified decoupler is designed by coefficient matrix (CM) method [16]. For 2 × 2 system,

$$G = \begin{bmatrix} g_{11} & g_{12} \\ g_{21} & g_{22} \end{bmatrix}$$

The cofactor of ***G*** and decoupler elements follows:

$$C = (adj\,G)^T = \begin{bmatrix} c_{11} & c_{12} \\ c_{21} & c_{22} \end{bmatrix} = \begin{bmatrix} g_{22} & -g_{21} \\ -g_{12} & g_{11} \end{bmatrix} \tag{16}$$

$$d_{21} = -\frac{c_{12}}{c_{11}} = \frac{-g_{21}}{g_{22}}, \quad d_{12} = -\frac{c_{21}}{c_{22}} = \frac{-g_{12}}{g_{11}} \quad \text{and} \quad D = \begin{bmatrix} 1 & d_{12} \\ d_{21} & 1 \end{bmatrix} \tag{17}$$

4.3 *Inverted Decoupler*

As per Gagnon et al. [23] it is possible to simplify the ideal decoupler as,

$$v_1 = u_1 - \frac{g_{12}}{g_{11}} u_2 \quad \text{and} \quad v_2 = u_2 - \frac{g_{21}}{g_{22}} u_1 \tag{18}$$

It gives an inverted decoupler which is similar to simplified decoupler, but apparent process model remains same as in case of ideal decoupler. These decoupler elements are connected back from one process input to the other controller output. However the performance may decrease with this decoupler because of implementation with lead-lag and delay function.

The decomposed individual loop controller is tuned using SIMC method [20] and PI controller parameters for FOPDT SISO processes given by

$$k_c = \left(\frac{1}{k}\frac{\tau}{\tau_c + \theta}\right) \quad \text{and} \quad \tau_I = \min\{\tau_1, 4(\tau_c + \theta)\} \tag{19}$$

5 Case Study

Consider the Vinnate and Luyben (VL) process transfer function model [24]

$$G(s) = \begin{bmatrix} \frac{-2.2}{7s+1}e^{-s} & \frac{1.3}{7s+1}e^{-0.3s} \\ \frac{-2.8}{9.5s+1}e^{-1.8s} & \frac{1.3}{9.2s+1}e^{0.35s} \end{bmatrix}$$

The RGA (1.6254), RNGA (1.5537) and NI (0.6152 > 0) suggests the pairing of input-output variable is 1–1/2–2. The ETF of the model is,

$$\hat{G}(s) = \begin{bmatrix} \frac{-1.3535}{(6.6910s+1)}e^{-0.9559s} & \frac{-2.0786}{(6.197s+1)}e^{-0.2656s} \\ \frac{4.4769}{(8.4103s+1)}e^{-1.5935s} & \frac{2.6455}{(8.7940s+1)}e^{-0.3345s} \end{bmatrix}$$

The decoupler elements designed by ideal, simplified and inverted are,

$$D_{ideal}(s) = \begin{bmatrix} \frac{89.97s+9.46}{25.116s^2+59.112s+5.82} & \frac{53.105s+5.59}{25.116s^2+59.112s+5.82} \\ \frac{-42.504s^2+52.052s+6.16}{25.116s^2+59.112s+5.82} & \frac{89.87s+9.46}{25.116s^2+59.112s+5.82} \end{bmatrix}$$

$$D_{simplified}(s) = \begin{bmatrix} 1 & 0.591 \\ \frac{0.651(9.2s+1)}{(9.5s+1)}e^{-1.45s} & 1 \end{bmatrix}$$

$$D_{inverted}(s) = \begin{bmatrix} 1 & 0.5909 \\ \frac{(59907s+0.6512)}{(9.5s+1)}e^{-0.75s} & 1 \end{bmatrix}; \quad N_x = e^{-0.7s}$$

The individual loop controller parameters, performance IAE (Integral Absolute Error), and ISE (Integral square Error) are listed in Tables 1 and 2 respectively and closed loop response of VL column is shown in Fig. 2.

Table 1 PI Controller parameters using SIMC for different decoupling schemes

Control loop	Proposed (simplified)		Inverted decoupler		Ideal decoupler	
	k_c	τ_i	k_c	τ_i	k_c	τ_i
u_1–y_1	−2.5858	6.6910	−1.5909	7.0	−1.5909	7.00
u_2–y_2	1.4945	8.7940	1.0189	8.4	1.0188	8.4

Table 2 Performance indices of different decoupling schemes (IAE_t is total IAE)

Performance indices		Proposed decoupler		Inverted decoupler		Ideal decoupler	
		Y_1	Y_2	Y_1	Y_2	Y_1	Y_2
IAE	Change in r_1	1.9320	0.0029	2.5910	1.1610	2.7180	1.7500
	Change in r_2	0.0661	2.1100	0.4454	1.9990	0.4929	2.0160
ISE	Change in r_1	1.4430	$2.27e^{-6}$	1.9030	0.2179	1.9660	0.5162
	Change in r_2	0.0008	1.4840	0.0349	1.1120	0.0457	1.2250
IAE_t		4.111		6.1964		6.9769	
ISE_t		2.9334		3.2678		3.7529	

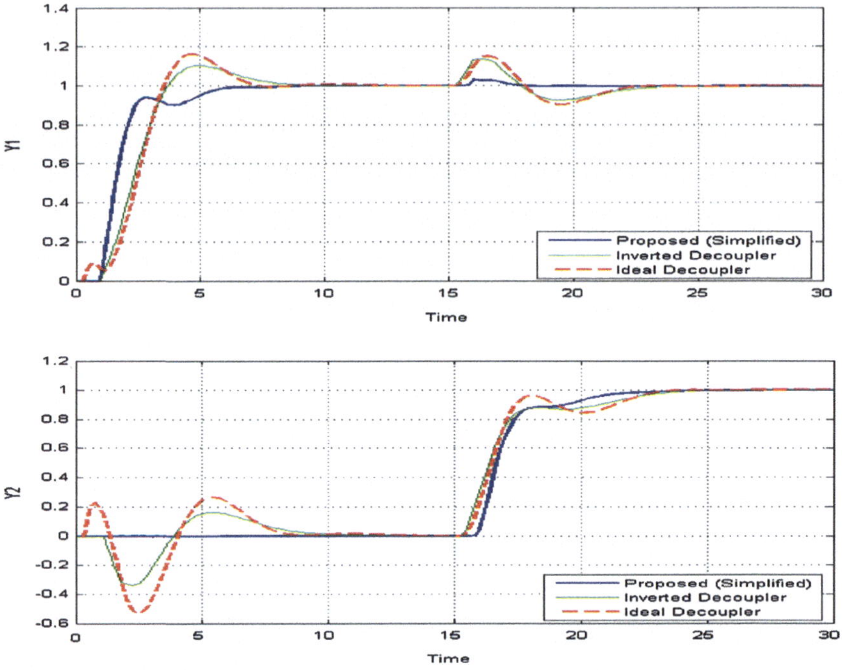

Fig. 2 Closed loop response of VL column for sequential set point changes at $r_1 = 0$ s and $r_2 = 15$ s

6 Conclusions

This paper aims to provide a comparison study of decoupler design techniques of Ideal, Inverted and cofactor matching based simplified decoupler. A case study (VL column) was incorporated to validate the efficacy of decoupler. The minimization of interaction among the loops and improvement of performance has been achieved by this approach. The IAE, ISE of proposed method for sequential changes in inputs are better than other two decoupler methods. This decoupler is straightforward, casual and stable.

References

1. Maciejowski JM (2002) Predictive control: with constraints. Pearson education
2. Luyben WL (1970) Distillation decoupling. AIChE J 16(2):198–203
3. Weischedel K, McAvoy TJ (1980) Feasibility of decoupling in conventionally controlled distillation columns. IEC Fund 19(4):379–384
4. Waller (Toijala) KVT (1974) Decoupling in distillation. AIChE J 20(3):592–594
5. Arkun Y, Manousiouthakis B, Palazoglu A (1984) Robustness analysis of process control systems. A case study of decoupling control in distillation. IEC Process Des Dev 23(1):93–101
6. Shinskey FG (1988) Process control systems: application, design and adjustment. McGraw-Hill, New York
7. Wang QG, Huang B, Guo X (2000) Auto-tuning of TITO decoupling controllers from step tests. ISA Trans 39(4):407–418
8. Wade HL (1997) Inverted decoupling: a neglected technique. ISA Trans 36(1):3–10
9. Tavakoli S, Griffin I, Fleming PJ (2006) Tuning of decentralised PI (PID) controllers for TITO processes. Control Eng Pract 14(9):1069–1080
10. Nordfeldt P, Hägglund T (2006) Decoupler and PID controller design of TITO systems. J Process Control 16(9):923–936
11. Shen Y, Cai WJ, Li S (2010) Normalized decoupling control for high-dimensional MIMO processes for application in room temperature control HVAC systems. Control Eng Pract 18(6):652–664
12. Shen Y, Cai WJ, Li S (2009) Multivariable process control: decentralized, decoupling, or sparse? Ind Eng Chem Res 49(2):761–771
13. Shen Y, et al. (2011) Partial decoupling control for multivariable processes. Ind Eng Chem Res 50(12):7380–7387
14. Maghade DK, Patre BM (2012). Decentralized PI/PID controllers based on gain and phase margin specifications for TITO processes. ISA Trans 51(4):550–558
15. Rajapandiyan C, Chidambaram M (2012) Controller design for MIMO processes based on simple decoupled equivalent transfer functions and simplified decoupler. Ind Eng Chem Res 51(38):12398–12410
16. Vu TNL, Lee M (2013) An extended method of simplified decoupling for multivariable processes with multiple time delays. J Chem Eng Jpn 46(4):279–293
17. Garrido J, Vázquez F, Morilla F (2011) An extended approach of inverted decoupling. J Process Control 21(1):55–68
18. Cai WJ, et al. (2008) Normalized decoupling a new approach for MIMO process control system design. Ind Eng Chem Res 47(19):7347–7356
19. He MJ, Cai WJ, Ni W, Xie LH (2009) RNGA based control system configuration for multivariable processes. J Process Control 19(6):1036–1042
20. Skogestad S (2003) Simple analytical rules for model reduction and PID controller tuning. J Process Control 13(4):291–309
21. Naik RH, Ashok Kumar DV, Anjaneyulu KSR (2014) Control configuration selection and controller design for multivariable processes using normalized gain. Int J Electr Comput Electron Commun Eng 8(10). World Academy of Science, Engineering and Technology
22. Morari M, Zafiriou E (1989) Robust process control. Prentice-Hall, Englewood Cliffs, NJ
23. Gagnon E, Pomerleau A, Desbiens A (1998) Simplified, ideal or inverted decoupling? ISA Trans 37(4):265–276
24. Luyben WL (1986) Simple method for tuning SISO controllers in multivariable systems. Ind Eng Chem Proc Des Dev 25:654–669

An Adaptive Hybrid Optimization Algorithm for OPF for Non-smooth Fuel Cost Functions with Facts Device

A. Immanuel and Ch. Chengaiah

Abstract This paper presents a Hybrid Particle Swarm Optimization with Differential Perturbed Velocity with adaptive acceleration coefficient (APSO-DV) to reduce generator fuel cost in Optimal Power Flow control with a powerful Flexible Alternating Current Transmission Systems (FACTS) device such as Unified power Flow Controller. The APSO-DV algorithm employs a strongly coupled differential operator acquired from differential evolution with adaptive acceleration coefficient in velocity update function of particle swarm Optimization. The strategic location of UPFC is found using Fuzzy approach by taking voltage magnitudes and voltage stability index (L-Index) as input parameters where L-Index is a real number which gives fair and consistent results for stability among different methods of voltage stability analysis. The feasibility of the proposed method has been tested on IEEE-30 bus system with three different objective functions that reflects fuel cost minimization, fuel cost with valve point effects and total system power loss. The test result shows the effectiveness of robustness of the proposed approach and provides superior results compared with the existing results.

Keywords OPF · Particle swarm optimization · Differential perturbed velocity · Adaptive acceleration · Fuzzy · UPFC · L-Index

1 Introduction

Optimal Power Flow (OPF) is a static nonlinear problem which optimizes a specific objective function while satisfying a set of physical and operational constraints imposed by equipment restrictions and security requirements. The latter is regarded

A. Immanuel (✉) · Ch. Chengaiah
Departments of Electrical & Electronics Engineering,
Sri Venkateswara University, Tirupati 517502, India
e-mail: anupalliimmanuel@gmail.com

Ch. Chengaiah
e-mail: Chinthapudisvu@gmail.com

K.R. Attele et al. (eds.), *Emerging Trends in Electrical, Communications and Information Technologies*, Lecture Notes in Electrical Engineering 394,
DOI 10.1007/978-981-10-1540-3_32

as the backbone tool that has been extensively researched since its first introduction by Carpentier in 1962 [1].

Over the last three decades, many successful methods [2–4] have been proposed i.e. reduced gradient method, successive linear programming, Newton method, P–Q decomposition, interior point method, Genetic Algorithm, Evolutionary Programming and Particle Swarm Optimization. In order to get a better search efficiency different hybrid algorithms have been developed such as hybrid evolutionary programming and tabu search [5] to resolve economic dispatch problem for non-smooth fuel cost functions.

From last two decades researchers have been developed many algorithms to solve optimal power flow incorporating FACTS devices. Still research is in progress to meet the present congestion management problem with help of FACTS devices effectively. Taranto et al. [6] have presented a decomposition method to solve OPF problem with FACTS devices. This method can deal with the representation of series compensators and phase shifters but this method didn't consider the specified line flow constraints. Gotham and Heydt [7] have developed the modeling of FACTS devices for power flow studies and deliberated the role of that modeling in the study of FACTS devices. Ambriz-Perez et al. [8] have solved OPF problem incorporating FACTS devices using Newton's method that leads a highly robust iterative solution. But it has been recognized that the OPF problem with series compensation may be a non-convex problem [6], which will lead the classical method to be struck at local minimum. To alleviate the above said problem numerous heuristic methods have been proposed and a few are: Chung and Li [9] have proposed genetic algorithm to find out the parameters of FACTS devices. Ongsakul and Bhasaputra [10] have proposed hybrid tabu search and simulated annealing (TS/SA) technique to solve OPF problem through FACTS devices.

This paper proposed the particle swarm optimization with differentially perturbed velocity with adaptive acceleration co-efficient to solve OPF problem with UPFC. The UPFC is located the lines closed to the weak nodes determined by using Fuzzy approach. The proposed Fuzzy-APSODV with UPFC is examined on IEEE-30 bus system with three objective functions and it contributes very remarkable results.

1.1 L- Index

In a transmission network consist of 'n' number of buses where 1, 2, 3, … g; generator buses, and the remaining g + 1…….. n load buses. For a given network operating condition, by using Load-flow results, the Voltage-Stability Index is determined as:

$$L_j = \left| 1 - \sum_{i=1}^{g} F_{ji} \frac{v_i}{v_j} \right| \tag{1}$$

where j = g +1,…., n. The values of F_{ji} are complex and are determined from the system Y-bus matrix.

$$\text{i.e.} \qquad F_{LG} = [Y_{LL}]^{-1}[Y_{LG}]^{-1} \tag{2}$$

$[Y_{LG}]$ and $[Y_{LL}]$ are the sectionalized parts of Y-bus system matrix. For voltage stability analysis, the L_j value should not be violated the maximum limit of 1 at any load bus j [11].

1.2 Power Flow Model of UPFC

As an advanced FACTS device UPFC can provide instantaneous control of voltage magnitude, real and reactive power flows. It is well located to overcome most of the issues related to power flow control while improving the considerable transient and dynamic stability. The equivalent circuit of UPFC power injection model is as shown in Fig. 1.

The two coordinated synchronous voltage sources of UPFC are:

$$V_{sh} = V_{sh}(cos\delta_{sh} + jsin\delta_{sh}) \tag{3}$$

$$V_{se} = V_{se}(cos\delta_{se} + jsin\delta_{se}) \tag{4}$$

where, V_{sh} = Voltage magnitude of shunt converter; δ_{sh} = Voltage angle of shunt converter; V_{se} = Voltage magnitude of series converter; and δ_{se} = Voltage angle of series converter. Based on the equivalent circuit and from Eqs. (3) and (4), the real and reactive power flow expressions are:

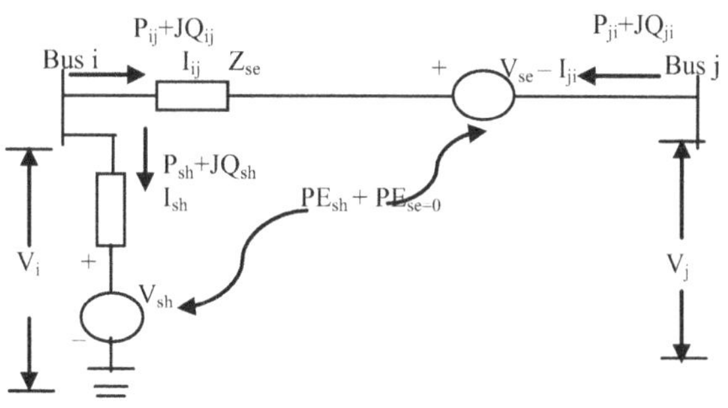

Fig. 1 UPFC equivalent circuit

$$P_{sh} = V_i^2 g_{sh} - V_i V_{sh}(g_{sh}\cos(\theta_i - \theta_{sh}) + b_{sh}\sin(\theta_i - \theta_{sh})) \quad (5)$$

$$Q_{sh} = -V_i^2 b_{sh} - V_i V_{sh}(g_{sh}\sin(\theta_i - \theta_{sh}) + b_{sh}\cos(\theta_i - \theta_{sh})) \quad (6)$$

$$P_{ij} = V_i^2 g_{ij} - V_i V_j(g_{ij}\cos\theta_{ij} + b_{ij}\sin\theta_{ij}) - V_i V_{se}(g_{ij}\cos(\theta_i - \theta_{se}) + b_{ij}\sin(\theta_i - \theta_{se})). \quad (7)$$

$$Q_{ij} = -V_i^2 b_{ij} - V_i V_j(g_{ij}\sin\theta_{ij} - b_{ij}\cos\theta_{ij}) - V_i V_{se}(g_{ij}\sin(\theta_i - \theta_{se}) + b_{ij}\cos(\theta_i - \theta_{se})). \quad (8)$$

$$P_{ji} = V_j^2 g_{ij} - V_i V_j(g_{ij}\cos\theta_{ji} + b_{ij}\sin\theta_{ji}) + V_j V_{se}(g_{ij}\cos(\theta_j - \theta_{se}) + b_{ij}\sin(\theta_j - \theta_{se})). \quad (9)$$

$$Q_{ji} = -V_j^2 b_{ij} - V_i V_j(g_{ij}\sin\theta_{ji} - b_{ij}\cos\theta_{ji}) + V_j V_{se}(g_{ij}\sin(\theta_j - \theta_{se}) + b_{ij}\cos(\theta_j - \theta_{se})). \quad (10)$$

where$g_{sh} + jb_{sh} = \frac{1}{z_{sh}}, g_{ij} + jb_{ij} = \frac{1}{z_{se}}, \theta_{ij} = \theta_i - \theta_j, \theta_{ji} = \theta_j - \theta_i$

2 OPF Problem Formulation

The solution of OPF aims to optimize a chosen objective function with best possible tuning of the power network control variables, by satisfying the number of equality and inequality constraints. The OPF problem can be formulated as:

$$\min \mathrm{J}(\mathrm{x}, \mathrm{u})$$

$$\text{Subject to:} \quad \mathrm{g}(\mathrm{x}, \mathrm{u}) = 0$$

$$\mathrm{h}_{\min} \le \mathrm{h}(\mathrm{x}, \mathrm{u}) \le \mathrm{h}_{\max}$$

where J = Objective function to be minimized, x = vector of dependent variables.

g = Equality constraints, h = operating constraints and u = vector of control variables such as:

1. Voltage magnitude of generators V_G at PV buses.
2. Real power output of generator PG at PV buses excluding at the slack bus PG_1.
3. Tap settings of Transformer T.
4. Shunt VAR compensators.

Therefore control vector can be represented as:

$$u^T = [PG_1 \ldots PG_{ng}, V_{G1} \ldots V_{Gng}, Qc_1 \ldots Qc_{nc}, T_1 \ldots T_{nt}]$$

where, nt = No. of the tap changing transformers and nc = No. of VAR compensators.

The UPFC is located to minimize the selective objective function and enhance the system performance while maintaining thermal limits and voltage constraints. The OPF problem after placing the UPFC can be formulated with the following three objective functions:

2.1 Smooth Cost Function Using Quadratic Form

The objective function 'f' is the total operating fuel cost function by imposing the constraints is:

$$f_1 = \left[\sum_{i=1}^{NG} \left(a_i P_{Gi}^2 + b_i P_{Gi} + c_i\right)\right] + KP\left(P_{G1} - P_{G1}^{lim}\right)^2 + KV\left(\sum_{i=1}^{NL} (V_i - V_{lim})^2\right) + KQ\left(\sum_{i=1}^{N} \left(Q_{G,i} - Q_{G,i}^{lim}\right)^2\right) + KS\left(\sum_{i=1}^{nl} abs\left(S_i - S_i^{lim}\right)^2\right) + KL\left(\sum_{j=1}^{NL} \left(L_j - L_j^{lim}\right)^2\right) \tag{11}$$

where, NG = Number of generating units, P_{Gi} = Generation of active power of ith generator, a_i, b_i and c_i are the cost coefficients of the ith generator and KP, KV, KQ, KS and KL are the penalty factors. NL = No. of PQ buses, nl = No. of transmission lines and Y^{lim} = limiting values dependent variable given as:

$$Y^{lim} = = \begin{cases} Y^{max}; & Y > Y^{max} \\ Y^{min}; & Y < Y^{min} \end{cases} \tag{12}$$

2.2 Fuel Cost with Valve-Point Loading Effects

A sine component is included into the cost of the generating units to apply the valve point loading effects which can be represented as:

$$f_2 = \left[\sum_{i=1}^{NG} \left(a_i P_{Gi}^2 + b_i P_{Gi} + c_i\right) + \left|d_i \sin\left(e_i\left(P_{Gi}^{\min} - P_{Gi}\right)\right)\right| \right] + KP\left(P_{G1} - P_{G1}^{lim}\right)^2$$
$$+ KV\left(\sum_{i=1}^{NL} (V_i - V_{lim})^2\right) + KQ\left(\sum_{i=1}^{N} (Q_{G,i} - Q_{G,i}^{lim})^2\right) + KS\left(\sum_{i=1}^{nl} abs(S_i - S_i^{lim})^2\right)$$
$$+ KL\left(\sum_{j=1}^{NL} \left(L_j - L_j^{lim}\right)^2\right) \tag{13}$$

where, d_i and e_i are the cost coefficients of the generators with valve-point loading.

2.3 Real Power Loss Minimization

The objective function for total power loss minimization can be expressed as follows:

$$f_3 = \sum_{i=1}^{nl} G_{ij}\left(V_i^2 + V_j^2 - 2V_iV_j\cos(\delta_i - \delta_j)\right) + KP\left(P_{G1} - P_{G1}^{lim}\right)^2 + KV\left(\sum_{i=1}^{NL} (V_i - V_{lim})^2\right)$$
$$+ KQ\left(\sum_{i=1}^{N} (Q_{G,i} - Q_{G,i}^{lim})^2\right) + KS\left(\sum_{i=1}^{nl} abs(S_i - S_i^{lim})^2\right) + KL\left(\sum_{j=1}^{NL} \left(L_j - L_j^{lim}\right)^2\right). \tag{14}$$

where G_{ij} is the conductance of the line i-j. The minimization problem is treated under the following constraints:

2.4 Equality Constraints

These are the set of nonlinear load flow expressions that regulate the power systems, i.e.

$$P_{Gi} - P_{Di} - \sum_{j=1}^{n} |V_i||V_j||Y_{ij}|\cos(\theta_{ij} - \delta_i + \delta_j) = 0 \tag{15}$$

$$Q_{Gi} - Q_{Di} + \sum_{j=1}^{n} |V_i||V_j||Y_{ij}|\sin(\theta_{ij} - \delta_i + \delta_j) = 0 \tag{16}$$

where, P_{Gi}, P_{Di} and Q_{Gi}, Q_{Di} are the real and reactive power generation and demands at bus-i respectively and $|Y_{ij}|$ are the elements of bus admittance matrix.

2.5 Inequality Constraints

The power network operational and security limits are represented as the set of inequality constraints, i.e.

1. Generators real and reactive power outputs.

$$P_{Gi}^{min} \leq P_{Gi} \leq P_{Gi}^{max},\ \mathrm{i} = 1, 2, \ldots\ldots, N_G \tag{17}$$

$$Q_{Gi}^{min} \leq Q_{Gi} \leq Q_{Gi}^{max},\ \mathrm{i} = 1, 2, \ldots\ldots, N_G \tag{18}$$

2. Voltage magnitudes of each bus

$$V_i^{min} \leq V_i \leq V_i^{max}, \mathrm{i} = 1, 2, \ldots\ldots.., N \tag{19}$$

3. Tap settings of Transformer

$$T_i^{min} \leq T_i \leq T_i^{max},\ \mathrm{i} = 1, 2, \ldots\ldots.., N_T \tag{20}$$

4. VAR injections by capacitor banks

$$Q_{Ci}^{min} \leq Q_C \leq Q_{Ci}^{max},\ \mathrm{i} = 1, 2, \ldots\ldots.., \tag{21}$$

5. Loading on Transmission lines

$$S_i \leq S_i^{max} \mathrm{i} = 1, 2, \ldots\ldots\ldots, N_L \tag{22}$$

6. Voltage stability index

$$Lj_i \leq Lj_i^{max}, \mathrm{i} = 1, 2, \ldots\ldots\ldots, N_{LD} \tag{23}$$

2.6 UPFC Constraints

UPFC Series injected voltage limits:

$$V_{se\ min} \leq V_{se} \leq V_{se\ max} \tag{24}$$

$$\theta_{se\ min} \leq \theta se \leq \theta_{se\ max} \tag{25}$$

UPFC Shunt injected voltage limits:

$$V_{sh\ min} \leq V_{sh} \leq V_{sh\ max} \tag{26}$$

$$\theta_{sh\ min} \leq \theta sh \leq \theta_{sh\ max} \tag{27}$$

The above constraints are controlled using APSO-DV technique which is discussed in subsequent section.

3 Overview of PSO-DV Algorithm

PSODV introduces a differential operator barrowed from differential evolution in the velocity update scheme of PSO. In this algorithm, for each particle 'i' in the swarm, the other two different particles, say 'j' and 'k' (i ≠ j ≠ k), are chosen randomly. The difference vector can be obtained as follows:

$$\overrightarrow{\delta_d} = \overrightarrow{Y_k} - \overrightarrow{Y_j} \tag{28}$$

The dth dimension velocity upgrade expression of ith target particle is:

$$\begin{aligned} V_{id}(k+1) &= \omega.V_{id}(k) + \beta\delta_d + C_{2}.\varphi_2.(P_{gid} - Y_{id})\text{If, rand } (0,1) \leq \text{CR} \\ &= V_{id}(k); \qquad \text{or else} \end{aligned} \tag{29}$$

where CR = Crossover probability, δ_d = difference vector d^{th} component and β = scaling factor between 0 and 1.

The cognitive part of the velocity upgrade expression is changed with the differential operator to generate further exploration capacity. If CR ≤ 1, a number of the velocity components will preserve their previous values. Now, a new trial position Tri is created for the particle by combining the upgraded velocity to the preceding position Y_i:

$$\overrightarrow{T_{ri}} = \overrightarrow{Y_i}(\text{k}) + \overrightarrow{V_i}(\text{k}+1) \tag{30}$$

The particle is positioned at this latest position only if the coordinates of the position gains a superior fitness. Therefore, if the minimum of an 'n' dimensional function $f(\vec{X})$ is required, then the target particle is repositioned as follows:

$$\overrightarrow{Y_i}(\mathrm{k}+1) = \overrightarrow{T_{ri}} \text{ if } (f(\overrightarrow{T_{ri}}) < f(\overrightarrow{Y_i}(k))$$

$$\overrightarrow{Y_i}(\mathrm{k}+1) = (\overrightarrow{Y_i}(k) \text{ otherwise} \tag{31}$$

Therefore, each time its velocity is modified, the particle either shifts to a superior location in the search space or holds to its preceding position. The recent position of the particle is the best position so far compared to the previous positions. On the other side, unlike the traditional PSO, in the present method, P_{lid} at all times equals Y_{id}. So the cognitive part of the algorithm involving $|P_{lid} - Y_{id}|$ is automatically removed. If the particle is stagnant at any position in the search space then the particle is moved to a random mutation to a new position. This procedure helps run away from local minima and also retain the swarm "moving":

$$\text{If } ((\overrightarrow{Y_i}(k) = \overrightarrow{Y_i}(k+1) = \overrightarrow{Y_i}(k+2) = \ldots\ldots\ldots = \overrightarrow{Y_i}(k+n))$$

$$\text{And } (f(\overrightarrow{Y_i}(\mathrm{k}+\mathrm{n})) \neq f^*) \text{ Then for } (\mathrm{r}=1 \text{ to } \mathrm{n})$$

$$Y_{ir}(k+n+1) = \mathrm{Y_{MIN}} + \mathrm{rand_r}(0,1)*(\mathrm{Y_{MAX}} - \mathrm{Y_{MIN}}) \tag{32}$$

where, f^* = Global minimum of the fitness function, n = Maximum No. of iterations up to which stagnation can be tolerated and (Y_{MAX}, Y_{MIN}) are the permissible bounds of the search space.

In this paper, the adaptive acceleration coefficient approach is applied for updating g the position of the particle for PSO-DV [12].

4 Proposed APSO-DV Hybrid Algorithm

The main objective of APSO-DV is to use the adaptive acceleration coefficient for updating g the position of the particle for PSO-DV to accelerate the search for global solution [12]. The APSODV is discussed as follows:

Step 1 *Initialization:*

The initial population is generated randomly and is given by:

$$Y_i^0 = Y_{i,\min} + rand().(Y_{i,\max} - Y_{i,\min}),\ i = 1, \ldots N_p \tag{33}$$

where rand() represents a uniformly distributed random number within the range of 0 to 1. This produces Np of individuals Y_i^0 randomly. During the initialization, the control variables, real power generations, generator voltages, transformer taps, shunt reactive power injections and velocities of the control variables are randomly generated within the allowable ranges.

Step 2 *Run the power flow and evaluate the fitness value of each individual*

Step 3 *Mutation Operation:*

To involve mutation operator in the velocity updating part of PSO, Two particles are selected randomly and the mutation operator is constructed as given below:

$$\overrightarrow{\delta_d} = \beta\left(\overrightarrow{Y_k} - \overrightarrow{Y_j}\right) \qquad i \neq j \neq k \tag{34}$$

Step 4 *Crossover Operation:*

In order to extend the diversity of further individuals at next generation, the perturbed individual of $\hat{Y}_i^{G+1}$ is generated from the present individual Y_i^G by adding differentially perturbed velocity to V_i^G. The crossover constant (CR) is used to determine if the newly generated individual is to be recombined. Each parameter's velocity j of the ith individual is reproduced from the perturbed individual velocity V_i^{G+1} is as follows:

$$V_{ij}^{G+1} = \begin{cases} \omega V_{ij}^G + \delta_d + C_2\varphi_2\left(P_{gj} - Y_{ij}^G\right), & if rand(0,1) < CR \\ V_{ij}^G, & Otherwise \end{cases} \tag{35}$$

where i = 1,…N_p; J = 1,…n; n = number of parameters.

The adaptive acceleration coefficient C_2 [12] is given below:

$$C_2 = \left(C_{2f} - C_{2i}\right)\frac{gen}{gen\max} + C_{2i} \tag{36}$$

where C_{2i} and C_{2f} are constants and the weighting factor is given by

$$\varpi = 1 - \frac{gen}{gen\max}$$

Step 5 *Estimation and selection:*

The fitness of the offspring is in competition with its parent. The parent is replaced by its offspring if the fitness of offspring is fitter than its parent. On the other hand, the parent retained in the next generation if the offspring is less fit than that of its parent. These two forms are presented as follows:

$$Y_i^{G+1} = \arg\max\{f(Y_i^G), f(Y_i^{G+1})\} \tag{37}$$

$$Y_b^{G+1} = \arg\max\{f(Y_i^{G+1})\} \tag{38}$$

where, arg max means the argument of the maximum. here, arg max is used because the fitness function, f = 1/OF where OF is the objective function to be minimized.
Step 6 *Repeat steps 2–5 until maximum generation quantity is reached*

5 Simulation Results and Discussions

The proposed method is tested on IEEE-30 bus system in MATLAB computing environment and the test system consists of six generators interconnected with 41 transmission lines with a total load of 283.4 MWand126.2 MVAR [13]. The shunt VAR compensators are provided at buses 10, 12, 15, 17, 20, 21, 23, 24 and 29 as given in [14]. The Weak nodes in the system are identified using Fuzzy by taking voltage magnitudes and L-Indices are the inputs and corresponding test results of top five weak nodes are tabulated in Table 1.

From the Table 1, the bus 27 has maximum severity considered as weakest node in the system and ranked according to the severity. The line between 29-30 is selected as most favorable location of UPFC by checking different possible locations in the vicinity of weak nodes. The APSODV-OPF results of the system with UPFC for fuel cost and fuel cost with valve point effects are shown in Tables 2 and 3 respectively.

From the above Table 2, It is observed that the optimal fuel cost in proposed method is reduced to 798.54 $/h compared to PSO-DV and APSO-DV is 800.16$/h and 799.53$/h respectively and the corresponding graphical representation is as shown in Fig. 2. It is also observed that, the L-Index value is reduced to 0.1301 compared to PSO-DV and APSO-DV is 0.1360 and 0.1342 respectively which indicates enhanced voltage stability. The voltage deviation also reduced to 0.6045

Table 1 Fuzzy severity of weak nodes

S. no	Bus no	Severity	Voltage (p.u)	L-Index	Rank
1	27	26.8642	1.0326	0.0827	1
2	22	26.2389	1.0318	0.0813	2
3	23	25.7536	1.0300	0.0842	3
4	29	25.0000	1.0248	0.1113	4
5	26	24.8783	1.0075	0.1041	5

Table 2 Comparison of APSO-DV& APSODV-UPFC results

Parameter	Limits		PSO-DV	APSO-DV	APSODV-UPFC
	Min	Max			
P_{G1}	0.5	2.0	1.7688	1.776	1.7805
P_{G2}	0.2	0.8	0.4844	0.4899	0.4922
P_{G5}	0.1	0.35	0.2149	0.2121	0.1989
P_{G8}	0.1	0.3	0.1212	0.1183	0.1218
P_{G11}	0.1	0.5	0.2137	0.2129	0.2095
P_{G13}	0.12	0.4	0.1200	0.1200	0.1200
V_{G1}	0.9	1.10	1.0829	1.0500	1.1000
V_{G2}	0.9	1.10	1.0667	1.0374	1.0841
V_{G5}	0.9	1.10	1.0442	1.0212	1.0633
V_{G8}	0.9	1.10	1.0161	1.0127	1.0180
V_{G11}	0.9	1.10	1.0341	1.0130	1.0594
V_{G13}	0.9	1.10	1.0829	1.0500	1.0616
T_{11}	0.9	1.10	1.0469	1.0163	1.0507
T_{12}	0.9	1.10	1.0240	1.0383	0.9868
T_{15}	0.9	1.10	0.9642	0.9786	1.0205
T_{36}	0.9	1.10	0.9396	0.9843	1.0312
Q_{C10}	0.0	0.2	0.2000	0.1544	0.0000
Q_{C12}	0.0	0.2	0.1305	0.0000	0.0331
Q_{C15}	0.0	0.2	0.0448	0.0680	0.1120
Q_{C17}	0.0	0.2	0.0567	0.0000	0.0000
Q_{C20}	0.0	0.2	0.0470	0.0375	0.0333
Q_{C21}	0.0	0.2	0.1445	0.0884	0.1883
Q_{C23}	0.0	0.2	0.0000	0.0296	0.0089
Q_{C24}	0.0	0.2	0.0340	0.0378	0.0323
Q_{C29}	0.0	0.2	0.0000	0.0407	0.1087
Cost ($/h)			**800.1666**	**799.5374**	**798.5683**
P_{loss} *(P.u)*			0.0890	0.0884	0.0857
L_j^{max}	0.0	0.5	0.1360	0.1342	0.1301
VD			0.895	0.6304	0.6045
V_{se}	0.0	0.2			0.0519
V_{sh}	0.9	1.1			0.9902

where as for PSO-DV and APSO-DV is 0.895 and 0.6304 respectively and the power loss also reduced to 0.0857 p.u from 0.0890 and 0.0884 respectively.

From the above Table 3, It is observed that the optimal fuel cost in proposed method is reduced to 928.60 $/h compared to PSO-DV and APSO-DV is 931.374$/h and 930.400$/h respectively where as the L-Index reduced to 0.1043 compared to PSO-DV and APSO-DV is 0.1150 and 0.1100 respectively which indicates

Table 3 Comparison of APSODV& APSODV-UPFC results

Parameter	Limits		PSO-DV	APSO-DV	APSO DV-PFC
	Min	Max			
P_{G1}	0.5	2.0	1.9761	1.9719	1.973
P_{G2}	0.2	0.8	0.3988	0.4001	0.4033
P_{G5}	0.1	0.35	0.1897	0.1933	0.1917
P_{G8}	0.1	0.3	0.1004	0.1000	0.1001
P_{G11}	0.1	0.5	0.1500	0.1500	0.1500
P_{G13}	0.12	0.4	0.1200	0.1202	0.1200
V_{G1}	0.9	1.10	1.0977	1.1000	1.0998
V_{G2}	0.9	1.10	1.0750	1.0778	1.0773
V_{G5}	0.9	1.10	1.0583	1.0570	1.0553
V_{G8}	0.9	1.10	1.0485	1.0654	1.0298
V_{G11}	0.9	1.10	1.0484	1.0502	1.0459
V_{G13}	0.9	1.10	1.0897	1.0840	1.0732
T_{11}	0.9	1.10	1.0035	0.9981	1.0704
T_{12}	0.9	1.10	1.0408	1.0753	0.9996
T_{15}	0.9	1.10	1.0151	0.9000	0.9000
T_{36}	0.9	1.10	0.9604	0.9920	1.0125
Q_{C10}	0.0	0.2	0.0000	0.0803	0.1853
Q_{C12}	0.0	0.2	0.1732	0.0000	0.0000
Q_{C15}	0.0	0.2	0.0693	0.0061	0.0897
Q_{C17}	0.0	0.2	0.2000	0.0000	0.0000
Q_{C20}	0.0	0.2	0.0635	0.0708	0.0730
Q_{C21}	0.0	0.2	0.1103	0.1997	0.0777
Q_{C23}	0.0	0.2	0.0000	0.0614	0.0000
Q_{C24}	0.0	0.2	0.0000	0.0207	0.0218
Q_{C29}	0.0	0.2	0.0329	0.0493	0.2000
Cost ($/h)			**931.3747**	**930.405**	**928.6**
P_{loss} *(P.u)*			0.1009	0.1006	0.1024
L_i *max*	0.0	0.5	0.1150	0.1100	0.1043
VD			1.7688	1.0116	1.0035
V_{se}	0.0	0.2			0.0519
V_{sh}	0.9	1.1			0.9768

enhanced voltage stability. The voltage deviation also reduced to 1.0035 where as for PSO-DV and APSO-DV is 1.7688 and 1.0116 respectively.

From the above Table 4, It is evident that the system power loss in proposed method is reduced to 5.9 MW compared to PSO-DV and APSO-DV is 7 MW and 6.86 MW respectively. From the results it is evident that the proposed method provides better results compared to the results available in the literature shown in Table 5.

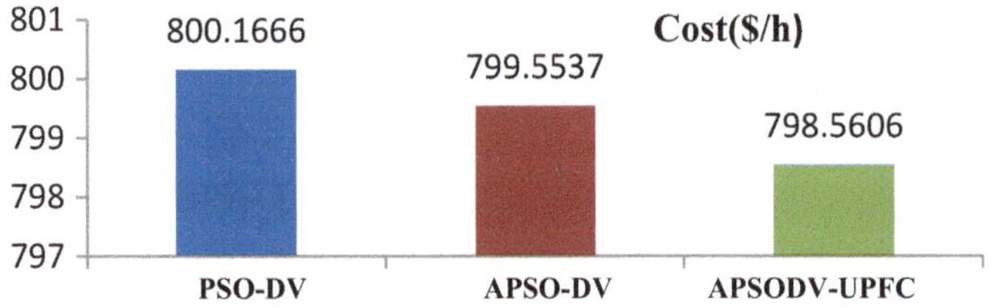

Fig. 2 Comparison of fuel cost for different OPF methods

Table 4 Comparison of APSODV& APSODV-UPFC with power loss as an objective

Parameter	Limits		PSO-DV	APSO- DV	APSO DV-UPFC
	Min	Max			
P_{G1}	0.5	2.0	1.3774	1.2857	1.2122
P_{G2}	0.2	0.8	0.4488	0.4318	0.4820
P_{G5}	0.1	0.35	0.1739	0.2395	0.2776
P_{G8}	0.1	0.3	0.2807	0.1912	0.1810
P_{G11}	0.1	0.5	0.3293	0.3988	0.4623
P_{G13}	0.12	0.4	0.2938	0.3514	0.2780
V_{G1}	0.9	1.1	1.0359	1.0264	1.0236
V_{G2}	0.9	1.1	1.0191	1.0131	1.0075
V_{G5}	0.9	1.1	0.9925	0.9983	0.9941
V_{G8}	0.9	1.1	1.0038	0.9814	0.9722
V_{G11}	0.9	1.1	0.9780	0.982	0.9745
V_{G13}	0.9	1.1	1.0118	1.0246	1.0409
T_{11}	0.9	1.1	0.9963	1.0683	1.0325
T_{12}	0.9	1.1	0.9447	1.0078	0.9933
T_{15}	0.9	1.1	0.9755	0.9701	0.9535
T_{36}	0.9	1.1	0.9146	0.9856	0.9679
Q_{C10}	0.0	0.2	0.1062	0.0431	0.1318
Q_{C12}	0.0	0.2	0.0000	0.0000	0.0482
Q_{C15}	0.0	0.2	0.1059	0.2000	0.1947
Q_{C17}	0.0	0.2	0.0983	0.0000	0.0000
Q_{C20}	0.0	0.2	0.0724	0.1288	0.0299
Q_{C21}	0.0	0.2	0.0336	0.0734	0.0568
Q_{C23}	0.0	0.2	0.0735	0.0836	0.0924
Q_{C24}	0.0	0.2	0.0801	0.0525	0.0512
Q_{C29}	0.0	0.2	0.0105	0.0448	0.0822
Cost ($/h)			833.5633	847.4463	858.19
P_{loss} (P.u)			**0.0700**	**0.0686**	**0.0590**
L_i *max*	0.0	0.5	0.1232	0.1322	0.1300
VD			0.7711	0.6670	0.4552
V_{se}	0.0	0.2			0.0518
V_{sh}	0.9	1.1			0.9949

Table 5 Cmparison of Fuel cost for IEEE-30 bus system

Method	Fuel cost ($/h)	Method	Fuel cost ($/h)
EP[15]	802.9070	SADE-AIM[16]	802.4040
TS[15]	802.5020	PSO[17]	800.4036
TS/SA [15]	802.7880	**PSO-DV**	**800.1666**
ITS[15]	804.5560	**PSO-DV**	**799.5374**
IEP[15]	802.4650	**APSODV-UPFC**	**798.5606**

6 Conclusions

In this paper, the proposed technique called APSO-DV with UPFC has been applied for solving the OPF problem with three objective functions such as both smooth and non-smooth generator fuel cost curves along system power loss with different equality and inequality constraints. UPFC is located in line connected to weak nodes in the systems which are determined by Fuzzy which effectively improved the system performance.

The proposed technique used to solves the OPF problem efficiently for the three objective functions and it eliminates the drawbacks canonical PSO. Simulation results shows that APSO-DV with UPFC outperforms the original PSO-DV and APSO-DV algorithms and it is effectively implemented to find the best possible settings of the control variables of the IEEE30-bus system. The comparison of the result shows the proposed method is effective and superior to find remarkable global solutions without any restrictions on the different type of fuel cost curves.

References

1. Carpentier J (1962) Contribution e létude do dispatching economique, Bull Soc Franc Elect, pp 431–447
2. Huneault M, Galiana FD (1991) A survey of the optimal power flow literature. IEEE Trans Power Syst 6(2):762–770
3. Momoh JA, El-Hawary ME, Adapa R (1999) A review of selected optimal power flow literature to 1993, Part I: Nonlinear and quadratic programming approach. IEEE Trans Power Syst 14(1):96–104
4. Momoh JA, El-Hawary ME, Adapa R (1999) A review of selected optimal power flow literature to 1993, Part II: Newton, linear programming and interior point methods. IEEE Trans Power Syst 14(1):104–111
5. LinWM Cheng F S, Tsay MT (2002) An improved tabu search for economic dispatch with multiple minima. IEEE Trans Power Syst 17(1):108–112
6. Taranto GN, Pinto LMVG, Pereira MVF (1992) Representation of FACTS devices in power system economic dispatch. IEEE Trans Power Syst 7(2):572–576
7. Gotham DJ, Heydt GT (1998) Power flow control and power flow studies for systems with FACTS devices. IEEE Trans Power Syst 13(1):60–65

8. Ambriz-Perez H, Acha E, Fuerte-Esquivel CR (2000) Advanced SVC model for Newton-Raphson Load Flow and Newton optimal power flow studies. IEEE Trans Power Syst 15(1):129–136
9. Chung TS, Li YZ (2001) A hybrid GA approach for OPF with consideration of FACTS devices. IEEE Power Eng Rev 21(2):47–50
10. Ongsakul W, Bhasaputra P (2002) Optimal power flow with FACTS devices by hybrid TS/SA approach. Electr Power Energy Syst 24:851–857
11. Kessel P, Glavitch H (1986) Estimating the voltage stability of a power system. IEEE Trans Power Deliv 1(3):346–354
12. Das S, Abraham A (2008) Amit konar,particle swarm optimization and differential evolution: technical analysis, applications, hybridization perspectives

Finite Set Model Predictive Current Control of Three Phase Neutral Point Clamped Inverter with Reduced Leg Count

Eedara Aswani Kumar, K. Chandra Sekhar and R. Srinivasa Rao

Abstract In this paper current control of two leg three phase Neutral Point Clamped (NPC) inverter feeding a resistance and inductance (RL) load using Model Predictive Control (MPC) method is presented. In this method decoupled currents are predicted using discretized load current equations obtained from application of Kirchhoffs voltage law. A cost function which is dependent on decoupled predicted currents and reference currents is evaluated for all the possible nine switching states of two leg NPC inverter. Switching state which gives minimum cost function is applied in the next sampling instant to the two-leg NPC inverter. The inverter and load are modeled using Simulink. MPC algorithm is coded in embedded MATLAB function. Simulation results confirm capability of algorithm to track reference and step change in reference currents.

Keywords Model predictive current control · Reduced leg count NPC converter · Two leg NPC converter

1 Introduction

Some of the possible ways of current control of power converter are non-linear hysteresis current control and linear proportional integral control [2, 3]. Predictive control strategy is implemented for current control of various power converters to

E. Aswani Kumar
Sasi Institute of Technology and Engineering, Tadepalligudem,
Andhra Pradesh, India
e-mail: aswaniatsasi@gmail.com

K. Chandra Sekhar (✉)
Department of EEE, RVR & JC College of Engineering, Guntur,
Andhra Pradesh, India
e-mail: cskoritala@gmail.com

R. Srinivasa Rao · E. Aswani Kumar
Department of EEE, Jawaharlal Nehru Technological University, Kakinada,
Andhra Pradesh, India
e-mail: srinivas.jntueee@gmail.com

K.R. Attele et al. (eds.), *Emerging Trends in Electrical, Communications and Information Technologies*, Lecture Notes in Electrical Engineering 394,
DOI 10.1007/978-981-10-1540-3_33

overcome the drawbacks of above methods in literature which is having advantages of no modulation stage requirement and no tuning of PI controllers [1, 5, 8–11]. In this paper a reduced leg count three phase Neutral Point Clamped (NPC) inverter is used instead of three phase three NPC inverter to reduce leg count to two and hence switch count to eight [6, 7]. It is called a three phase two leg eight switch NPC inverter. Model predictive control algorithm is applied to control currents in all phases of load.

2 Three Phase Two Leg NPC Inverter

Three phase two leg NPC inverter feeding a three phase load is shown in Fig. 1 (Table 1).

Switching states of inverter are defined in (1).

$$S_u = \begin{cases} 2 & \text{if } S_{1u}, S_{2u} \text{ are ON} \\ 1 & \text{if } \overline{S}_{1u}, S_{2u} \text{ are ON} \quad \text{where } u \epsilon \{R, Y\} \\ 0 & \text{if } \overline{S}_{1u}, \overline{S}_{2u} \text{ are ON} \end{cases} \tag{1}$$

Pole voltages with respect to DC source neutral N based on switching states are defined as

$$v_{uN} = \begin{cases} \frac{V_{dc}}{2} * (S_u) & \text{if } u \epsilon \{R, Y\} \\ \frac{V_{dc}}{2} & \text{if } u \epsilon \{B\} \end{cases} \tag{2}$$

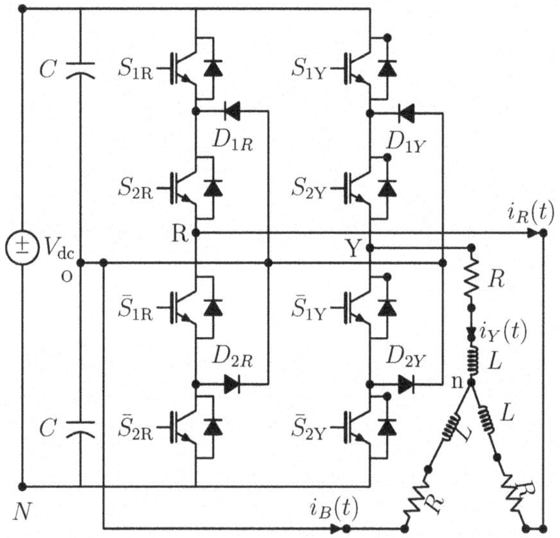

Fig. 1 Three phase two leg NPC inverter

Table 1 Voltage vectors of three phase two leg NPC Inverter [6]

$\vec{V}$	S_R	S_Y	v_α	v_β	$\lvert\vec{V}\rvert\angle\theta^\circ$	CMV
V_0	1	1	0	0	$0\angle 0^\circ$	0
V_1	2	1	$\frac{V_{dc}}{3}$	0	$\frac{V_{dc}}{3}\angle 0^\circ$	$\frac{V_{dc}}{6}$
V_2	2	2	$\frac{V_{dc}}{6}$	$\frac{V_{dc}}{2\sqrt{3}}$	$\frac{V_{dc}}{3}\angle 60^\circ$	$\frac{V_{dc}}{3}$
V_3	1	2	$-\frac{V_{dc}}{6}$	$\frac{V_{dc}}{2\sqrt{3}}$	$\frac{V_{dc}}{3}\angle 120^\circ$	$\frac{V_{dc}}{6}$
V_4	0	2	$-\frac{V_{dc}}{2}$	$\frac{V_{dc}}{2\sqrt{3}}$	$\frac{V_{dc}}{\sqrt{3}}\angle 150^\circ$	0
V_5	0	1	$-\frac{V_{dc}}{3}$	0	$\frac{V_{dc}}{3}\angle 180^\circ$	$-\frac{V_{dc}}{6}$
V_6	0	0	$-\frac{V_{dc}}{6}$	$-\frac{V_{dc}}{2\sqrt{3}}$	$\frac{V_{dc}}{3}\angle 240^\circ$	$-\frac{V_{dc}}{3}$
V_7	1	0	$\frac{V_{dc}}{6}$	$-\frac{V_{dc}}{2\sqrt{3}}$	$\frac{V_{dc}}{3}\angle 300^\circ$	$-\frac{V_{dc}}{6}$
V_8	2	0	$\frac{V_{dc}}{2}$	$-\frac{V_{dc}}{2\sqrt{3}}$	$\frac{V_{dc}}{\sqrt{3}}\angle 330^\circ$	0

Application of Kirchhoff's voltage law for N-u-n-N closed paths gives

$$\begin{bmatrix} v_{RN} \\ v_{YN} \\ v_{BN} \end{bmatrix} = L\frac{d}{dt}\begin{bmatrix} i_R \\ i_Y \\ i_B \end{bmatrix} + R * \begin{bmatrix} i_R \\ i_Y \\ i_B \end{bmatrix} + \begin{bmatrix} v_{nN} \\ v_{nN} \\ v_{nN} \end{bmatrix} \tag{3}$$

Clarke's three phase to two phase transformation is defined as

$$h_\alpha + j * h_\beta = \frac{2}{3} * \begin{bmatrix} 1 & 1\angle 120^\circ & 1\angle 240^\circ \end{bmatrix} * \begin{bmatrix} h_R \\ h_Y \\ h_B \end{bmatrix} \tag{4}$$

where $h\epsilon\{v, i\}$. Pre multiplying (3) with $\frac{2}{3} * \begin{bmatrix} 1 & 1\angle 120^\circ & 1\angle 240^\circ \end{bmatrix}$ gives

$$\begin{aligned} \frac{2}{3} * \begin{bmatrix} 1 & 1\angle 120^\circ & 1\angle 240^\circ \end{bmatrix} * \begin{bmatrix} v_{RN} \\ v_{YN} \\ v_{BN} \end{bmatrix} &= L\frac{d}{dt} * \frac{2}{3} * \begin{bmatrix} 1 & 1\angle 120^\circ & 1\angle 240^\circ \end{bmatrix} * \begin{bmatrix} i_R \\ i_Y \\ i_B \end{bmatrix} \\ &+ R * \frac{2}{3} * \begin{bmatrix} 1 & 1\angle 120^\circ & 1\angle 240^\circ \end{bmatrix} * \begin{bmatrix} i_R \\ i_Y \\ i_B \end{bmatrix} \\ &+ \frac{2}{3} * \begin{bmatrix} 1 & 1\angle 120^\circ & 1\angle 240^\circ \end{bmatrix} * \begin{bmatrix} v_{nN} \\ v_{nN} \\ v_{nN} \end{bmatrix} \end{aligned} \tag{5}$$

Last term in (5) is zero as $1 + 1\angle 120° + 1\angle 240° = 0$. So (5) is simplified by using (4) as

$$v_\alpha + j * v_\beta = L\frac{d}{dt}(i_\alpha + ji_\beta) + R * (i_\alpha + ji_\beta) \tag{6}$$

as (6) is similar in α and β can be formulated as

$$v_\gamma = L\frac{di_\gamma}{dt} + R * i_\gamma \text{ where } \gamma \epsilon \{\alpha, \beta\} \tag{7}$$

Voltage vector $\overline{V}$ in terms of switching states can be obtained by using (2) & (4) as

$$v_\alpha + j * v_\beta = \tfrac{2}{3} * \begin{bmatrix} 1 & 1\angle 120° & 1\angle 240° \end{bmatrix} * \begin{bmatrix} 0.5V_{dc} * (S_R) \\ 0.5V_{dc} * (S_Y) \\ 0.5V_{dc} \end{bmatrix}$$

$$\Rightarrow v_\alpha + j * v_\beta = \frac{V_{dc}}{6}(2S_R - S_Y - 1) + j\frac{V_{dc}}{2\sqrt{3}}(S_Y - 1) \tag{8}$$

Common Mode Voltage (CMV) [4] of three phase two leg NPC inverter is defined as

$$CMV = \frac{\sum\limits_{u=R,Y,B} v_{uo}}{3}$$

$$CMV = \frac{1}{3} * \{\frac{V_{dc}}{2} * [S_R - 1] + \frac{V_{dc}}{2} * [S_Y - 1] + 0\}$$

$$CMV = \frac{V_{dc}}{6} * (S_R + S_Y - 2) \tag{9}$$

3 Predictive Current Control Scheme

According to Euler's formula of forward difference

$$\frac{di_\gamma}{dt} = \frac{i_\gamma(m+1) - i_\gamma(m)}{T_s} \tag{10}$$

Currents can be predicted by substituting (9) in (7)
$v_\gamma = \frac{L}{T_s} * [i_\gamma(m+1) - i_\gamma(m)] + R * i_\gamma(m)$

$$i_\gamma^p(m+1) = \frac{T_s}{L} * v_\gamma(m) + (1 - \frac{R * T_s}{L}) * i_\gamma(m) \text{ where } \gamma \epsilon \{\alpha, \beta\} \tag{11}$$

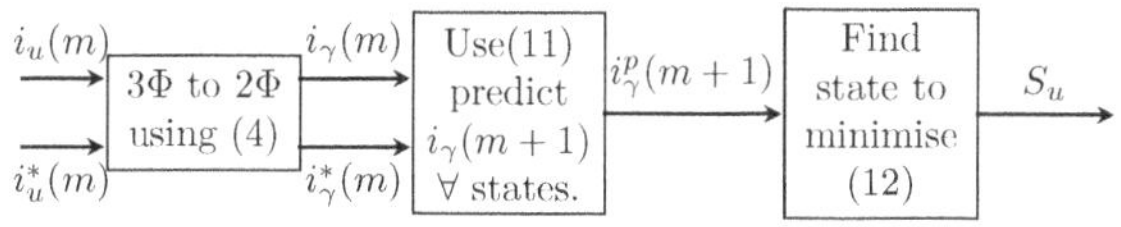

Fig. 2 Model predictive current controller block diagram

Subscript p is used to denote predicted current. Algorithm for the implementation of predictive current control scheme [11] is explained below:

1. Store reference and measured currents at mth instant converted in to $\alpha\beta$ frame of reference.
2. Predict currents corresponding to all the nine vectors using (11).
3. Evaluate cost function for all the predicted currents using (12).
4. Find voltage vector giving minimum cost function value.
5. Select switching state corresponding to the voltage vector giving minimum cost function value.
6. Apply this switching state for next sampling instant.
7. Go to first step (Fig. 2).

$$CF = \sum_{\gamma=\alpha,\beta} |i_\gamma^*(m) - i_\gamma^p(m+1)| \qquad (12)$$

4 Simulation Results

The simulation parameters are $V_{dc} = 100$ V, $R = 1.5\ \Omega$, $L = 1.5$ mH, $T_s = 25\ \mu$s, power frequency=50 Hz. A step change in reference current vector amplitude is created from 5 A to 8 A at $t = 0.07$ s. Figure 3 shows that the measured load currents are in track with reference currents and step change in load current is tracked with

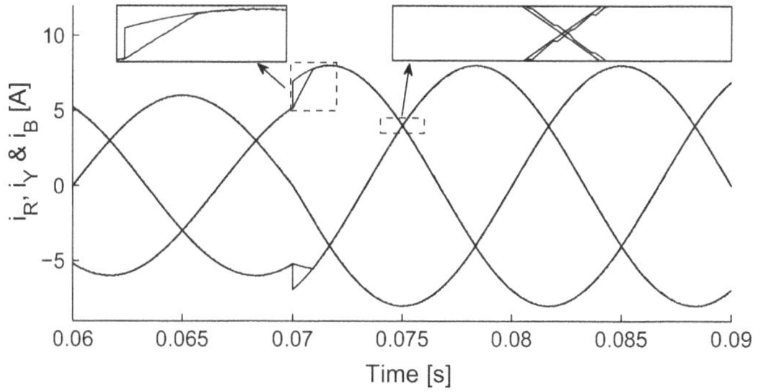

Fig. 3 3Φ measured and reference load currents

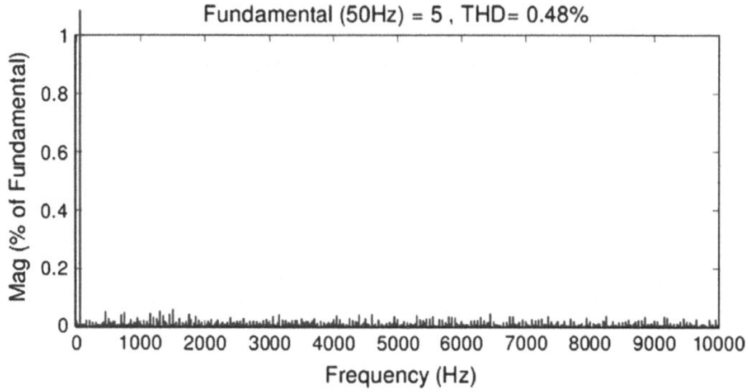

Fig. 4 Spectrum of load current

Fig. 5 Pole to DC neutral(N) voltages

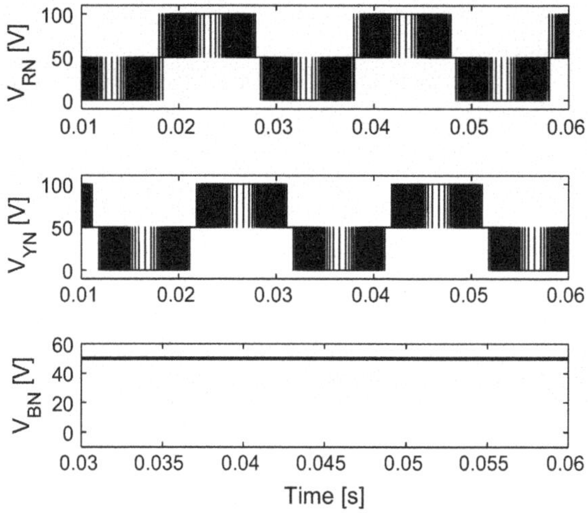

fast dynamics because of MPCC Algorithm. Current spectrum of R-phase load current as shown in Fig. 4 indicates that total harmonic distortion is 0.48% only with fundamental current as 5 A. Steady state pole voltages as shown in Fig. 5 indicates that V_{RN}, V_{YN} are of three level and V_{BN} is of single level as B is clamped to $\frac{V_{dc}}{2}$. Steady state three phase line to line load voltages as in Fig. 6 shows that V_{RY} is of five level and V_{YB}, V_{BR} are of three level as pole R and Y are of three level and B is single level. Figure 7 showing line to neutral voltages of load indicates that V_{Rn}, V_{Yn} are of seven level and V_{Bn} is of three level.

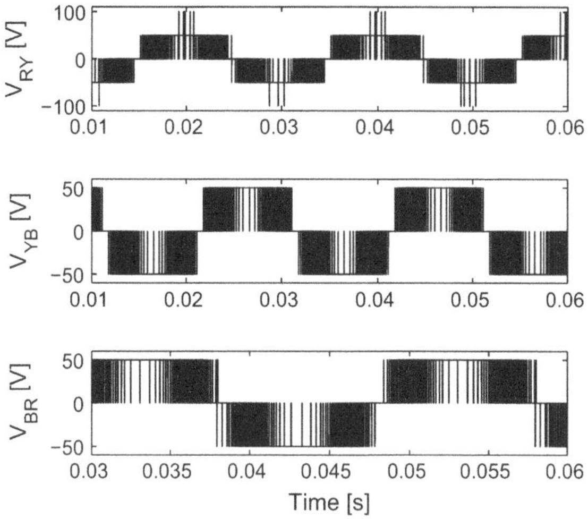

Fig. 6 Line to line load voltages

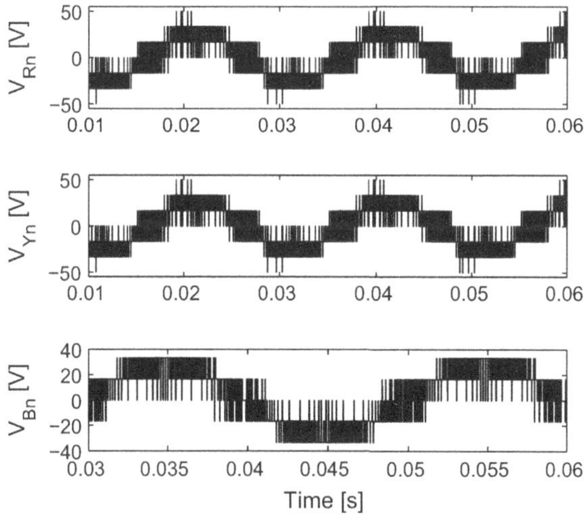

Fig. 7 Line to neutral load voltages

5 Conclusions

A three phase two leg NPC inverter is analyzed to obtain voltage vectors and modelled for predictive current control implementation. The modeled system is simulated in MATLAB/Simulink for step change in load current. From simulation results it is clear that the predictive current control algorithm predicts one step ahead switching state to be applied to the inverter in next instant, reference currents and change in reference currents are tracked well by using finite set model predictive current control scheme with fast dynamics for the three phase two leg NPC inverter feeding a RL load.

Acknowledgments The first author would like to acknowledge Mr. B. Ravi Kumar, vice chairman, Prof. N. Venkat Rao, dean academics, Prof. K. Bhanu Prasad, principal, Prof. R.B. Choudary, HOD ME Dept. and his colleagues at Sasi Institute of Technology & Engineering for consistent encouragement.

References

1. Cortés P, Kazmierkowski MP, Kennel RM, Quevedo DE, Rodríguez J (2008) Predictive control in power electronics and drives. IEEE Trans Ind Electron 55(12):4312–4324
2. Holtz J (1994) Pulsewidth modulation for electronic power conversion. Proc IEEE 82(8):1194–1214
3. Kaźmierkowski MP, Krishnan R, Blaabjerg F (2002) Control in power electronics: selected problems. Academic Press
4. Kimball JW, Zawodniok M (2011) Reducing common-mode voltage in three-phase sine-triangle pwm with interleaved carriers. IEEE Trans Power Electron 26(8):2229–2236
5. Kouro S, Cortés P, Vargas R, Ammann U, Rodríguez J (2009) Model predictive control and simple and powerful method to control power converters. IEEE Trans Ind Electron 56(6):1826–1838
6. Lin BR, Wei TC (2004) Analysis and implementation of a three-phase two-leg neutral point clamped converter based on space vector pwm for power factor correction. IEE Proc-Electr Power Appl 151(1):38–46
7. Lin BR, Wei TC (2004) Space vector modulation strategy for an eight-switch three-phase npc converter. IEEE Trans Aerosp Electron Syst 40(2):553–566
8. Rodriguez J, Cortes P (2012) Predictive control of power converters and electrical drives, vol 40. Wiley
9. Rodríguez J, Pontt J, Silva C, Correa P, Lezana P, Cortés P, Ammann U et al (2007) Predictive current control of a voltage source inverter. IEEE Trans Ind Electron 54(1):495–503
10. Uddin M, Mekhilef S, Rivera M (2014) Experimental validation of minimum cost function-based model predictive converter control with efficient reference tracking. IET Power Electron 8(2):278–287
11. Vargas R, Cortés P, Ammann U, Rodríguez J, Pontt J (2007) Predictive control of a three-phase neutral-point-clamped inverter. IEEE Trans Ind Electron 54(5):2697–2705

Coordination of Energy Storage Devices in Hybrid Power Systems

Aayush Sharma and K. Jamuna

Abstract Energy storage becomes a vital and potent factor in the economic development with extensive use of electricity. It hard to succeed the ever growing demand of electrical energy with the use of conventional available electrical network. Although new technologies based on solar, wind etc. have been introduced to feed this very purpose but their uncertainty nature make its application inflexible with limited infrastructure. Past studies have come out with the idea of energy storage to balance the supply and demand of electricity and cope up load challenges. The design of self-reliant future energy network can be achieved by the use of Energy Storage Devices (ESD) that have potential to toil for load balancing, continuous energy supply, conquering fluctuation and creating reserve for emergency. This paper deals with the coordination among different energy storage devices in the hybrid power system environment. The solar-photo voltaic and wind power plant are considered as renewable energy resources. The simulation is performed in MATLAB-Simulink environment. The results are discussed in detail.

Keywords ESD (energy storage devices) · Buck boost converter · Solar module · Wind module

1 Introduction

Electricity has fascinated human clan since they witnessed lighting. Since then electricity has become the most vital part of industrialization and modernization of the society. With the depletion of conventional sources, the attention is moved into a new era of non-conventional energy (Renewable energy) sources. Conventional electrical power transmission and distribution system are operated in the unidirectional mode from remote and large power plants to energy consumers. With the use of renewable in the grid, it is possible to transmit the power in both directions.

A. Sharma (✉) · K. Jamuna
School of Electrical Engineering, VIT University, Chennai Campus, Chennai, India
e-mail: s.aayush11@yahoo.com

K.R. Attele et al. (eds.), *Emerging Trends in Electrical, Communications and Information Technologies*, Lecture Notes in Electrical Engineering 394,
DOI 10.1007/978-981-10-1540-3_34

But the uncertainty nature of alternative sources may lead to system collapse during peak times. The distribution of the electricity is expressed by a typical load curve. The load curve of the demand of electricity by the consumer is non-linear. The demand changes from day to day, week to week and season to season. In order to satisfy the demand, a reliable, potential and complex electronic infrastructure equipped with computer software, power electronic devices (converters, inverter etc.) and energy storage devices are required to make an efficient power transmission [1]. Establishment of such energy storage devices network at load end yields an energy reserve for arising demands at the energy utility. This reserve can be obtained by storing electrical energy during times when production exceed consumption and then using these reserves when consumption exceeds production. According to a research, it has been found that an electric system should have 15–20 % of reserve capacity to realize any consumer request. The inadequacy of reserve capacity would result in sudden stress at load side, when the generation surpasses consumer demands or vice versa leading to voltage fluctuation at consumer end causing the performance decline of home appliances [2]. In order to avoid such inconsistence of power supply and system collapse due to renewable plants, the needy devices can provide back up at the time of voltage sags and on sudden deviation of voltages [2]. Transient stability means the ability of a load to restore its steady state after overcoming the sudden disturbance (sudden change in load, loss of generation, fault etc.) [3]. Frequency stability means the load operate under steady frequency when there is a sudden disturbance at its side [4]. The addition of different topologies of Enery Storage Systems (ESS) in electric network can perform the frequency stabilization function [5, 6]. The presence of ESS at a heavily loaded line can deal with sudden raise of voltage by charging and fall of voltage by discharging of electricity [6]. Furthermore the sensitivity of solar and wind power plant on climatic parameters (solar irradiance, wind speed) make their connections to grid a tough task which can be handled more precisely and cost effectively by ESS than single-purpose mitigation measures as stated by [6]. The characteristic of donating and absorbing electric energy from the system at the time of sudden upset makes ESS to be utilized as virtual inertia [7]. This paper particular stresses on the use of hybrid ESD's in the electric network and also discusses the improvement of power quality to achieve frequency, transient and dynamic stability of electric network. This paper has the following sections. Section 2 explains about the type of energy storage devices. The Sect. 3 describes the simulation model of both solar and wind power plant. The ESD models are designed and the results are discussed. Finally the Sect. 4 summaries the paper as a conclusion

2 Types of Energy Storage Devices

Electrical energy is in form of AC which cannot be stored directly. However this AC energy can be transformed and stored in the form of kinetic, electromagnetic (SMES, Super capacitor) electrochemical (Battery), or as potential energy. All type of energy storage device has a power conversion unit to convert the energy from one form to another [1, 2, 8]. There are some types of ESD's: SMES (superconducting magnetic energy storage system), Super capacitor energy storage system, Battery energy storage system. The selection of energy storage system depends on "**time response**" Time response depends on the physical characteristics of these ESD's which they exhibit and their mode of application required by an energy utility [8, 9]. Devices with such capability are useful for the stabilization of the electric network. The aim of using these devices together is to check how the devices are operated together to compensate the change in the output power. So in this paper, the hybrid energy storage devices has been utilized and depicted in Fig. 1. The ESD's receive energy from hybrid system [Solar + Wind]. The energy generated from solar and wind power plants are fed as an input to converter, where the wind input is converted to DC from AC then fed to load and energy storage devices. When demand is less than supply, then charging of ESD's take place and when demand is greater than supply discharging take place. The system is simulated in MATLAB Simulink environment and its model is shown in Fig. 2. In simulation part it is has been explained how this hybrid system of ESS can coordinate in overcoming voltage fluctuation and annihilating micro grid challenge. All three ESD's are connected in parallel and are operated, and coordinated by buck boost converter [10–14]. The output voltage of the ESD's is applied across the "load" (Figs. 3 and 4). The Fig. 2 shows the MATLAB Simulink model of the proposed hybrid system: Fig. 2 consists of solar module (3), wind module (4), buck boost dc-dc convertor (5) with compensator, SMES, super-capacitor and battery connected with DC bus.

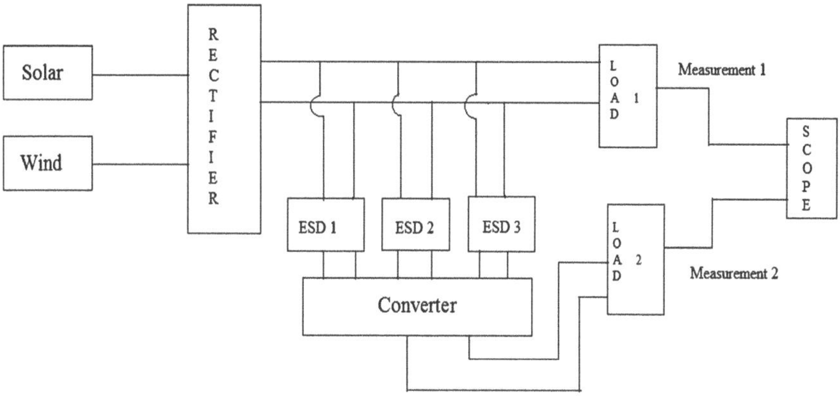

Fig. 1 Block diagram of interconnection of energy storage devices in hybrid power system

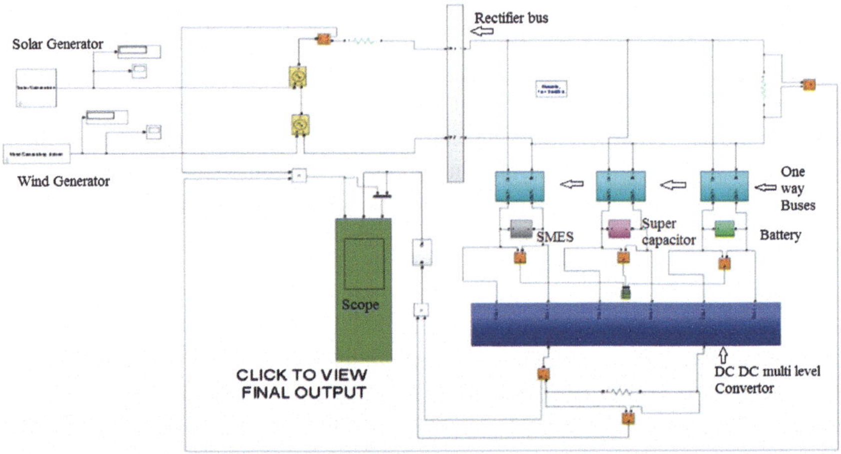

Fig. 2 Hybrid module connected to ESDs through DC bus

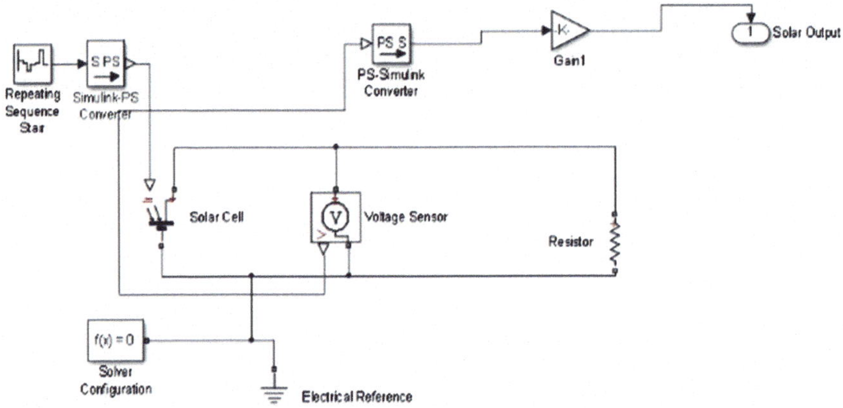

Fig. 3 PV/solar module

The solar module is designed in the MATLAB environment which uses irradiance value as input and generates direct current as output. This output is supplied to DC bus and then to ESDs. As the output generated by this PV module is direct current, hence no conversion is required. The variable wind speed is applied to the wind turbine and then output is feed to permanent magnet synchronous machine and the AC output is then converted to DC output through rectifier and feed to DC bus. The Fig. 5 depicts one of the topology of the buck boost dc-dc converter [10–14] presented in [10], that is used to connect and coordinate the charging and discharging of hybrid energy storage devices in a single system to meet the load demand, improve power quality, assure stability & reliability. The value of inductance & capacitance used in the above circuit are 20 mH and 47 μF. This buck

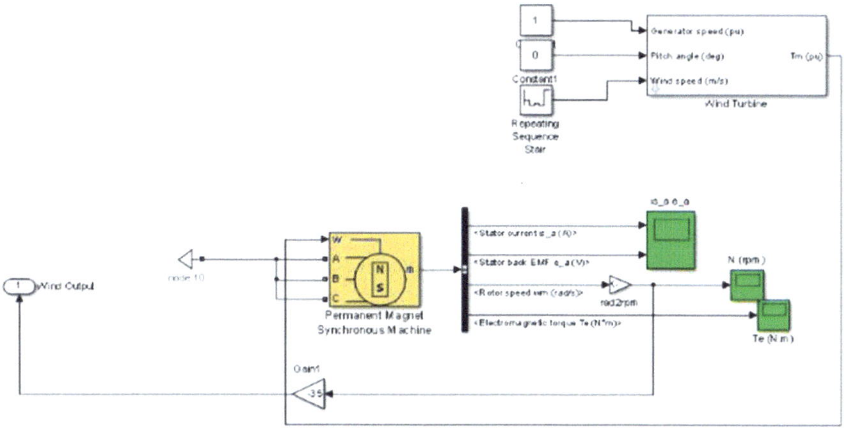

Fig. 4 Wind module

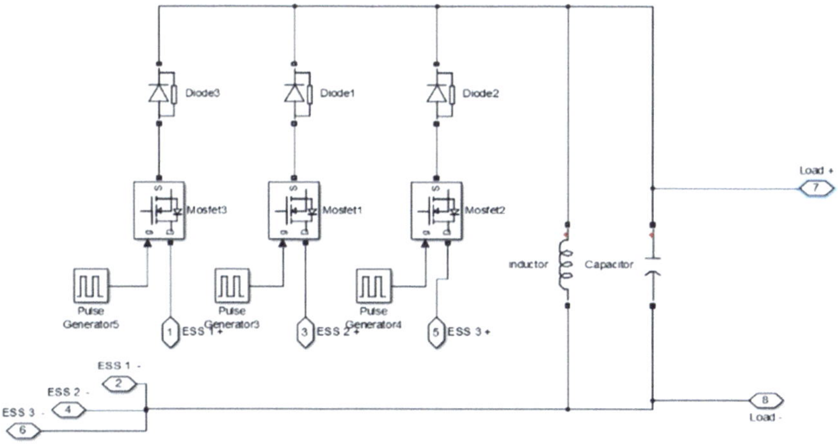

Fig. 5 Buck boost converter [10–14]

boost works in continues conduction mode. If the output value of the hybrid system is less than the reference value then the converter start discharging otherwise switch to charging process (Table 1).

ESD in the circuit shown in Fig. 5 operate at a frequency of 60 Hz. Hence as per three devices, one cycle of 1/20 s yields 60 Hz on each device. The entire cycle is

Table 1 Technical specification for the hybrid system

Battery rating	12 V and 10A
Super capacitor	200 F and 16 V
Buck boost converter	47 μf and 20 mH
Load	10 kW
SMES coil	500 H
Rectifier	4 F

divided in 3 parts first 33.33 % of the cycle to SMES, from 33.33 to 66.66 % to battery and rest to super capacitor.

The efficiency of energy circle as per [1] at load due to ESS can be written as:

$$\eta_e = \frac{\eta_{in}}{\frac{1}{\eta_{out}} + \frac{T}{\tau_s} \cdot \frac{E_s}{E_{out}}}, \quad \text{where } |\tau_s = E_s / P_{idl} \tag{1}$$

Pidl (losses at ESS), E_s (energy stored at ESS), T (single charging or discharging time cycle), η_{in} (conversion efficiency from load to ESS), η_{out} (conversion efficiency from ESS to load), The amount of energy loss is very small than energy storage which means Pidl can be assumed as constant [1]. This makes T/τ_s a very small quantity that can be neglected and the expression [1] can be deduced to

$$\eta_e = \eta_{in}\eta_{out}. \tag{2}$$

3 Simulations

This section explains the different simulation results of the hybrid system. The plots of input solar voltage, input wind voltage, input power after ESS and comparative study with respect to time (t) are shown in this section.

Case 1 As the solar irradiance is increased from 1000 to 4000 W/m^2, the voltage generated by solar module is increased from 222 to 248 V which is shown in Fig. 6. The value of wind speed is assumed to be increased from 40 to 80 kmph, and then the voltage (Vw) generated by wind module is increased from 230 to 600 V and displayed in Fig. 7. The input power (Pin) (solar + wind) fluctuates from 16 to 30 kW, but after using ESS the output power (Pout) fluctuates between 25–30 kW which is shown in Fig. 8.

Case 2 When the solar irradiance value is randomly decreased from 7500 to 4500 W/m^2 and thus output voltage (Vs.) decreases from 240 to 230 V and the wind speed decreased from 75 to 45 kmph so is voltage (Vw) from 590 to 240 V. The input power (Pin) (solar + wind) obtained fluctuates between 15–28 kW, and output (Pout) from 25 to 30 kW is shown in Fig. 9.

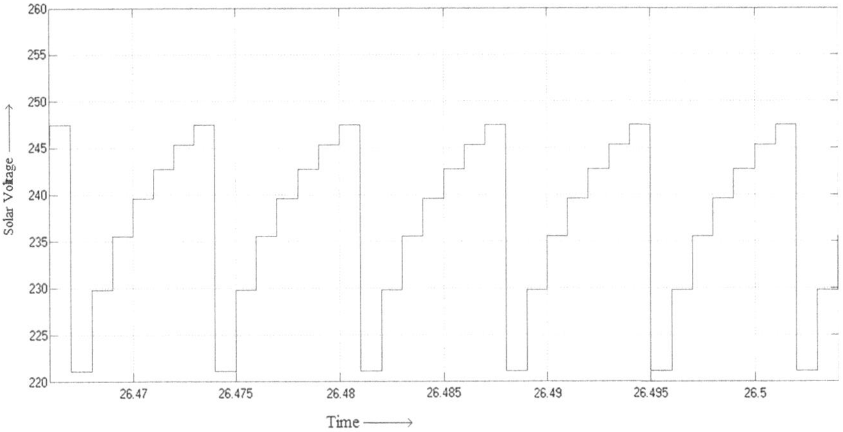

X-axis- time, Y-axis-Solar voltage (V)

Fig. 6 Output voltage of solar module with respect to time (Vs versus t)

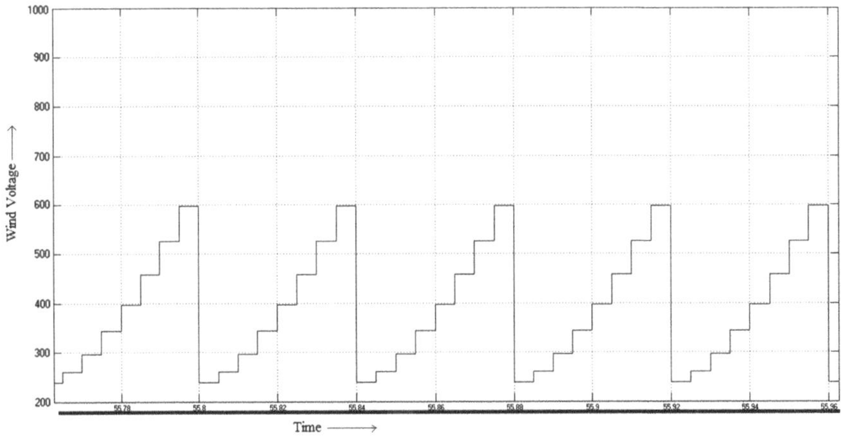

X-axis- time, Y-axis-wind voltage (V)

Fig. 7 Output voltage of wind module (Vw/t)

Case 3 The input solar irradiance values from 1000 to 15000 W/m^2 is fed to generate voltage (Vs.) of between 222–257 V. The input wind speed values from 30 to 80 kmph are taken to generate voltage V from 250 to 930 V. The input power (Pin) fluctuates between 15–40 kW (solar + wind). But after the use of ESS, the obtained output power (Pout)

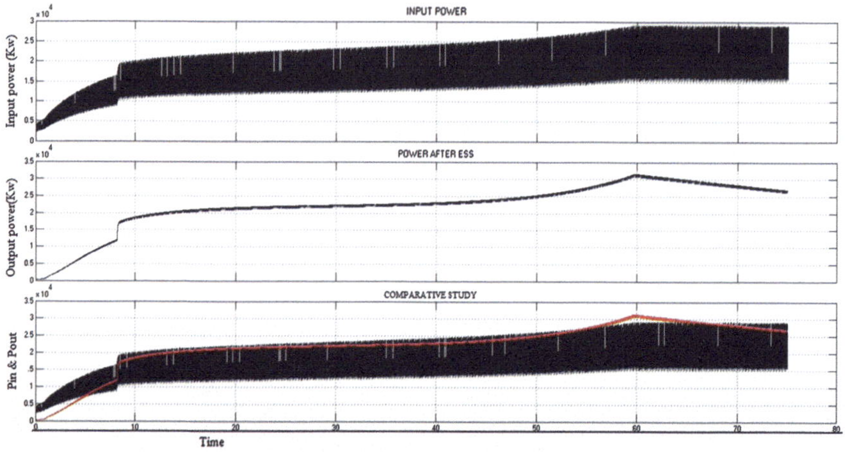

X-axis- time, Y-axis-Input power (kW), Output power (kW), Pin Pout

Fig. 8 Comparative study of input power, output power (with ESS) and comparison b/w input and output power with respect to time

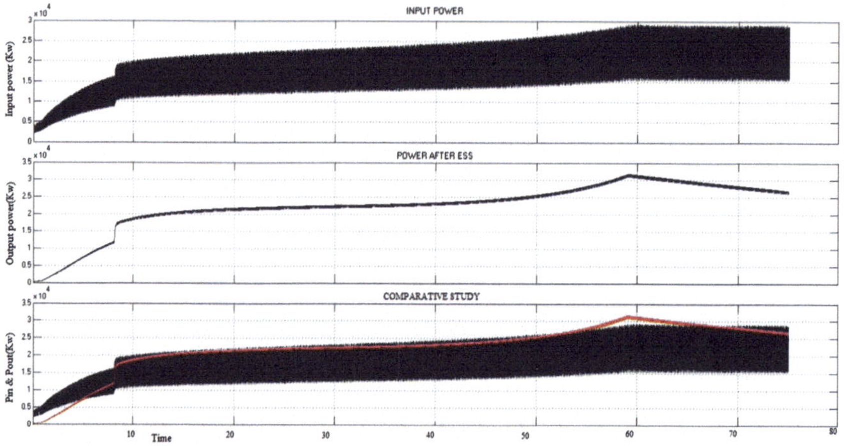

X-axis- time, Y-axis-Input power (kW), Output power (kW), Pin Pout

Fig. 9 Comparative study of input power, output power (with ESS) and comparison b/w input and output power with respect to time

from 30 to 33 kW satisfies 10 kW load demand and remaining power is stored in batteries which are shown in Fig. 10. Same improvement of power is achieved between 30–35 kW when solar irradiance increase and wind speed decrease and vice versa.

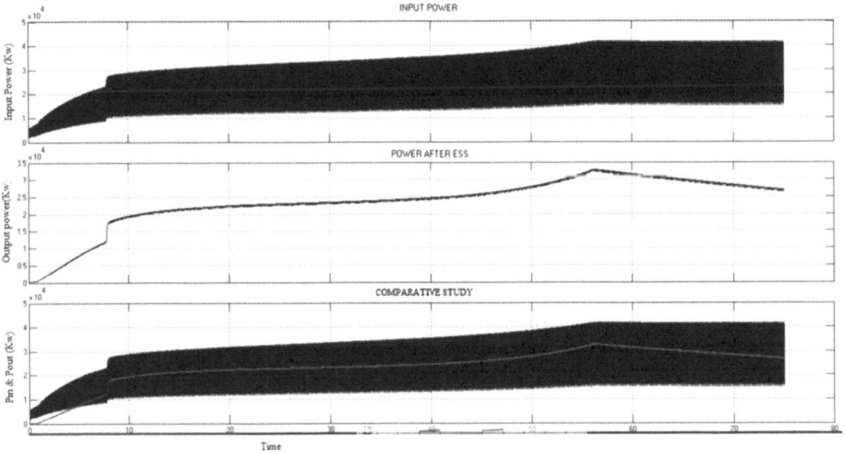

X-axis- time, Y-axis-Input power (kW), Output power (kW),Pin Pout

Fig. 10 Comparative study of input power, output power (with ESS) and comparison b/w input and output power with respect to time

4 Conclusion

In this study a simulation model of ESS along with solar and wind hybrid system is carried out. Due to periodic nature of solar and wind power plant, three cases featuring different irradiance and wind speed values have been shown in this paper. These set of values are given as input to hybrid system. The power obtained from solar and wind hybrid system (when both modules working like case 3) fluctuates between 15–40 kW, but after using coordinated hybrid ESS network, the power supply ranging from 30 to 33 kW is obtained. Similarly in case 1 and 2, the power deviation is effectively reduced. It is understood that the use of multiple energy storage devices help to improve power efficiency and provide stable power supply to network. Furthermore the application of hybrid ESS gives flexibility and potential infrastructure to renewable plants to withstand system collapse at peak times.

References

1. Kondoh* J, Ishii I, Yamaguchi H, Murata A, Otani K, Sakuta K, Higuchi N, Sekine S, Kamimoto M Electrical energy storage systems for energy network. http://www.dept.aoe.vt.edu/~cdhall/courses/aoe4065/OtherPubs/energystorage.pdf
2. Gupta R, Sharma NK, Tiwari* P, Gupta A, Nigam N, Gupta A (2011) Application of energy storage devices in power systems. Int J Eng Sci Technol 3(1):289–297
3. Saadat H (1999) Book: power system analysis. WCB McGraw-Hill International Editions

4. Kundur P (1994) Power system stability and control. McGraw Hill. Electrical Energy storage by International Electro technical Commission. http://www.iec.ch/whitepaper/pdf/iecWP-energystorage-LR-en.pdf
5. Tana X, Li[b] Q, Wang H (2013) Advances and trends of energy storage technology in Microgrid. Int J Electr Power Energy Syst 44(1):179–191
6. Electrical Energy storage by International Electro technical Commission. http://www.iec.ch/whitepaper/pdf/iecWP-energystorage-LR-en.pdf
7. Benidris M, Elsaiah S, Sulaeman S, Mitra J (2012) Transient stability of distributed generators in the presence of energy storage devices. In: North American Power Symposium (NAPS), pp 1–6
8. Kondoh* J, Denholm P, Ela E, Kirby B, Milligan M The role of energy storage with renewable electricity generation. http://www.nrel.gov/docs/fy10osti/49396.pdf
9. Mohiuddin* SM, Sheikh MRI (2014) Stabilization of solar-wind hybrid power system by using SMES. Int J Electr Comput Eng 4(3):351–358
10. Krishna YP, Ferdowsi M, Corzine K New double input DC-DC converters for automotive applications. Power Electronics and Motor Drives Laboratory, University of Missouri-Rolla
11. Gummi K, Ferdowsi M (2010) Double-input DC–DC power electronic converters for electric-drive vehicles—topology exploration and synthesis using a single-pole triple throw switch. IEEE Trans Ind Electr 57(2):617–623
12. Napoli AD, Crescimbini F, Rodo S, Solero L (2002) Multiple input DC-DC power converter for fuel-cell powered hybrid vehicles. In: IEEE power electronics specialists conference, vol 4, pp 1685–1690
13. Solero L, Lidozzi A, Pomilio JA (2005) Design of multiple-input power converter for hybrid vehicles. IEEE Trans Power Electron 20:1007–1016
14. Dobbs BG, Chapman PL (2003) A multiple-input DC-DC converter topology. Power Electr Lett 1:6–9

Design of Closed Loop Controller for DC-DC Converter by Using K-Factor Method Used For Renewable Energy Applications

K.M. Ravi Eswar and D. Elangovan

Abstract Renewable Energy resources are an important aspect of sustainability. The world's energy needs will reduce by one-third by 2050 if individuals and corporations save energy and depend on renewable energy sources. We need to control energy from renewable sources by interfacing particular type of converter which is suitable for the renewable source. If we consider solar application, the output of solar panel (i.e. input to converter) depends upon solar radiation. The energy from renewable sources are fluctuating because of nature. So, they give output also fluctuating. But for our applications there should not be any voltage variations even when load changes which is possible with the help of closed loop control of Converter. The control design can be done with the help of K factor method. This paper describes closed loop control design and operation of dc-dc converter (buck, boost and buck-boost model) used for renewable applications.

Keywords Renewable energy source · Closed loop control · Buck · Boost and buck-boost converter · K-factor method

1 Introduction

Fossil-based energy resources causes the largest emission of greenhouse gas in the world that leads to climate change and pollution which results in negative impact on environment and creatures. Therefore encouragement for using nonpolluting and more efficient resources and technologies has been needed as a solution for the major environmental problems and the high prices of fossil energy resources. Renewable energy sources, with capacity varying from 100 kW to a few MW, are increasingly present in electrical networks as shown in Table 1. This RE systems

K.M. Ravi Eswar (✉) · D. Elangovan
School of Electrical Engineering, V.I.T University, Vellore, Tamilnadu, India
e-mail: ravieswar.km@gmail.com

K.R. Attele et al. (eds.), *Emerging Trends in Electrical, Communications and Information Technologies*, Lecture Notes in Electrical Engineering 394,
DOI 10.1007/978-981-10-1540-3_35

Table 1 Total Renewable Energy Installed Capacity (31 Dec 2014) [1]

Source	Total installed capacity (MW)
Bagasse cogeneration	2,800.35
Biomass power	1,365.20
Small hydro power	3,990.83
Solar power	3,062.68
Total	33,791.74
Waste to power	107.58
Wind power	22,465.03

can be used in small, decentralized power plants or in large ones. The utilization of RE resources is efficiently usable and financially feasible for electricity generation because of the rapid development of the new technologies. Electrical power can be generated from a variety of resources such as sunlight, wind, biomass, and others.

From many previous studies, the utilization of RE resources for electricity generation can be proved that they could provide many advantages in many aspects such as reliability enhancement, loss reduction, and grid expansion postponement. However, RE-based electricity generation sources (abbreviated as "RE source") may bring about the problem to the connecting power system. For example, the uncertainty of RE resources will cause the uncertain electricity generation that can lead to voltage fluctuation and may cause the voltage profile to exceed the voltage limit. The over-voltages may occur when output from RE sources are high, while the under voltages are vice versa.

To solve the above problem the duty of converter should keep on adjusting automatically with changes in input voltage so that output voltage profile can be maintained in steady state even there is a load change. Therefore closed loop control operation of dc-dc converter is necessary. This paper explains about controller design technique for buck, boost and buck-boost model using K-factor method. A simple, straight forward method exists for achieving optimum control system performance without trial-and-error. The way of designing controller can be done by Ziegler-Nichols method; however it has some disadvantages. It yields an aggressive gain and overshoot that can be overcome by loop shaping method. K-factor method is a type of loop shaping technique. It is widely used in industry for design of dc-dc controller.

Some of advantages of loop shaping method are

1. Easy, intuitive, step by step design procedure
2. Controller can be designed accurately for a specified phase margin and cross over frequency
3. Accurate

2 Designing of Feedback Controllers Using K Factor Method

In this loop shaping method we shape the transfer function of the open loop gain specifically bode plot of loop gain to make sure that it has desired characteristics based on the required control specification. Design procedure starts with following steps as follows. [2].

1. Determine desired crossover frequency (bandwidth)
   (From specifications on dynamic performance, switching frequency limits)—It is desired to take 1/10th of switching frequency because feedback signal needs bandwidth at most 1/10th–1/5th of switching frequency
2. Integrator (pole at origin) for infinite dc gain- steady state error depends on the dc gain of loop gain. So higher the gain smaller the error which is required for dc-dc converter.
3. Zeroes at appropriate locations to achieve desired phase margin.
4. Adjust constant gain to achieve required cross over frequency.
   There are three types of K-factor based controllers.
   Type 1 controller $G_C(s) = \frac{K_c}{s}$ Type 2 controller $G_C(s) = \frac{K_c}{s} * \frac{1+\frac{s}{\omega_z}}{1+\frac{s}{\omega_p}}$
   Type 3 controller $G_C(s) = \frac{K_c}{s} * \frac{\left(1+\frac{s}{\omega_z}\right)^2}{\left(1+\frac{s}{\omega_p}\right)^2}$

Type 1 controller is simple integrator with gain Kc. Type 2 has same integrator Kc/s in addition it has one zero at ω_z and one pole at ωp such that $\omega_z < \omega_p$ then only we can get phase boost. Type 3 is same as previous, in addition it has 2 zeroes and 2 poles. Both of zeroes at same location, similarly both the poles are also at same location and $\omega_z < \omega_p$ to have phase boost.

3 Choice of Controller Type

1. After determine the desired crossover frequency ωc and desired phase margin obtain the phase of the plant ϴsys at the cross over frequency. This can be obtained directly from simulation (PLECS software) or by substituting ωc in plant transfer function [3].
2. Calculate the phase boost
   Φ_{boost} = PM (desired) – ϴsys – 90
   Where PM = desired phase margin (deg) i.e., 180 plus phase at crossover frequency
   ϴsys = Phase of system without controller
3. From φ_{boost} choice of controller is obtained as follows in Table 2.

Table 2 Controller type based on phase boost

Required φ_{boost}	Controller type
0°	Type 1
Less than 90°	Type 2
Greater than 90°	Type 3

Type 1 controller has only integrator and is suitable when no phase boost is required; so it's not widely used in typical dc-dc converters. Transfer function of type 1 controller is given as

$$G_c(s) = Kc/s$$

3.1 Design of Type 1 Controller

At desired cross over frequency (ωc) the magnitude of entire loop gain should be one.

$|G_{OL}(J\omega_C)| = \frac{K_C}{J\omega_C}|G_P(J\omega_C)| = 1$, Where $|G_{OL}(J\omega_C)|$ = Magnitude of entire loop gain, $|G_P(J\omega_C)|$ = Plant gain. Design of type 2 and 3 controller can be explained with the help of converters.

3.2 Design of Controller for Buck, Boost and Buck-Boost Converter

Converter topologies are shown in Fig. 1. From Small signal model transfer function of buck, boost and buck-boost converter is given as shown in Table 3. [4]. Where $D^1 = 1 - D$ and D is the duty cycle

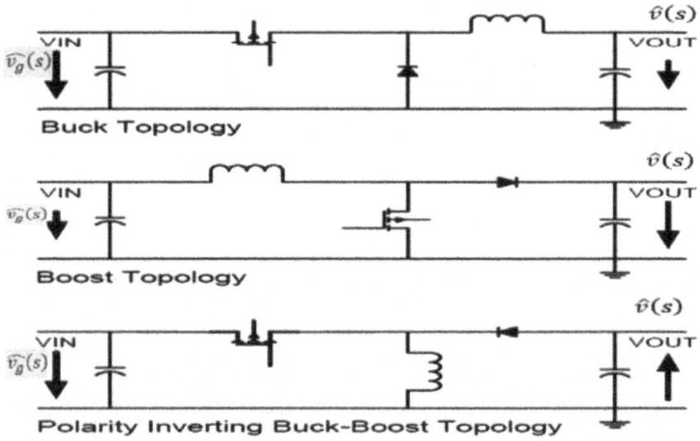

Fig. 1 Different DC-DC converter topologies and their transfer functions

Table 3 Generalized transfer function terms of dc-dc converters

Converter	G_{go}	G_{do}	ω_o	Q	ω_z
Buck	D	$\frac{V}{D}$	$\frac{1}{\sqrt{LC}}$	$R\sqrt{\frac{C}{L}}$	∞
Boost	$\frac{1}{D^1}$	$\frac{V}{D^1}$	$\frac{D^1}{\sqrt{LC}}$	$D^1 R\sqrt{\frac{C}{L}}$	$\frac{(D^1)^2 R}{L}$
Buck-boost	$\frac{-D}{D^1}$	$\frac{V}{DD^{12}}$	$\frac{D^1}{\sqrt{LC}}$	$D^1 R\sqrt{\frac{C}{L}}$	$\frac{(D^1)^2 R}{DL}$

Where the transfer functions are written in the standard forms [5].

$$G_{vd}(s) = G_{do} * \frac{1 - \frac{s}{\omega_z}}{1 + \frac{s}{Q\omega_o} + \left(\frac{s}{\omega_o}\right)^2} \quad G_{vg}(s) = G_{go} * \frac{1}{1 + \frac{s}{Q\omega_o} + \left(\frac{s}{\omega_o}\right)^2}$$

$$\text{Where } G_{vg}(s) = \frac{\hat{v}(s)}{v_{\hat{g}}(s)} \text{ At } \partial(s) = 0,\ G_{vd}(s) = \frac{\hat{v}(s)}{\partial(s)} \text{ At } v_{\hat{g}}(s) = 0$$

4 Design Procedure

1. After determining the phase of plant at crossover frequency, calculate phase boost from the above equation.
2. According to the obtained phase boost value choose appropriate type of controller.
3. Calculate k, ω_z and ω_p. The value of k mentioned below is for type 2 controller

$$K = \tan\left(\frac{\varphi\,\text{boost}}{2} + 45\right) \qquad \omega_Z = \frac{\omega c}{K} \qquad \omega_p = K * \omega_c$$

For type 3 controller the value of k given as

$$K = \tan\left(\frac{\varphi\,\text{boost}}{4} + 45\right)$$

4. Obtain the magnitude of the loop gain, at cross-over frequency with the designed controller with Kc of the controller set initially at 1.
5. Therefore Kc equal to inverse of the magnitude obtained in previous step

$$K_C = \frac{1}{\text{magnitude of the loop gain, at cross-over frequency with the designed controller}}$$

5 Simulation Circuits and Results of Closed Loop DC-DC Converter

5.1 Buck Converter

Circuit parameters were considered as L = 24 μH, C = 500 μF, ESR of C = 0.08 Ω, R load = 4 Ω.

Consider an input voltage of 20 v, with the variations in input voltage the output voltage changes as observed in simulation for buck, boost and buck-boost converter, Therefore by designing closed loop controller the output remains constant even with changes in input as observed in Figs. 2, 3 and 4. Controller design parameters of converters obtained by following procedure given above. Controller design parameters for buck converter are f = 400 kHz, Kc = 4551, ω_z = 6492, ω_p = 608108.

5.2 Boost Converter

Controller design parameters for boost converter are K_c = 7420.25, ω_z = 25992.57, ω_p = 151883.4, f = 400 kHz.

5.3 Buck-Boost Converter

Controller design parameters for buck-boost converter are Kc = 5689.76, ω_z = 25992.57, ω_p = 151883.4, f = 400 kHz. With the obtained control design values,

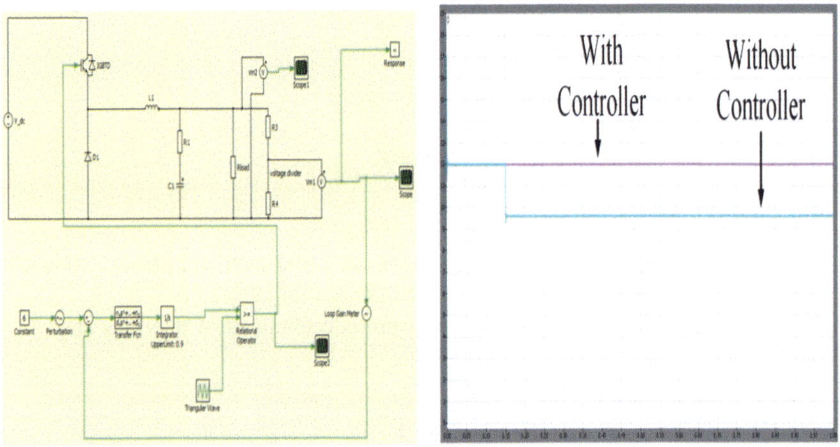

Fig. 2 Buck converter closed loop design circuit and simulated output voltage waveform with and without controller

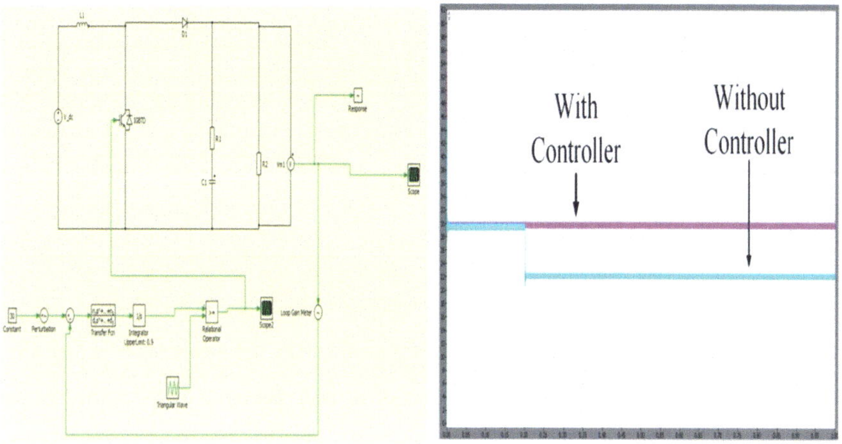

Fig. 3 Boost converter closed loop design circuit and simulated output voltage waveform with and without controller

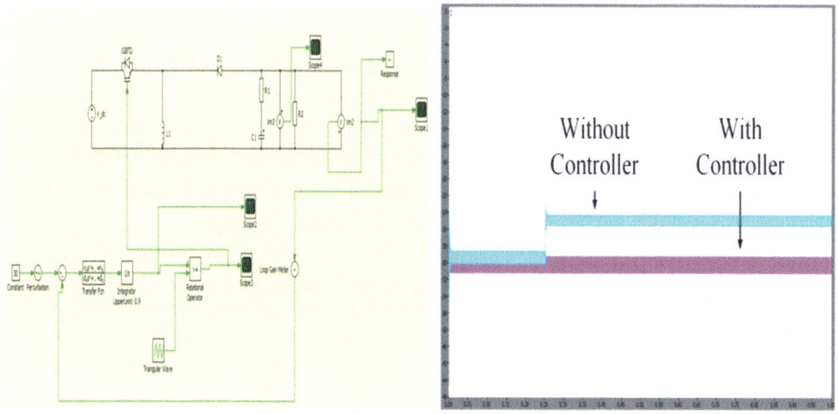

Fig. 4 Buck-boost converter closed loop design circuit and simulated output voltage waveform with and without controller

a practical operational amplifier circuit can be designed and simulated as shown in Fig. 5. The values of resistor and capacitor can be obtained by comparing obtained transfer function of controller with generalized form of controller. Therefore op-amp circuit controller can be designed. [6, 7].

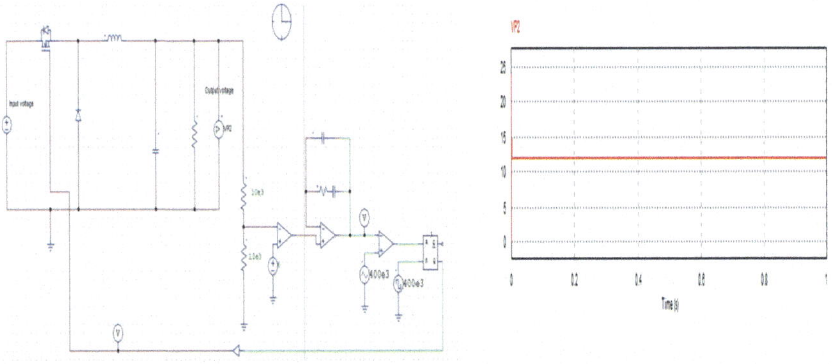

Fig. 5 Practical op-amp circuit closed loop control of buck converter and simulated output voltage waveform

6 Conclusion

India has vast supply of renewable energy resources. With the help of DC-DC converters it is possible to use efficiently them. As there is supply variations in renewable source, to make supportable with our applications we need to implement controller part in converters for successful operation and to improve efficiency. This paper has given complete procedure for designing controller for converter applications using K-Factor method. The above results in this paper depicts that even with supply variations in input side of converter the output remains constant. This can be done by properly choosing a controller for the converter. Even there is a load variation, controller can be able to stabilize the output voltage.

References

1. Renewable energy From Wikipedia, the free encyclopedia
2. Venable HD The k-factor: A new mathematical tool for stability analysis and synthesis. In: Proceedings of Powercon 10. www.venable.biz
3. PLECS (Piecewise Linear Electrical Circuit Simulation). www.plexim.com/
4. RASHID MH (2001) Power Electronics Handbook
5. Tang W, Lee FC, Rid4ey RB (1993) Small-signal modeling of average current-mode control. IEEE Trans Power Electr 8(2):112–119
6. Lee SW Demystifying Type II and Type III Compensators Using Op-Amp and OTA for DC/DC Converters—(Texas Instruments Application Report July 2014)
7. Czarkowski D, Kazimierczuk MK (1992) Static and dynamic circuit models of PWM buck-derived DC-DC converters. IEE Proc Part G, Circuit Devices Syst 139(6):669–679

Application of Bio-Inspired MPPT Techniques for Photovoltaic System

Jagadish Kumar Patra, Soumya Bhanu Mohanty, H.M. Tania, D. Elangovan and G. Arunkumar

Abstract The need of photovoltaic generation system is getting elevated day by day since it has no involved fuel cost, eco-friendly and it has less maintenance. In case of partial shaded condition there are many peaks thus we use maximum power point tracking (MPPT) method to get maximum power. The proposed bioinspired technique has the ability to track maximum power point under any environmental condition i.e. high change in insolation and partially shaded condition. This paper reflects the implementation of MPPT method for photovoltaic system by making use of particle swarm optimisation (pso) technique. Simulation results have been provided to prove correctness of proposed model and results were compared with Hill Climbing method.

Keywords Particle swarm optimisation · Hill climbing · MPPT

1 Introduction

Now a days with the growing world energy demand and rapid depletion of fossil fuels the interest of harvesting power from renewable sources has increased. In recent years there has been major growth in the field of photovoltaic power generation. We need to extract maximum power from PV source since its initial cost is high and efficiency of energy conversion is low. To maximize the use of PV system we need to use it efficiently. Thus we use MPPT which can be interfaced with inverters or dc-dc converter. MPPT tracks maximum power generated by PV under varying environmental conditions. To meet the load demand power we connect PV panels in series and parallel. As environmental condition varies the maximum power point also changes. To track this maximum power point many methods are

J.K. Patra · S.B. Mohanty (✉) · H.M. Tania · D. Elangovan · G. Arunkumar
School of Electrical Engineering, V.I.T University, Vellore, Tamilnadu, India
e-mail: soumyabhanu.mohanty@gmail.com

J.K. Patra
e-mail: jags606@gmail.com

K.R. Attele et al. (eds.), *Emerging Trends in Electrical, Communications and Information Technologies*, Lecture Notes in Electrical Engineering 394,
DOI 10.1007/978-981-10-1540-3_36

there like P & O, incremental conductance, short circuit current, open circuit voltage, hill climbing and fuzzy logic control. Each of these above mentioned techniques vary in speed, accuracy and simplicity [1, 2]. In this proposed model we use particle swarm optimization(PSO) method since its structure is simple, simple computational steps and has high maximum power point tracking efficiency. Simulations are done using MATLAB/Simulink and the results are compared with existing MPPT methods.

2 Modeling of PV Cell

A single diode equivalent circuit of photovoltaic cell is shown in figure. Mathematical equation governing the equivalent circuit is given as

$$I = Iph - IO[\exp(\frac{\mathrm{V} + \mathrm{RsI}}{\mathrm{Vta}}) - 1] - \frac{V + RsI}{Rp}$$

where,

Iph is PV generated current
I0 is diode saturation current
Vt = NskT/q is the thermal voltage
Ns is the number of cell connected in series
A is diode ideality constant
Rs is he series resistance
Rp is the parallel resistance.

The PV generated current Iph is given by following equation

$$Iph = \frac{G}{Gn}[Iph,\ n + Ki(T - Tn)]$$

where:

Iph, n is the PV generated current at the nominal condition
(25 °C and 1000 W/m2);
G is the irradiance;
Gn is the nominal irradiance;
T is the PV cell temperature;
Tn is the nominal cell temperature;
Ki is the short-circuit current/temperature coefficient.

The equations for diode saturation current I0 is given by

$$IO = IOn(\frac{T}{Tn})^3 \exp[\frac{qEg}{aK}(\frac{1}{Tn} - \frac{1}{T})]$$

where:

IOn is the nominal diode saturation current;
q = 1.602 10–19 C is the electron charge;
K = 1.380 10–23 J/K is the Boltzmann constant;
Eg = 1.12 eV is the bandgap energy.

The nominal diode current I0, n can be described by following equation

$$Ion = \frac{Iscn}{\left[\exp\left(\frac{Voc,\, n}{\propto Vt,\, n}\right)\right] - 1}$$

Where:

Voc, n is the nominal open-circuit voltage;
Vt, n is the nominal thermal voltage of the cell;
Isc, n is the short-circuit current at the nominal condition (25 °C and 1000 W/m^2).

A practical PV array is the combination of PV cell connected in combination of series and parallel, and its represented by the following equation

$$I = NppIpv - NppIO[\exp\left(\frac{V + IRs\left(\frac{Nss}{Npp}\right)}{Vt \propto Ns}\right) - 1]$$

In above equation Npp represent the number of PV modules connected in parallel;

Nss represents the number of PV modules connected in series.

Manufacturer specifies the most of parameter described in the above equation in datasheet.

3 MPPT Techniques

The maximum output produced by PV panel depends upon the presence of irradiation and temperature. The temperature and irradiation are not constant but varies with atmospheric condition. The present scenario demands the maximum energy for utilization. Various MPPT techniques were introduced to harvest maximum power and to reduce the hardware size. The MPPT techniques are classified on basis of control variables such as: (1) Current (2) voltage and (3) Duty. In method (1) Vmp or Imp is observed with Voc or Isc. The accuracy in this techniques is not guaranteed the tracked power is below MPP. In second method obtain the actual information of power. The points are made to update with climatic condition. In the P&O the present power is compared with previous power [2, 5]. Inspite of its simplicity it faces serious drawbacks such as slow tracking speed and oscillation. The other algorithm is incremental conductance which compares the derivative of conductance with instant. PSO and HC algorithms will be discussed in the forthcoming part.

3.1 Hill Climbing

This method is the perturbation of the duty ratio of the power converter to which the PV array is connected. As the duty of PV array perturbs it in turn pertubs both PV array current and PV array voltage from Fig. 1 it can be observed that, there is increment or decrement in voltage. This increment or decrement, increases or decreases the power when the operating point is situated on the left side of the Maximum Power Point(MPP) and decreases or increases the power when the operating point is on the right side of the MPP. Until MPP is achieved this process is continuously repeated. It will start oscillating around the MPP. By reducing the perturbation step size the oscillations can be minimized. Small perturbation size results in slowing down the MPPT. If variable perturbation size is used that reduces towards MPP, then the above problem can be solved. This is a two stage algorithm that where tracking is faster in the first stage. Under frequently changing atmospheric conditions this fails (Figs. 2, 3, 4 and 5).

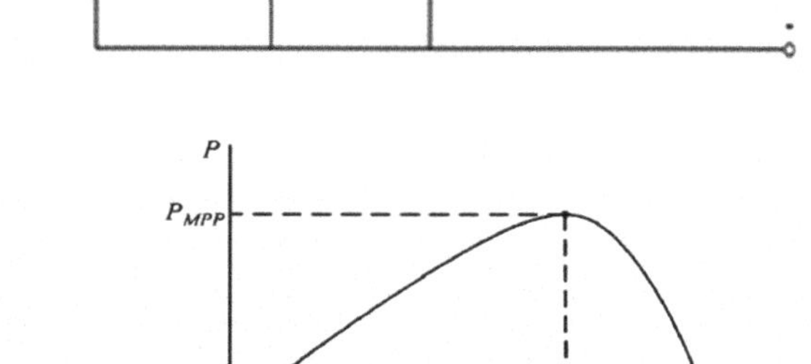

Fig. 1 Equivalent circuit of PV cell

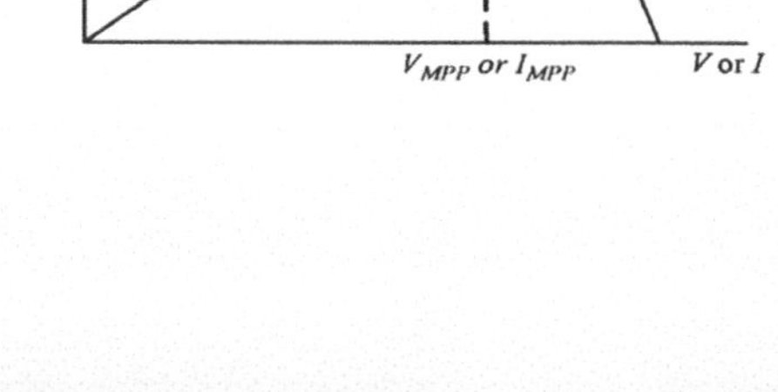

Fig. 2 Characteristic PV array power curve

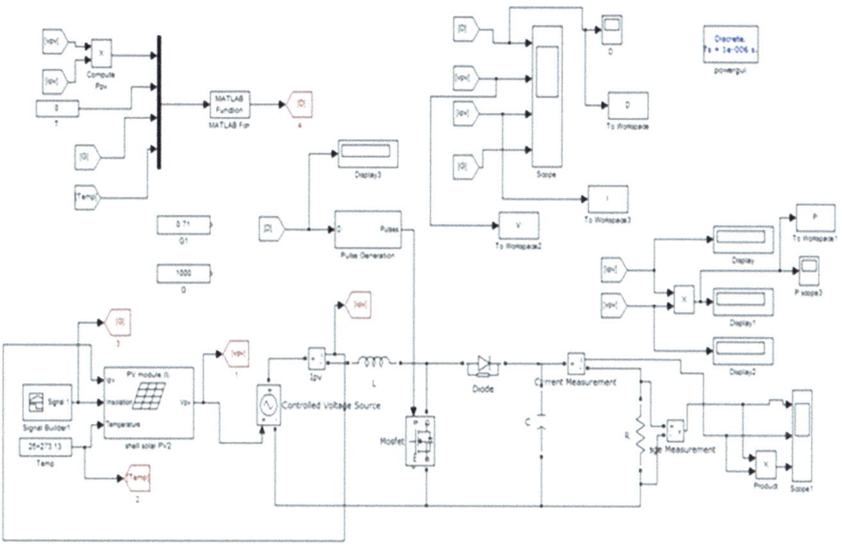

Fig. 3 Simulink model

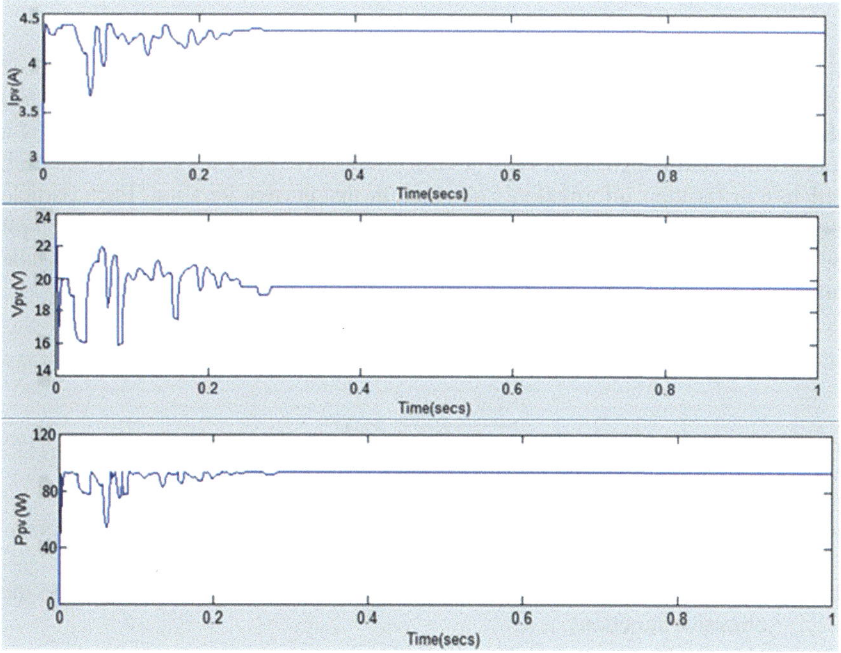

Fig. 4 PSO-current, voltage and power waveform

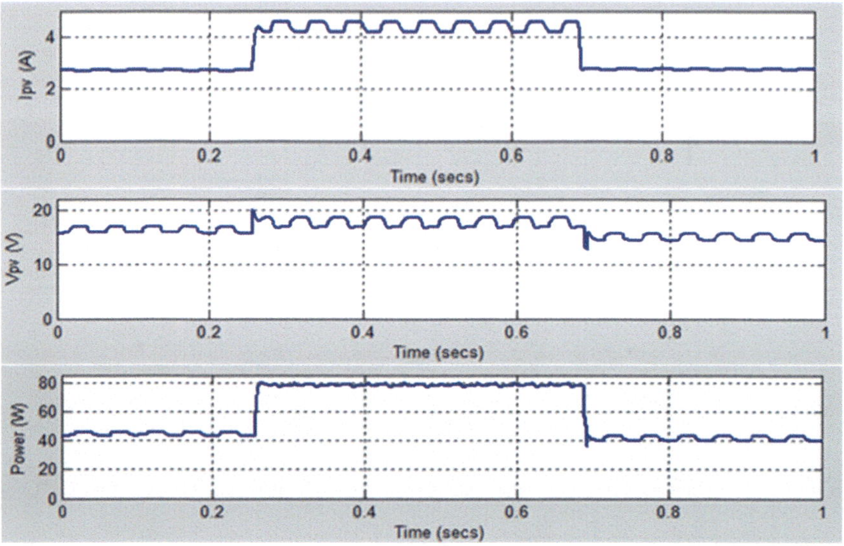

Fig. 5 Hill Climbing-current, voltage and power waveforms

3.2 PSO Algorithm

This is developed by Eberhart and Kennedy in 1995 which is a swarm intelligence optimization algorithm. This has been derived from birds flocking and fish schooling. For best solutions in an n-dimensional space, PSO is used which is a global optimization algorithm. Here several cooperative particles are used and each particle transfer their information's obtained in that current iteration. Each particle obey two simple rules. First one is to follow the particle which is performing best and the second one is to move towards the best conditions found by the same particle. As a result each particle reaches close to the optimal solution [2, 4].

$$|Vs + 1| < \Delta V$$

$$\frac{Psi + 1 - Psi}{Psi} > \Delta P$$

Different steps involved in PSO are as follows:-

Step 1 Initialization of PSO-In this step the particles are initialized randomly. Random values of velocities are also taken

Step 2 Evaluation of fitness-This is done by giving the candidate solution to the objective function

Step 3 Updating local and global values-Here the local and global values are updated by comparing with the newly calculated values

Step 4 Updating velocity and position position-The velocity and position of each particle gets updated

Step 5 Checking of convergence-If the convergence criterion is satisfied, then the process can be terminated

4 Results and Discussions

The implementation of HC has less complexity compared to PSO algorithm but the PSO has improved performance in transient state. Thus steady state can be achieved faster compared to HC. It can be observed from figure that PSO algorithm reaches steady state at around 0.3 s, while HC does not and HC takes longer time to settle. When compared to PSO we conclude from the figure that HC oscillates in steady state but PSO does not oscillate once it reaches steady state.PSO can respond to varying environmental conditions but in case of HC if environmental condition changes i.e. irradiance changes then its operating point also changes which diverges from the required operating point. In HC the duty cycle varies continuously with a constant step size.

5 Conclusion

Modeling of PV system was accomplished successfully. Validation of the same was done using MATLAB environment. Comparison between PSO and HC was carried out. It was found that PSO is superior in many ways than HC. PSO is an accurate, dynamic and efficient method for PV systems. PSO has high tracking efficiency and convergence rate is also high. In all these ways PSO has a upper hand on HC algorithm.

References

1. Peddapati S, Palani S, MPPT of PV systems under partial shaded conditions through a colony of flashing fireflies. IEEE Trans Energy Convers
2. Salam Z (2012) Member, IEEE, Muhammad Amjad, and Saad Mekhilef, Member, IEEE an improved particle swarm optimization (PSO)–Based MPPT for PV with reduced steady-state oscillation. IEEE Trans Power Electron 27(8) Aug 2012
3. Omar AM (2011) Modeling and simulation of photovoltaic (PV) array and maximum power point tracker (MPPT) for grid-connected PV system. In: 2011 3rd International Symposium & Exhibition in Sustainable Energy & Environment. Malaysia, pp 1–3, June 2011

4. Phimmasone V, Endo T, Kondo Y, Miyatake M, Improvement of the Maximum Power Point Tracker for Photovoltaic Generators with Particle Swarm Optimization Technique by Adding Repulsive Force among Agents
5. Miyatake M, Veerachary M, Toriumi F, Fujii N, Ko H (2011) IEEE Trans Aerosp Electron Syst 47(1) Jan 2011

Active Power Loss Minimization in Radial Distributed Micro Grid Incorporating Distribution Generators

S. Angalaeswari and K. Jamuna

Abstract Now-a-days, Micro grid is getting more and more attractive due to its relative advantages. In this paper, the integration of distribution Generation (DG) into the radial distribution network is taken for analysis. Normal load flow analysis cannot be applied to the radial distribution network since the R/X ratio is very high. Hence Forward/Backward sweep load flow analysis is implemented to IEEE 33 bus system. Different distributed energy sources are introduced in different location with same ratings. The results prove that the introduction of DGs considerably reduces the active power loss in the test system and improves the voltage profile.

Keywords Micro grid · Radial distribution network · Forward/backward sweep · DG

1 Introduction

In today scenario, the consumption of energy is increased more. The energy world is slightly moving towards renewable or non-conventional energy sources usage, due to the source availability, pollution free, and eco-friendly nature. The micro grid comprises of the distributed sources which is connected nearer to the load in order to increase the reliability, to reduce the losses, to improve the quality of supply. When the renewable sources are major part of the micro grid, the continuous supply is ensured. The sources like wind energy, Energy from solar panel, micro turbine, Fuel cell energy, diesel generator are the major components of Micro grid.

S. Angalaeswari (✉) · K. Jamuna
School of Electrical and Electronics Engineering, VIT University Chennai Campus, Chennai, Tamilnadu, India
e-mail: angalaeswari.s@vit.ac.in

K. Jamuna
e-mail: jamuna.k@vit.ac.in

K.R. Attele et al. (eds.), *Emerging Trends in Electrical, Communications and Information Technologies*, Lecture Notes in Electrical Engineering 394,
DOI 10.1007/978-981-10-1540-3_37

1.1 Radial Distribution Network

After transmission, the energy has to be distributed among various consumers. It can be divided into Primary and secondary distribution network. Primary is the one which supplies power to big consumers with medium voltage level. Secondary distribution mainly supplies the power to consumers for household appliances with voltage level of 415 V for 3Φ and 230 V for 1Φ. Any distribution network can be considered as

1. Radial distribution network.
2. Ring distribution network.
3. Interconnected distribution network.

Radial distribution network is employed only when the power is generated at low voltage and the substation is located at the center of the load. The system is very simple and the initial cost is too low. But the radial distribution network is considered as ill conditioned system due to the following reasons-high R/X ratio, distributed generation and distributed load. In order to determine the power flow in the radial distribution network, Newton raphson and Fast decoupled load flow methods are inefficient. Hence algorithms like forward/backward sweep method, vector based distribution load flow method, Primitive impedance distribution load flow method, current injection based load flow method and Ladder network theory etc. can be used for load flow analysis.

1.2 Literature Survey

In forward/backward sweep method also, three methods can be employed: Current Summation method, Power summation method and the admittance summation method. The forward/backward sweep load flow for radial distribution network using Power summation method has been tested on IEEE 33 bus system using MATLAB [1]. The current summation method of forward/backward sweep load flow analysis is implemented in IEEE 15 and IEEE 33 bus radial distribution system [2]. The real and reactive line losses are also evaluated. A new algorithm is proposed for large distribution networks by formulating numbering scheme [3]. The results proved that the algorithm is efficient for all types of systems. The three methods of forward/backward sweep are clearly explained [4]. The forward sweep which is started from the last node and move towards the source node is basically a current summation one. It uses the initial voltages for the current calculation. Whereas the backward sweep which is started from the source node and move towards the last or end node is voltage calculation method from the updated current values [4].

The authors mentioned that the backward sweep started from last or end node to the source node and the forward sweep calculates the voltages at each node from source node to the end node [5]. A new and efficient method for the evaluation of receiving end voltages is proposed [6]. The method has good and fast convergence characteristics and the results are compared with the other existing methods. The method calculates the number the branches beyond the considered node and the currents in the branches are calculated accordingly. The method [6] has been implemented on a Meteor 400 VT with a 66 MHz clock.

Srinivasa [7] explains a brief review about the various methods used for load flow analysis in radial distribution systems. An efficient branch and nodal numbering scheme [8] with variable step size factor and next linked branch matrix is employed in practical and complex radial distribution networks. The proposed method is simple; flexible, hence it is suitable for multi type DGs [8]. The authors use the power summation method to find the power in forward sweep and then finding voltages in the backward sweep.The current summation method of forward/backward sweep has been implemented with data structure to increase the speed of the algorithm [9].

In this paper, the current summation method of forward/backward sweep algorithm is chosen for analysis. The base work is started with the simple algorithm and tested in IEEE 33 bus test system. The work is carried out in MATLAB coding for three cases with and without integration of DGs. This paper is organized as follows: Sect. 2 presents the mathematical formulation. Sect. 3 deals with the forward and backward sweep method of load flow. Section 4 presents the results of the IEEE 33 bus system with and without DG and finally conclusion discusses in Sect. 5.

2 Mathematical Formulation

The following assumptions are made in the considered system: Three phase radial distribution networks are balanced; the system can be represented by the equivalent single line diagram; Line charging capacitances are neglected at the distribution voltage levels.

Figure 1 shows the equivalent circuit of transmission line Where, V_i is the voltage at the sending end with an angle of δ_i. The voltage at the receiving end is V_j with the angle of δ_j. Z_{ij} is the series impedance of the transmission line. R_{ij} is the resistance of the line and X_{ij} is the reactance of the line.

Fig. 1 Equivalent circuit of transmission line

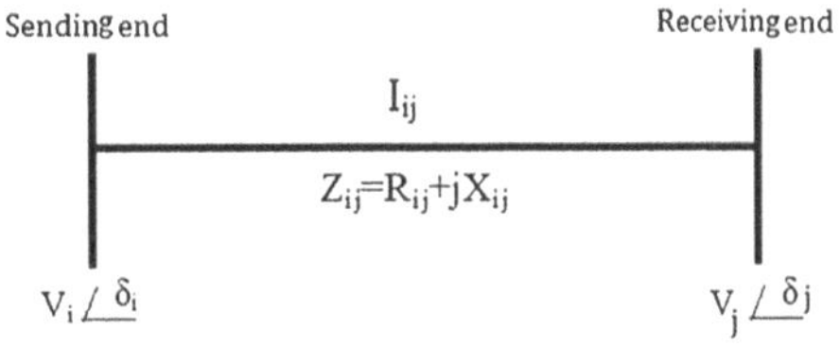

The current flow in the transmission line can be calculated as

$$I_{ij} = \frac{V_i\angle\delta_i - V_j\angle\delta_j}{Z_{ij}}. \tag{1}$$

The real and reactive power losses can be determined from the following equations:

$$P_{Lij} = r_{ij}\frac{P_{ij}^2 + Q_{ij}^2}{V_j^2}. \tag{2}$$

$$Q_{Lij} = x_{ij}\frac{P_{ij}^2 + Q_{ij}^2}{V_j^2}. \tag{3}$$

Pi_j and Qi_j are the real and reactive powers flowing in the line at the receiving end. When load at bus j and transmission power loss in the line between i and j are known, then the load at bus i can be calculated easily. This process continues till the voltage is computed. The phase angle can be calculated by the formula:

$$\delta_j = \delta_i - cos^{-1}\left\{\sqrt{1 - \left(\frac{P_{ij}x_{ij} - Q_{ij}r_{ij}}{V_iV_i}\right)^2}\right\}. \tag{4}$$

3 Algorithm for Forward/Backward Sweep Load Flow Method

1. Assume the flat initial voltages at all nodes for the first iteration.
2. Start with end node and the node current can be computed using the following equation:

$$I_{ij} = \left(\frac{S_i}{V_i}\right)^*. \tag{5}$$

3. By applying KCL, the branch current from node i to node j can be calculated using the following equation:

$$I_{i,i+1} = I_{i+1} + \sum \text{Currents in branches flowing away from node i } + 1. \tag{6}$$

4. Thee voltage at i[th] bus(node) is computed by the following equation:

$$V_i = V_{i+1} + (I_{i,i+1} \times Z_{i,i+1}). \tag{7}$$

5. The load current is updated with the new voltages.
6. Repeat the above steps till the voltage difference between the successive iteration is less than the tolerance level.

4 Simulation Results

For analysis purpose, three cases are considered. The simulation is done in IEEE 33 radial distribution network as shown in Fig. 2 [2]:

Case 1 Without integration of DG in the IEEE 33 system
Case 2 A wind turbine of 1KW is integrated at bus 4
Case 3 A wind turbine of 1KW at bus 4 and solar panel of 1KW at bus 25

Case 1: The total real and reactive power of the system is measured as 3924 KW and 2442KVAR. Table 1 gives the voltage magnitude in p.u and phase angle in degrees. The real and reactive power loss are measured as 209.68 KW and 142.23 KVAR. The minimum voltage occurs at bus no 18 as 0.9038 p.u.

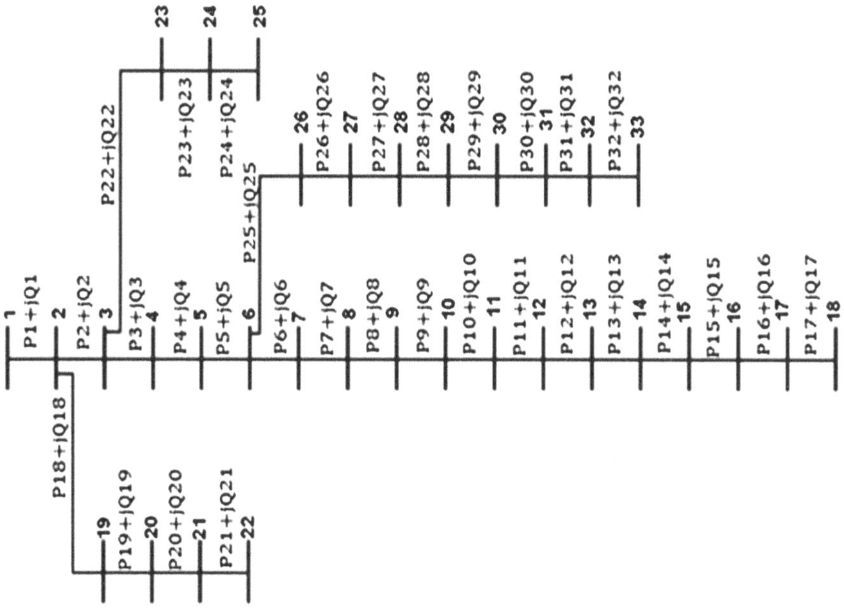

Fig. 2 IEEE 33 radial network

Table 1 Voltage profile of IEEE 33 bus system without DG—Case 1

Bus no	Vol mag	Phase angle	Bus no	Vol mag	Phase angle	Bus no	Vol mag	Phase angle
1	1.0000	0.0000	12	0.9178	−0.0065	23	0.9793	0.0011
2	0.9970	0.0002	13	0.9116	−0.0081	24	0.9726	−0.0004
3	0.9829	0.0017	14	0.9093	−0.0095	25	0.9693	−0.0012
4	0.9754	0.0028	15	0.9079	−0.0101	26	0.9476	0.0030
5	0.9680	0.0040	16	0.9065	−0.0106	27	0.9450	0.0040
6	0.9495	0.0024	17	0.9044	−0.0119	28	0.9336	0.0055
7	0.9460	−0.0017	18	0.9038	−0.0121	29	0.9253	0.0068
8	0.9323	−0.0044	19	0.9965	0.0000	30	0.9218	0.0087
9	0.9260	−0.0057	20	0.9929	−0.0011	31	0.9176	0.0072
10	0.9201	−0.0068	21	0.9922	−0.0015	32	0.9167	0.0068
11	0.9193	−0.0067	22	0.9916	−0.0018	33	0.9164	0.0067

Table 2 Voltage profile of IEEE 33 bus system with wind plant at bus 4—Case 2

Bus no	Vol mag	Phase angle	Bus no	Vol mag	Phase angle	Bus no	Vol mag	Phase angle
1	1.0000	0.0000	12	0.9244	−0.0033	23	0.9831	0.0030
2	0.9976	0.0005	13	0.9182	−0.0049	24	0.9765	0.0015
3	0.9867	0.0036	14	0.9160	−0.0062	25	0.9732	0.0007
4	0.9816	0.0059	15	0.9145	−0.0069	26	0.9539	0.0061
5	0.9742	0.0071	16	0.9132	−0.0073	27	0.9514	0.0071
6	0.9559	0.0054	17	0.9111	−0.0087	28	0.9400	0.0085
7	0.9524	0.0014	18	0.9105	−0.0088	29	0.9319	0.0098
8	0.9388	−0.0012	19	0.9971	0.0003	30	0.9284	0.0117
9	0.9325	−0.0025	20	0.9935	−0.0008	31	0.9242	0.0102
10	0.9267	−0.0036	21	0.9928	−0.0012	32	0.9233	0.0098
11	0.9259	−0.0035	22	0.9922	−0.0015	33	0.9230	0.0097

Case 2: A wind turbine of 1 KW is included at bus no 4. Table 2 shows the voltage magnitude and phase angle after integration of wind turbine. The results prove that the real and reactive power losses are greatly reduced by the inclusion of Distribution energy sources and also the voltage profile is improved from 0.9754 to 0.9816 at bus4. It is not only improves at bus4, all the bus voltage profile has been improved.

Case 3: In this case, a wind turbine of 1 KW and a solar panel of 1 KW is included. The wind turbine is assumed as giving constant power output of 1 KW and is connected at bus 4. A solar power output of 1 KW is connected at bus no.25.

Table 3 Voltage profile of IEEE 33 bus system with wind plant at bus 4 and solar power at bus 25—Case 3

Bus no	Vol mag	Phase angle	Bus no	Vol mag	Phase angle	Bus no	Vol mag	Phase angle
1	1.0000	0.0000	12	0.9284	−0.0013	23	0.9898	0.0068
2	0.9982	0.0008	13	0.9223	−0.0029	24	0.9889	0.0099
3	0.9905	0.0054	14	0.9200	−0.0043	25	0.9912	0.0136
4	0.9854	0.0077	15	0.9186	−0.0049	26	0.9578	0.0080
5	0.9780	0.0089	16	0.9172	−0.0054	27	0.9553	0.0089
6	0.9598	0.0073	17	0.9152	−0.0067	28	0.9440	0.0103
7	0.9563	0.0033	18	0.9146	−0.0069	29	0.9359	0.0117
8	0.9428	0.0007	19	0.9977	0.0006	30	0.9324	0.0135
9	0.9365	−0.0006	20	0.9941	−0.0005	31	0.9283	0.0120
10	0.9307	−0.0017	21	0.9934	−0.0009	32	0.9273	0.0116
11	0.9299	−0.0015	22	0.9928	−0.0012	33	0.9271	0.0115

Table 3 shows the results for the third case. By the inclusion of the additional DG, the voltage profile has been improved further.

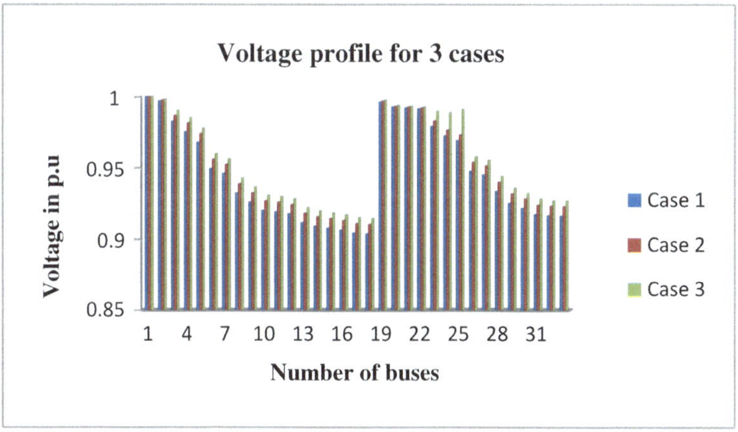

Table 4 shows that the comparision of real, reactive power losses for the three cases. By the introduction of the DG, the line losses are reduced considerably and the voltage profile are improved effictively.

Table 4 Comparison of real and reactive power losses in all cases

Quantity	Case 1	Case 2	Case 3
Real power loss in KW	209.68	176.1	154.37
Reactive power loss in KVAR	142.23	124.64	111.94
Minimum voltage	0.9038	0.9105	0.9146

5 Conclusion

In this paper, the load flow analysis to determine the voltage magnitude and the phase angle is done. For radial distribution system, the forward/backward sweep load flow is used here. To validate the algorithm, IEEE 33 bus system is considered. The results of standard test system are almost closer to the expected results. Three different cases with the addition of wind and solar are taken for analysis. It is proved that the real and reactive power losses are reduced by DG's. But in this paper, the location of DG is taken by trial and error method. In future, any optimization algorithm can be implemented to locate the optimum location of DG to minimize the losses.

References

1. Michline Rupa JA, Ganesh S (2014) Power flow analysis for radial distribution system using backward/forward sweep method. Int J Electr Comput Electron Commun Eng 8(10):540–1544
2. Rana AD, Darji JB, Pandya M (2014) Backward/Forward sweep load flow algorithm for radial distribution system. IJSRD 2(01):398–400
3. Ravi Babu P, Kumar MPVVR, Hemachandra VS, Vanamali MPR (2010) A novel power flow solution methodology for radial distribution systems. In: IEEE Region 8 SIBIRCON, Irkutsk Listvyanka, Russia, pp: 507–512, 11–15 July 2010
4. Balamurugan K, Srinivasan D (2011) Review of power flow studies on distribution network with distributed generation. In: IEEE PEDS. Singapore, pp: 411—417, 5–8 Dec 2011
5. Mishra S, Das D, Paul S (2014) A simple algorithm for distribution system load flow with distributed generation. In: ICRAIE-May 09–11 (2014)
6. Ghosh S, Das D (1999) Method for load-flow solution of radial distribution network. IEEE Proc Gener Transm Distrib 146(6):641–648
7. Srinivas MS (2000) Distribution load flows: a brief review. IEEE Proc 942–945
8. Zhang L, Tang W, Guan H (2009) The Back/Forward sweep-based power flow method for distribution networks with DGs. In: 2nd International Conference on Power Electronics and Intelligent Transportation System. pp. 145–149(2009)
9. Sianipar GHM, Siahaan F (2014) An efficient data structure for radial distribution load flow. In: 2nd IEEE Conference on Power Engineering and Renewable Energy, ICPERE. pp. 126–129

Loop Interaction and It's Influence in Multivariable Process Control

R. Hanuma Naik, D.V. Ashok Kumar and K.S.R. Anjaneyulu

Abstract The principal problem encountered in controlling of multivariable processes is the interaction amid the loops because of non-zero diagonal elements present in the system. The modest interaction between the control loops may lead the system unstable. In order to achieve the desired performance and guarantee robust stability margin, it is important to know which output is driving by which input. Hence the pairing is formed; consequently the controller is designed for resultant '*n*' independent loops. Thus improves the set point tracking and disturbance rejection property of a control system. This paper deals with loop interaction measurement, control configuration selection and thereby controller is designed based on reduced apparent process models. The case study is included to illustrate the efficacy of proposed method.

Keywords Interaction · Control loop pairing · Multivariable processes · Niederlinski's index · Independent loops · Integrity

1 Introduction

Even, an array of control schemes like model predictive, have been proposed and implemented for controlling the multivariable processes over the period, still many of processes using conventional proportional –integral-derivative (PID) controllers. It is because of its simplicity in design and easy to understand by field engineers. Basically two approaches are widely used in multivariable process control. First, the centralized control which is fully cross coupled and needs to tune $3n^2$ parameters, where '*n*' is dimension (assume square process) of processes [1]. It gives an improved performance but some control loops are loosely tuned to maintain the

R.H. Naik (✉) · D.V.A. Kumar
RGM College of Engineering and Technology, Nandyal, A.P, India
e-mail: rhnaik.1717@gmail.com

K.S.R. Anjaneyulu
JNTUA College of Engineering, Anantapur, A.P, India

K.R. Attele et al. (eds.), *Emerging Trends in Electrical, Communications and Information Technologies*, Lecture Notes in Electrical Engineering 394,
DOI 10.1007/978-981-10-1540-3_38

overall stability of closed loop system. Sometimes it may lead the system unstable. However, this approach remains complexity when any sensor/actuator fails in the loop.

Second, the decentralized control which controls the multivariable system by converting it into multiple single loops. This approach is widely using in industries because of lesser parameters to tune, tolerance to sensor/actuator failures, easy identification of failed loops, proficient maintenance and simple design. It needs to tune only *3n* parameters in controller's settings. However, it improves the performance of system significantly when and only the proper loop pairing is done. Improper pairing of variable may lead the control system poor. Hence the primary step in design of decentralized control is control configuration (pair of manipulated variable and controlled variable) selection. This is done by the measurement of interaction existing amid the variables of square open loop stable processes.

This paper focuses on control configuration selection by analyzing the interaction existing among different loops of square processes. Thereby the improper and unstable pairing is avoided by integrity of Niederlinski's theorem. The equivalent transfer function model of corresponding effective open loop transfer function (apparent process model) is approximated as first order plus dead time (FOPDT) model [2]. Therefore any existing single loop tuning method can be easily exploited and controller settings are determined. Thus remains the overall stability of system as in case of single input and single output process.

2 Loop Interaction Measurement

Consider open loop stable multivariable processes shown in Fig. 1 where '*r*' is input vector, '*u*' control vector and '*y*' is output vector. The $G_c(s) = diag(g_{c1}(s), g_{c2}(s) \ldots g_{cn}(s))$ is diagonal controller and $G_p(s)$ is process transfer function matrix.

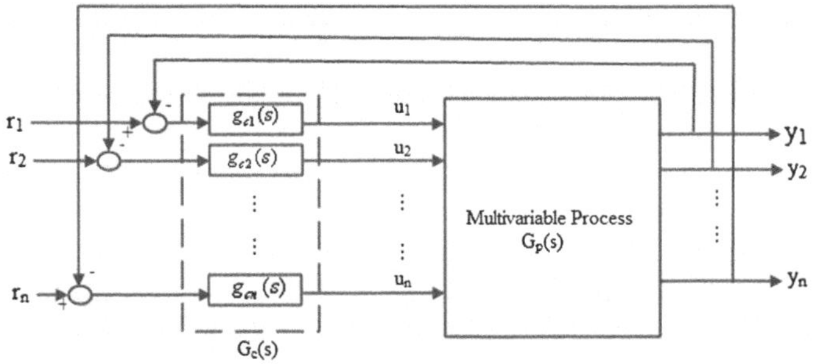

Fig. 1 Decentralized control system for multivariable process

Assume that, fully centralized process transfer function matrix is,

$$G_p(s) = \begin{bmatrix} g_{p11}(s) & g_{p12}(s) & \dots & g_{p1n}(s) \\ g_{p21}(s) & g_{p22}(s) & \dots & g_{p2n}(s) \\ \dots & \dots & \dots & \dots \\ g_{pn1}(s) & g_{pn2}(s) & \dots & g_{pnn}(s) \end{bmatrix} \tag{1}$$

and each element of processes is of the form,

$$\mathrm{g_{p,\,ij}(s)} = \frac{k}{\tau \mathrm{s} + 1} \mathrm{e}^{-\theta \mathrm{s}} \tag{2}$$

An interaction amid the control loops because of non-zero diagonal elements of MIMO processes is determined as follows:

RGA: The relative gain array (RGA) of n × n processes can be determined using steady state gain of the process as,

$$\Lambda_{ij} = G_p(0) \otimes G_p^{-T}(0) = \begin{bmatrix} \lambda_{11} & \lambda_{12} & \cdots & \lambda_{1n} \\ \lambda_{21} & \lambda_{22} & \cdots & \lambda_{2n} \\ \cdots & \cdots & \cdots & \cdots \\ \lambda_{n1} & \lambda_{n2} & \cdots & \lambda_{nn} \end{bmatrix} \tag{3}$$

RNGA: The normalized gain *($K_{N,\,ij}$)* for a particular transfer function element is defined as,

$$K_{N,ij} = \frac{k_{ij}}{\sigma_{ij}} = \frac{k_{ij}}{\tau_{ij} + \theta_{ij}} \tag{4}$$

where $\sigma_{ij} = \tau_{ij} + \theta_{ij}$ is the average residence time (ART) and signifies the speed of the manipulated variable to controlled variable. The relative normalized gain array is expressed using normalized gain matrix, and hence, the RNGA (ϕ) can be obtained as,

$$\phi = K_N \otimes K_N^{-T} \tag{5}$$

and, $\phi = \begin{bmatrix} \phi_{11} & \phi_{12} & \cdots & \phi_{1n} \\ \phi_{21} & \phi_{22} & \cdots & \phi_{2n} \\ \cdots & \cdots & \cdots & \cdots \\ \phi_{n1} & \phi_{n2} & \cdots & \phi_{nn} \end{bmatrix}$

The loop pairing is done from RNGA elements whose value is closest to unity and positive [3]. The pairing which directs unstable is keep away by Niederlinski's Index.

Niederlinski's Index (NI): The Niederlinski's index (NI) for the complex control configuration is denoted by $N(G)$ and defined as,

$$N(G) = \frac{|G_p(0)|}{\pi g_{ij}} \qquad i,j = 1,2,3,\ldots..n \tag{6}$$

Where $|G_p(0)|$ denotes the determinant matrix of $G_p(0)$ and πg_{ij} denotes product of diagonal elements of $G_p(0)$ for a centralized control system [4]. For a stability of intricate nonlinear system, NI must be larger than zero. Consequently, better pairing can be selected by the use of RGA-RNGA-NI rules [5].

The Equivalent transfer function (ETF) of each selected diagonal pairing is approximated as first order plus dead time model using RGA and RARTA [2]. The maximum weighted parameters are considered for controller parameter design. Consider each diagonal controller is in the form of

$$g_{c,ij}(s) = k_{c,ij} + \frac{1}{\tau_{i,ij}s} \tag{7}$$

For better trade-off between performance and robustness, SIMC method is used for determination of controller settings based on each corresponding process ETF's as [6],

$$g_{cii}(s) = \frac{1}{k_{ii}}\frac{\tau_{ii}}{2\theta_{ii}} \quad and \quad \tau_I = min(\tau_{ii}, 8\theta_{ii}) \tag{8}$$

3 Case Study

Consider the process ISPR (Industrial scale polymerization reactor) proposed by Chien et al. is [7],

$$G_p(s) = \begin{bmatrix} \frac{22.89}{4.572s+1}e^{-0.2s} & \frac{-11.64}{1.807s+1}e^{-0.4s} \\ \frac{4.689}{2.174s+1}e^{-0.2s} & \frac{5.80}{1.801s+1}e^{-0.4s} \end{bmatrix}$$

The RGA (Λ), RNGA (ϕ), NI values for main diagonal and off diagonal pairing of is given in Table 1.

According to the RGA-RNGA and NI rules, the recommended pairing is 1-1/2-2, and corresponding approximated Equivalent Transfer Function based on their maximum weights obtained is,

Table 1 The RGA, RNGA, NI values of diagonal pairings

S. No	Control configuration	RGA (Λ)	RNGA (ϕ)	NI
1	1-1/2-2	0.7087	0.5482	1.4111
2	1-2/2-1	0.2913	0.4518	−3.4323

$$\hat{g}_{11}(s) = \frac{32.3003}{4.572s+1}e^{-0.2s},\quad \hat{g}_{22}(s) = \frac{8.1844}{1.801s+1}e^{-0.4s}$$

By simplification, the apparent processes of diagonal elements (1-1/2-2) are,

$$\begin{aligned} g_1(s) &= \frac{22.89}{4.572s+1}e^{\;0.2s} + \frac{(16.947s+9.4103)}{(3.9284s^2+3.981s+1)}e^{\;0.2s} \\ g_2(s) &= \frac{5.8}{1.801s+1}e^{-0.4s} + \frac{(10.89s+2.384)}{(3.9284s^2+3.981s+1)}e^{-0.4s} \end{aligned}$$

To further verify the approximated model, Nyquist plot of $\hat{g}_{11}(s), g_1(s)$ and $\hat{g}_{22}(s), g_2(s)$ are shown in Figs. 2 and 3.The Nyquist plot indicates the dynamical

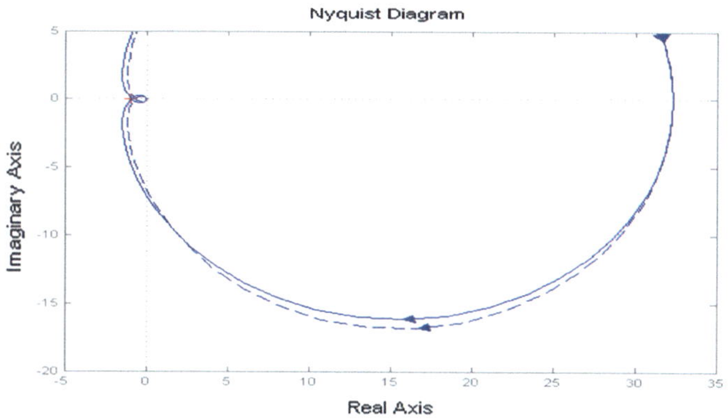

Fig. 2 The Nyquist plot of $\hat{g}_{11}(s), g_1(s)$ (*solid line*: original process, *dash line*: approximated model)

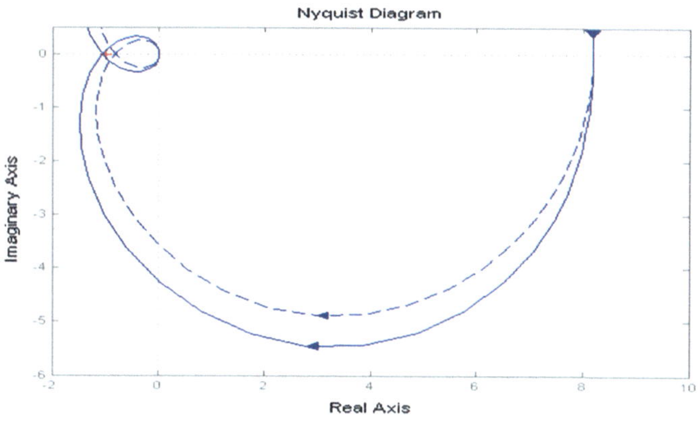

Fig. 3 The Nyquist plot of $\hat{g}_{22}(s), g_2(s)$ (*solid line*: original process, *dash line*: approximated model)

characteristics of approximated model are similar over a frequency range to an original process.

The decentralized controller parameters for main diagonal and off-diagonal are given in Table 2, and their performance is shown in Fig. 4.

The pairing recommended by RGA-RNGA-NI is 1-1/2-2 and same is preferred for designing of decentralized control. To show the efficacy of proposed pairing the decentralized controller is also designed for 1-2/2-1(off diagonal) pairing and resultant performance is shown in Fig. 4 and Table 3. The integral absolute error (IAE) of off-diagonal is very large compared to the main diagonal pairing.

Table 2 SIMC PI controller settings for both diagonal pairing

Control loop	Diagonal pairing		Off-diagonal pairing	
	k_c	τ_I	k_c	τ_I
1	0.3538	1.6	0.0565	2.8022
2	0.2750	1.801	0.3376	2.4816

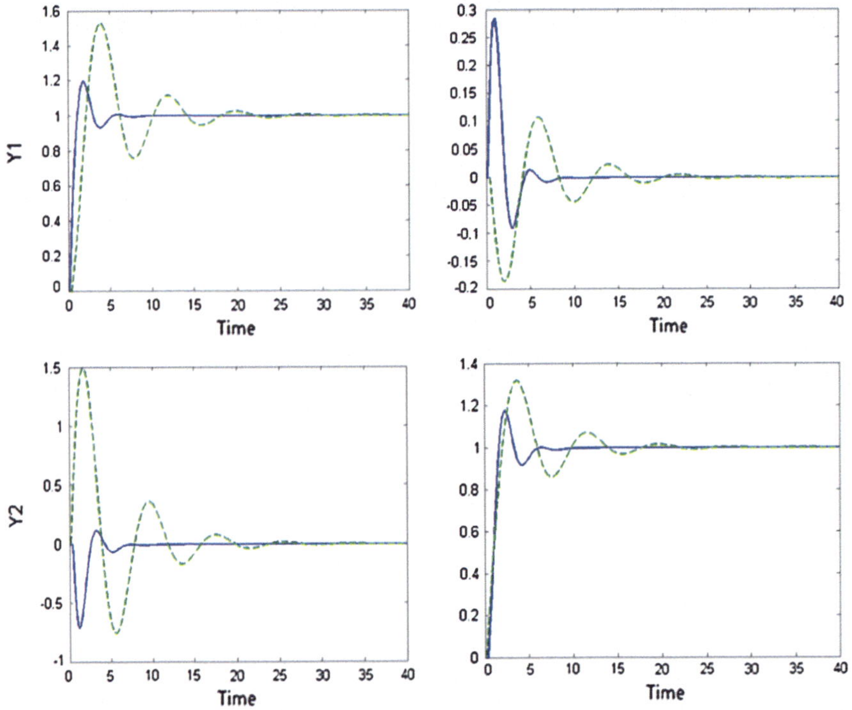

Fig. 4 performance of decentralized control for both diagonal pairing (*solid line*: main diagonal, *dash line*: off-diagonal)

Table 3 Performance indices of both diagonal pairings (IAE_t is sum of IAE)

Performance indices		Main diagonal pairing		Off-diagonal pairing	
		Y_1	Y_2	Y_1	Y_2
IAE	Change in r_1	0.9468	0.4835	3.841	0.9297
	Change in r_2	1.1899	1.37	7.03	2.481
ISE	Change in r_1	0.4874	0.0809	1.776	0.0920
	Change in r_2	0.5209	1.37	5.481	0.9675
IAE_t		3.9902		14.2817	
ISE_t		2.4592		8.3165	

Consequently, off diagonal pairing provides poor response in control system because of modest interaction in between the variables.

The overall performance of 1-1/2-2 combination is significantly better than the combination 1-2/2-1.

4 Conclusions

This paper aims in study the interaction and assortment of pairing in favor of design a controller for multivariable processes. Once interaction is measured and pairing is formed then individual loop controller parameters can be easily design using any single loop technique. In this paper, it is used the RNGA, RGA and NI for selection pair of loops. Consequently, diagonal processes are approximated as FOPDT using RGA and RARTA. Finally, controller parameters for corresponding ETFs are designed using SIMC tuning technique. The case study was included for analyzing the performance of main and off diagonal pairing.

References

1. Tavakoli S, Griffin I, Fleming PJ (2006) Tuning of decentralized PI (PID) controllers for TITO processes. Control Eng Pract 14(9):1069–1080
2. Rajapandiyan C, Chidambaram M (2012) Controller design for MIMO processes based on simple decoupled equivalent transfer functions and simplified decoupler. Ind Eng Chem Res 51 (38):12398–12410
3. He MJ, Cai WJ, Ni W, Xie LH (2009) RNGA Based Control System Configuration for Multivariable Processes. J Process Control 19(6):1036–1042
4. Naik RH, Kumar DV, Anjaneyulu KSR (2012) Controller for multivariable processes based on interaction approach. Int J Appl Eng Res 7.11(2012):1203–1213

5. Naik RH, Kumar DVA, Anjaneyulu KSR (2014) Control configuration selection and controller design for multivariable processes using normalized gain. World Acad Sci Eng Technol Int J Electr Comput Electron Commun Eng 8(10)
6. Skogestad S (2003) Simple analytical rules for model reduction and PID controller tuning. J Process Control 13(4):291–309
7. Chien IL, Huang HP, Yang JC (1999) A simple multi-loop tuning method for PID controllers with no proportional Kick. Ind Eng Chem Res 38(4):1456–1468

Four Level Boost Converter for Linear Loads

H.M. Tania, Jagadish Kumar Patra, Vinson John, D. Elangovan and G. Arunkumar

Abstract In renewable energy systems, the output voltage obtained will be of low level. In order to bring the voltage to nominal level, we need a step up DC-DC converter. Conventional boost converters has disadvantages like switch voltage stress, diode reverse recovery losses and high duty cycle losses. This paper presents a Multilevel DC-DC Boost Converter. To obtain 4 level output voltage, required number of switches, inductors, diodes and capacitors are one, one, seven and seven respectively. This converter has the ability to maintain equal voltage magnitudes in all the 4 output levels and also has a command over the input current. In this paper, a four level boost converter was simulated and implemented in hardware for an input and output specifications of 20 volts and 220 volts respectively.

Keywords DC-DC converter · Multi-level converter (MLC) and boost converter

1 Introduction

The need for Renewable Energy resources is increasing day by day due to the depletion of fossil fuels. Along with the researches going in the area of renewable energy sources, a lot of importance is being given to distributed generation (DG) which can resolve power related issues. Renewable energy sources like solar, wind can be used for generation of power as they are pollution free and inexhaustible [1]. Fuel cell is gaining popularity in many fields as a renewable energy source. Fuel cell provides high efficiency, reliability, low emission and also co-generation [2]. Single. Fuel cell gives a voltage output of around 1.2 v. By connecting several fuel cells in series—parallel combination, we form a stack which gives a considerable level of voltage [3]. The need of high output voltage cannot be

H.M. Tania · J.K. Patra · V. John (✉) · D. Elangovan · G. Arunkumar
School of Electrical Engineering, V.I.T University, Vellore, Tamil Nadu, India
e-mail: winsonbethesda@gmail.com

J.K. Patra
e-mail: jags606@gmail.com

K.R. Attele et al. (eds.), *Emerging Trends in Electrical, Communications and Information Technologies*, Lecture Notes in Electrical Engineering 394,
DOI 10.1007/978-981-10-1540-3_39

achieved by series connection of stacks keeping cost in consideration. So, to obtain high voltage level fuel cell stack has to be interfaced with power converters to make the system efficient in terms of various aspects which includes cost and size. Non-Isolated Converters are used to convert DC voltage from one level to another. The voltage level can either be stepped up or stepped down. The conventional topologies in use are Buck Converter, Boost Converter, Buck—Boost Converter [4]. Boost Converters are used for various voltage step up applications. But they have few inherent drawbacks [5]. Cascading boost converters to obtain a high voltage gain proves to be disadvantageous due to the losses associated with it. Isolated DC-DC converters can also be used to get a high voltage but the transformers and inductors used in this converter makes it costly and bulky. Power density will be low in these cases. CUK and SEPIC topologies can only give a normal conversion ratio even at higher duty cycle. MLC topologies can overcome these drawbacks thereby achieving a high step up ratio with low duty cycle and better efficiency. The output obtained from MLC can be raised by increasing the stages or levels of multiplier unit which is independent of input voltage given to the converter. Hence, modularity is one of its advantage which makes it unique over other available topologies. In MLC's, the stress on switch is lesser when compared to conventional converter topologies. MLC's have characteristics like lower voltage across switches, lower switching losses and also low voltages across the capacitors and diodes which makes them superior to other topologies. In addition to these, the converter can draw a continuous input current. Cascaded multi cell, Diode clamped (Neutral point) and capacitor clamped (flying capacitor) are types of MLCs. MLC's can be used to provide a dc link which serves as the input to a Voltage source inverter. It can also be connected to a multilevel inverter directly. The proposed converter topology can be a boon in fields where multiple voltage levels are desired.

2 DC-DC Boost Converter

It is a type of switching converter in which the switch is periodically opened and closed such that the output voltage is greater than the input voltage. However the boosting ratio is limited by the Inductor's internal parasitic resistance R_L.

Figure 1 shows the general circuit diagram of a conventional Boost Converter. The duty cycle, D is defined by the equation

Fig. 1 Conventional boost converter

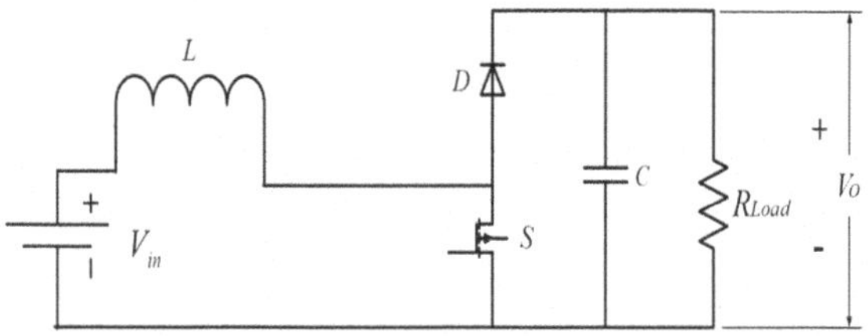

$$D = \frac{T_{on}}{T} \tag{1}$$

where, T_{on}—On Time of the switch and T—Total Time Period

In steady state analysis, by keeping the current in the inductor and voltage in the capacitors as constant, the equation of output voltage, V_O, is given by

$$V_o = \frac{V_{in}}{(1-D)} \tag{2}$$

where, V_{in}—Input Voltage, D—Duty Cycle

3 Proposed DC-DC MLC

In this paper, a four level DC-DC Boost Converter is implemented. This converter is implemented by using one switch, seven capacitors and seven diodes. One of the main advantage of this topology is that the output level can be increased by adding further capacitors and diodes. Also, as only one switch is used, the switching losses can be effectively reduced as compared to other MLC's that uses multi switches. The lower portion of Fig. 2 is the conventional DC-DC boost converter. In this Multilevel Boost Converter, the output is N times V_C. Here, the output voltage, V_o is given by the equation,

$$V_o = N \times V_c = \frac{N \times V_{in}}{(1-D)} \tag{3}$$

Where, V_c—Voltage across each Capacitor, N—Number of Converter's Level, D—Duty cycle of the switch.

When the switch (S) is Close (turn ON state), if the voltage across the capacitance C_6 (Fig. 2) is smaller than that of C_7, then C_7 clamps the voltage of C_6 through the diode D_6 and the Switch. Also at the same time, if the voltage across C_4 and C_6 is lesser than that across the capacitances C_5 and C_7, then C_5 and C_7 clamps the voltage across C_4 and C_6 through the diode D_4 and the Switch. Similarly, C_3, C_5 and C_7 clamps the voltage across C_2, C_4 and C_6. When the switch is open (turn OFF state), the inductor current closes the diode D_7 and thereby charging the capacitor C_7. When D_7 closes, both the input and the inductor voltage and C_6 will clamp the voltage across C_5 and C_7 through the diode D_5. Also, the voltage across the inductor and V_{in}, C_4 and C_6 will clamp the voltage across C_3, C_5 and C_7 through the diode D_3. Finally, V_{in} and inductor voltage along with C_2, C_4 and C_6 clamp the voltage across C_1, C_3, C_5 and C_7. Also it can be observed that the diodes D_1, D_3, D_5 and D_7 switches in a synchronous manner, complemented by the diodes D_2, D_4, D_6 and the switch.

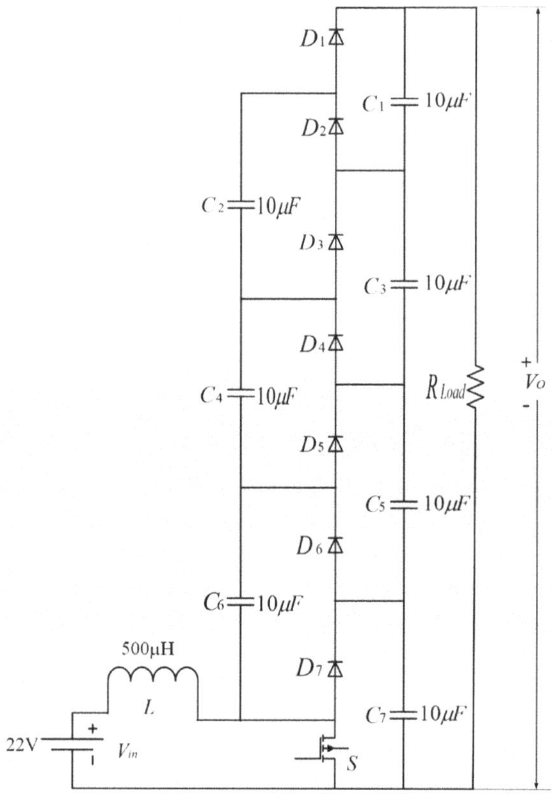

Fig. 2 Proposed DC-DC MLC

4 Simulation Result

The proposed topology has been simulated in MATLAB by using the parameters given in Table 1. The circuit was simulated for open loop operation and the output waveforms of various voltages and currents were observed in the scope.

The output voltage waveform is shown in Fig. 3. The initial voltage shoots up to 300 V but slowly the voltage settles down to the steady state value of 218 V. Similarly, the output current shoots up to 16 A and settles down at a steady state value of 3.5 A. The output waveform is shown in Fig. 4. The switch current was also observed. The gate pulse, for 56.37 % Duty, given to the switch and the

Table 1 Parameter values

V_{in}	20 V
V_O	220 V
Switching frequency, f_{sw}	100 kHz
Inductor, L with R_L	500 μH with $R_L = 0.35\ \Omega$
Capacitor, C	10 μF
D	63.6 %

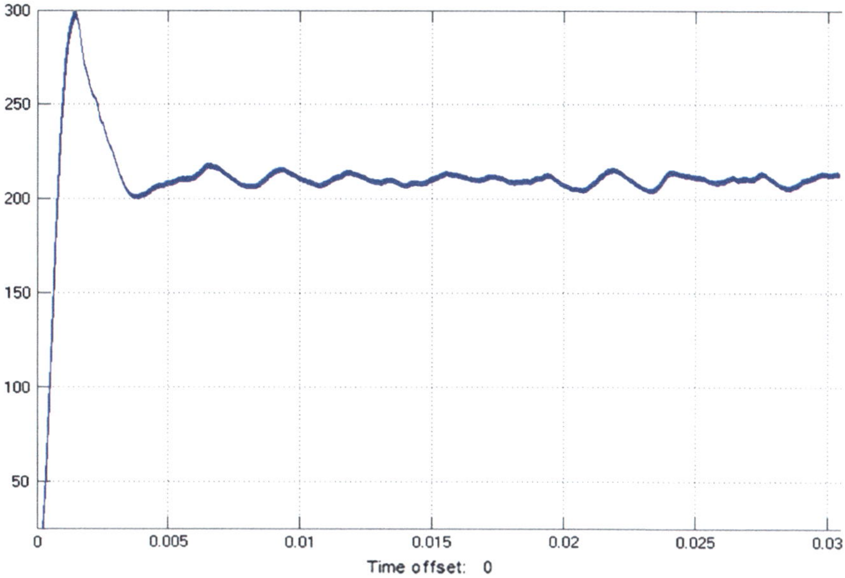

Fig. 3 Output voltage waveform

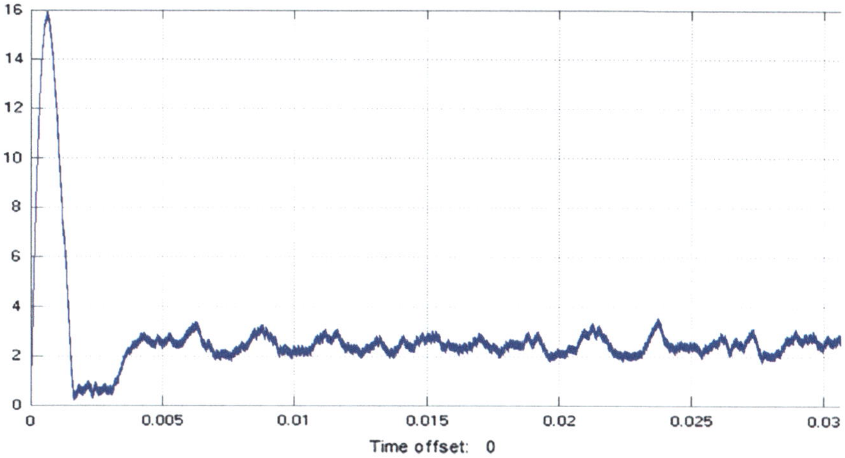

Fig. 4 Output current waveform

voltage across the switch is shown in Fig. 5. After carefully analysing each waveform, it is clear that the simulated circuit can be implemented as hardware. As the initial shoot up in the output voltage is within the range of the voltage rating of 10μF power capacitors available in the market, the cost for implementing in hardware is relatively cheaper.

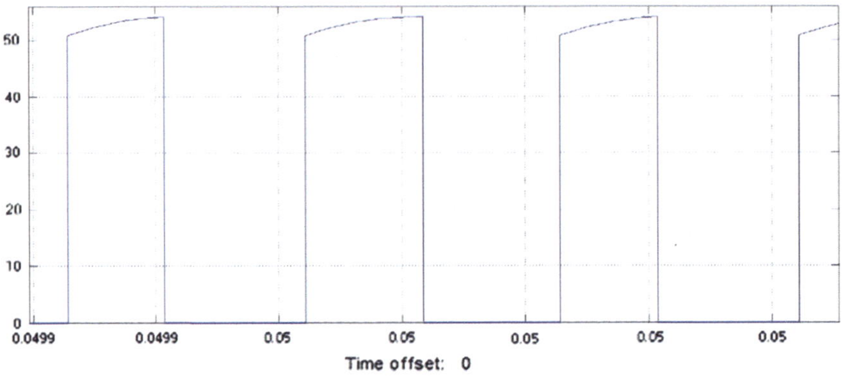

Fig. 5 Voltage across switch

5 Hardware Implementation

The Simulated circuit has been implemented in hardware with the same parameter values. Initially, gate pulse was generated using function generator with a voltage level of 5 V. Switching pulses were generated in a pattern to get a duty of 63.6 %. Output of the function generator was given to HCPL 3120 based driver circuit. This driver circuit provides output gate pulse with voltage level of 15 V which is the minimum requirement to drive the switch. When the switch gets turned on by applying the gate pulse, the switch current rises and when it is turned off, the swich current falls. The waveform observed in the DSO can be observed in Fig. 6. Input voltage is given to the circuit by using programmable regulated supply, to test under steady state conditions. Input voltage of 20 V was given and maintained constant for a resistive load of 60 watts. The output voltage obtained during steady state was

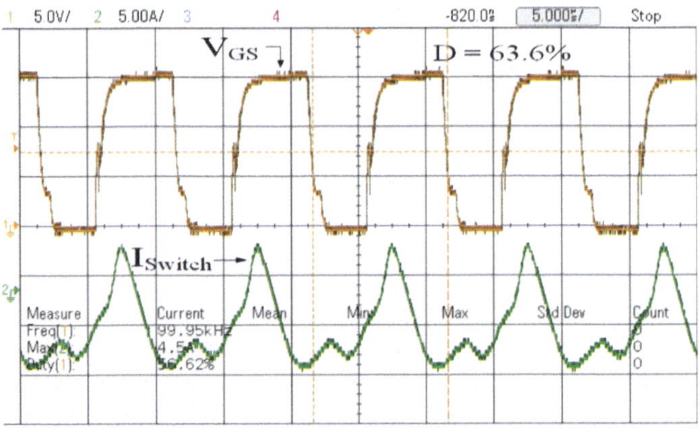

Fig. 6 Gate pulse and switch current

216.55 V. The desired output of 220 V is not obtained because of the internal resistance of the components used. The input and output voltage and current waveform obtained is shown in Fig. 7. A picture of the implemented hardware circuitry is shown in Fig. 8. MOSFET is used as the switch because of its low conduction losses.

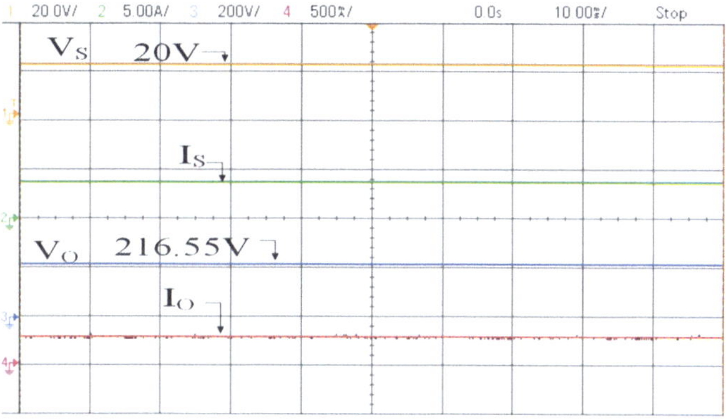

Fig. 7 Input and output voltage and current waveform

Fig. 8 **a** Gate drive input power circuit. **b** MOSFET. **c** Capacitors. **d** Diodes

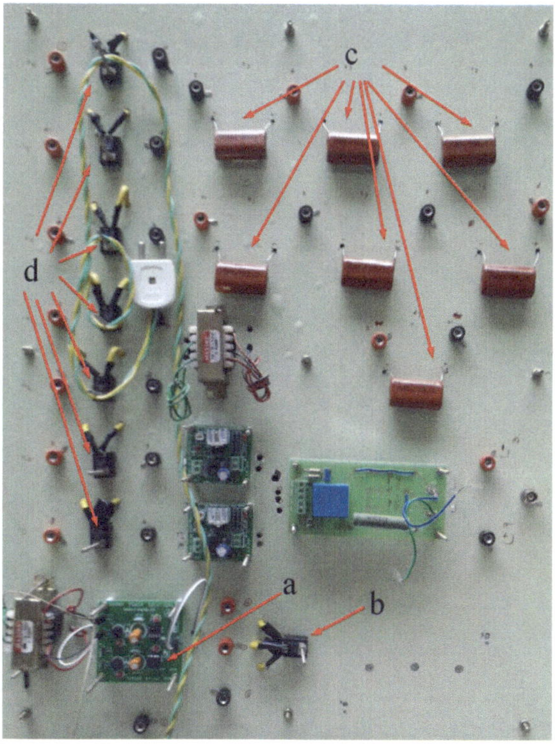

6 Conclusion

The proposed multilevel DC-DC Converter was simulated and implemented as hardware. The output was tested with linear loads having power levels ranging from 60 watts to 180 watts. With an input voltage of 20 V and switching frequency of 100 kHz, an output voltage of 216.55 V was obtained for a 60 W resistive load. Since, it is an open loop system, the output obtained was not exact as the desired output. The same system can be implemented as a closed loop system to get the desired voltage level exactly. Thus it proves that the proposed MLC is better than normal topologies like conventional Boost, SEPIC, CUK converters and it also suits better for Renewable Energy applications like Solar PV and Fuel Cells.

References

1. LaBella T, Lai J-S (2014) A Hybrid resonant converter utilizing a bidirectional GaN AC switch for High–Efficiency PV applications. IEEE Trans Ind Appl 50(5), Sep/Oct 2014
2. Kim H, Yoon C, Choi S (2010) An improved current-fed ZVS isolated boost converter for fuel cell applications. IEEE Trans Power Electron 25(9), Sep 2010
3. Thounthong P (2012) Control of a three—level boost converter based on a differential flatness approach for fuel cell vehicle applications. IEEE Trans Veh Technol 61(3), Mar 2012
4. Du X, Zhou L, Tai H-M (2009) Double-frequency buck converter. IEEE Trans Ind Electron 56 (5), May 2009
5. Hwu KI, Peng TJ A novel buck-boost converter combining KY and buck converters. IEEE Trans Power Electron 27(5), May 2012

Field Failure Rate Reduction Through ESS with MATLAB Based GUI

K. Susanna, Swarna Bai Arniker, K. Sita Rama Rao and M. Anka Rao

Abstract The increased vulnerability of the components to various electrical, thermal, mechanical, chemical and electromagnetic stresses poses a big threat in attaining the reliability required for various mission critical applications. Stress screening is a process in which imposed vibration, thermal and electrical stresses are applied for sufficient periods of time to precipitate defects and early life failure characteristics. In this paper, different stress screens are selected and their relative screen strengths are evaluated. The resulting average failure rate in field is calculated and presented for different components of PCB of different quality grades. A comparison of field failure rate without stress screening and with random vibration and temperature cycling stress screenings are carried out by using MATLAB based Graphical User Interface (GUI) tool.

1 Introduction

Ensuring reliable operation over an extended period of time is one of the biggest challenge facing present day electronic systems. Decrease the failure rate, ESS method is proposed [1]. It refers to the process of exposing a newly manufactured

K. Susanna (✉)
Reliability Engineering, JNTUACEA, Anantapur, India
e-mail: susannakeela@gmail.com

S.B. Arniker (✉)
SSQD, RCI, Hyderabad, India
e-mail: arniswarna@rcilab.in

K. Sita Rama Rao (✉)
R&QA, RCI, Hyderabad, India
e-mail: sitarama.rao@rcilab.in

M. Anka Rao (✉)
EEE, JNTUACEA, Anantapur, India
e-mail: ankaraomogili@gmail.com

K.R. Attele et al. (eds.), *Emerging Trends in Electrical, Communications and Information Technologies*, Lecture Notes in Electrical Engineering 394,
DOI 10.1007/978-981-10-1540-3_40

or repaired product or component (typically electronic) to stresses such as temperature cycling and random vibration in order to force latent defects to manifest themselves by permanent or catastrophic failure during the screening process [2].

2 Proposing ESS Method on PCB

2.1 Printed Circuit Board

The missile system consists of different assemblies. As the power supply assembly is the major part in any hardware, there is a need to observe failures. PCB is the main module in power supply assembly. Each assembly is build with huge number of PCBs and each one consists of different MIL grade components with different quality levels. It plays a major role in the missile system in which different environments will be considered i.e., missile launch, missile flight etc. The components present in the board are microcircuits, diodes, transistors, relays, switches, resistors, capacitors, inductors and connectors. As it plays important role in missile system there is need to check and reduce the defects present in it.

2.2 ESS

ESS is a process or series of processes in which environmental stimuli, such as rapid thermal cycling and random vibration, are applied to electronic systems in order to precipitate latent defects to early failure. An equally important and inseparable aspect of the screening process is the testing which is done as part of the screen, so as to detect and properly identify the defects which have been precipitated to failure. The precipitation and testing process is basically a search for defects. The latent defects are precipitated by applying random vibration and temperature cycling stress screens on the components of PCB. Latent defects and failure rate are reduced by applying ESS method. Manufacturing techniques for modern electronic hardware consists of hundreds of individual operations and processes through which defects can be introduced into the product.

3 Failure Rate Analyses on PCB

Failure rate of PCB is calculated without applying any stress and applying stress by using ESS method.

3.1 Collecting Data of PCB Module

Power supply assembly of a missile system with different quality levels is considered.

The different quality levels of PCB module which is present in power supply assembly is considered as case study. The data of PCB module which is of two types i.e., time-to-failure data and field data for different components with different quality levels is taken from Ref. [1].

3.2 Failure Analysis on PCB

Every product or process has modes of failure. Several systematic methodologies have been developed to quantify the effects and impacts of failures. In this view, the failure modes compete as to which causes the failure for each particular item. Failure analysis segregates the analyses of failure modes and then combines the results to provide an overall model for the product. The main objective of failure analysis is to calculate the initial number of defects, defects removed, defects remaining and failure rate from the data of PCB.

3.3 Failure Rate Without Stress Screening

Failure rate can be calculated without applying any stress by normal inspection method with the initial number of defects. Data for initial number of defects for different components of PCB is taken from Ref. [1]. By using defect density and the number of components present in the PCB initial number of defects and failure rates are calculated.

$$\mathrm{FRfield} = \frac{Din\,field}{T} * (1 - e^{-kT}). \tag{1}$$

where Din field = initial number of field defects

T = operating time

k = constant = 1/500.

3.4 Failure Rate by Applying Stress Screening

Failure rate of PCB is calculated by applying two different types of stresses. They are Random vibration stress screening and Temperature cycling stress screening.

3.4.1 Random Vibration Stress Screening

Random Vibration is one of the more common types of vibration testing services performed by vibration test labs. Random vibration is characterized as stationary process for estimating the failure rate. To apply random vibrations firstly, screening strength is calculated which is the product of precipitation efficiency and detection efficiency. The data for calculating precipitation efficiency and detection efficiency is taken from MIL-HDBK-344A. By applying random vibrations on the components of PCB failure rate is calculated.

Dremoved = Din × SS.

Dremaining = Din − Dremoved.

$$\text{FRfield} = \frac{Dremaining}{T} * (1 - e^{-kT}). \tag{2}$$

where Dremoved = defects removed

Dremaining = defects remaining.

3.4.2 Temperature Cycling Stress Screening

Temperature cycling is the process of cycling through two temperature extremes, typically at relatively high rates of change. It is an environmental stress test used in evaluating product reliability as well as in manufacturing to catch early-term, latent defects by inducing failure through thermal fatigue. Different number of cycles with specific temperature at certain rate of temperature screening strength is calculated. Failure rate is calculated by applying different temperature cycling's on PCB.

4 Test Results Using GUI

Initially the data is collected for calculating number of defects, screening strength and failure rate. The program is developed in GUI by using collected data in order to calculate the field failure rate by applying ESS method. The test results in GUI by failure rate with and without stress screening is shown in Figs. 1, 2 and 3 respectively.

4.1 Results of Failure Rate Without Stress Screening

The failure rate for total assembly is obtained without stress screening from GUI is shown Fig. 1.

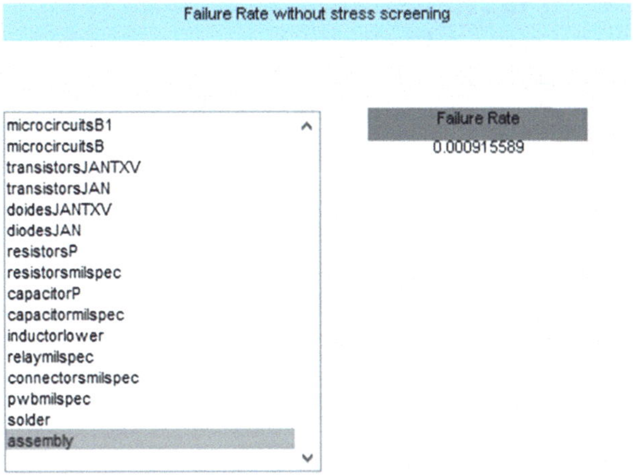

Fig. 1 Failure rate without stress screening

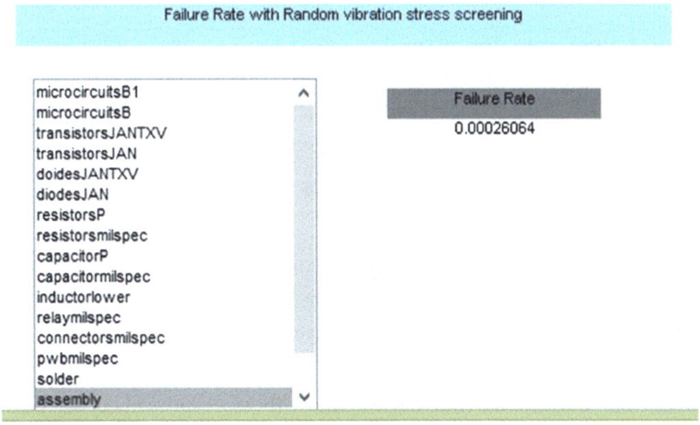

Fig. 2 Failure rate by random vibration stress screening

4.2 Results of Failure Rate by Applying Random Vibration Stress Screening

The results of failure rate for assembly are obtained by applying random vibration stress screening is shown in Fig. 2.

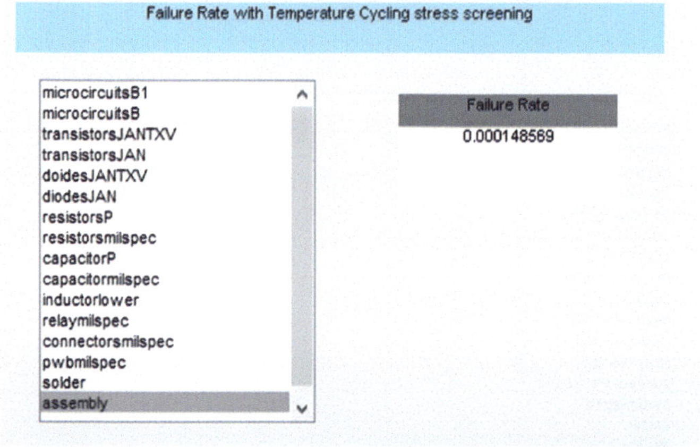

Fig. 3 Failure rate by temperature cycling stress screening

4.3 Results of Failure Rate by Applying Temperature Cycling Stress Screening

The results of failure rate for assembly are obtained by applying Temperature Cycling stress screening is shown in Fig. 3.

4.4 Analogy of Failure Rate Values

The comparison of failure rates of different components of PCB with and without stress screening is observed. The Failure rate results without screening and with screening are obtained from GUI and are shown in Table 1.

Table 1 Failure rate values for all components and PCB Assembly without screening and with Random vibration and Temperature cycling screening

Components	Without stress screening	Random vibration stress screening	Temperature cycling stress screening
Microcircuits B	9.1517e-007	6.3508e-006	3.620e-006
Transistors JAN	5.8746e-005	4.1364e-005	2.3578e-004
Diodes JAN	1.7833e-005	1.2429e-005	1.0851e-005
Resistors Milspec	5.0375e-006	3.511e-006	2.0014e-006
Capacitor Milspec	0.0001097	0.0000764	0.0000436
Inductor Lower	0.000428	0.0002983	0.00017005
Relay Milspec	8.2523e-005	1.9331e-005	1.019e-005
Connector Milspec	1.2985e-005	9.09510e-006	5.91592e-006
PCB Assembly	0.000915589	0.00026064	0.000148569

5 Conclusion

In this paper, failure rate analysis of PCB is carried out. From the field data initial number of defects of different components for all quality levels is calculated. To decrease the defects and failure rate ESS method is applied on PCB.

Random vibration and temperature cycling stress screenings are applied on the components of PCB to reduce the field failure rate. Screening strengths are evaluated and latent defects are reduced. This resulting average failure rate is calculated and presented for different components of different quality grades. A comparison of field failure rate without stress screening and with stress screening is carried out and results are analyzed and discussed. After calculating the failure rates, temperature cycling stress screening is observed as the best one because the failure rate got reduced as compared with the random vibration stress screening.

This complete case study of PCB is implemented in a MATLAB based GUI tool. The Failure rate is reduced after applying the ESS method as a result the Reliability is increased.

References

1. MIL-HDBK 344A, Environmental stress screening of electronic equipment
2. MIL-HDBK 217F, Reliability prediction of electronic equipment

A Novel Space Vector Approach Using Shoot Through State for Three Level Z Source Inverter

B.M. Manjunatha, D.V. Ashok Kumar and M. Vijaya Kumar

Abstract This paper presents a novel approach of space vector pulse width modulation techniques (SVPWM) used for two and three level Z Source Inverter (ZSI). The ZSI is a single stage converter and capable of doing buck and boost operation. Boost facility is obtained by adding extra state named as shoot through state and additionally DC bus utilization can be improved by proper placing of shoot through states in all the sectors. The voltage unbalance is reduced by proper use of pivot vectors. It has been observed that the proposed SVPWM offers better results in comparison to standard sine PWM techniques. Simulated results are presented for validating the analysis.

Keywords Z source inverter · Shoot through · Space vector modulation · Two level inverter · Three level inverter · Multilevel inverter

1 Introduction

A lot of research took place in power conversion technique due to the development of power semiconductor devices. Among all the power converting circuits, ZSI has lot of advantages. As it is capable of doing both buck and boost operation, eliminates the need of DC–DC converter, as it is a single stage power converter and miss gating will no longer destroys the circuit [1]. The power circuit of two level ZSI is shown in Fig. 1.

At the input side of the inverter, two capacitors and two inductors are connected in X shape and diode will prevents the reverse flow of current. The boost facility

B.M. Manjunatha (✉) · D.V. Ashok Kumar
RGMCET, Nandyal, AP, India
e-mail: manjunath.bmeee@gmail.com

M. Vijaya Kumar
JNTUA, Anantapur, AP, India

K.R. Attele et al. (eds.), *Emerging Trends in Electrical, Communications and Information Technologies*, Lecture Notes in Electrical Engineering 394,
DOI 10.1007/978-981-10-1540-3_41

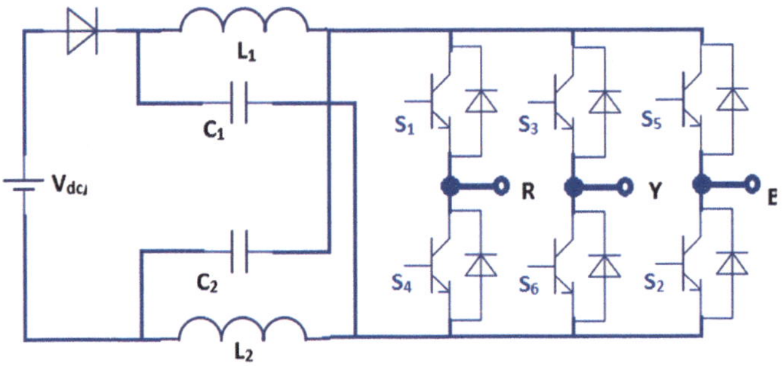

Fig. 1 Power circuit of two level ZSI

can be enabled by turning on all the switches in any one or two or three legs. Without disturbing the active states, then shoot through state is placed during the null states. The traditional two level voltage source inverter is having six active states and two null states. In ZSI beside with these two states, the extra state known as shoot through state must be considered. Simple, maximum constant and maximum boost techniques are used to generate shoot through pulses [2] along with the conventional PWM for Z source inverter.

This paper deals with: SVPWM for two level ZSI is discussed in Sect. 2. Section 3 presents SVPWM for three level ZSI, MATLAB implementation is discussed in Sect. 4, simulation results are presented in Sect. 5 and conclusions are made in Sect. 6.

2 SVPWM for Two Level ZSI

Conventional VSI has eight switch states, includes six active states (V_1–V_6) and two null or zero states (V_7 and V_8). In ZSI along with eight states, the additional state named shoot through state $(V_1^{'})$ need to be inserted. The shoot through state should be symmetrically distributed in all the switching periods to avoid the uneven stress on the switches [3]. Shoot through state in ZSI and null state in VSI appears same at the load side. Hence, without disturbing the active states the shoot through states are inserted during zero state. In order to achieve shoot through state, two switches in leg ($V_1^{'}$ or $V_2^{'}$ or $V_3^{'}$) or four switches in two legs ($V_{12}^{'}$ or $V_{23}^{'}$ or $V_{31}^{'}$) or six switches in three legs ($V_{123}^{'}$) are turned on. Figure 2 represents the switching states in which shoot through is inserted in all three legs.

The two-level switching states are indicated based on the phase connections to the dc bus. The first letter in the state name corresponds to the leg 1, the second letter corresponds to the leg 2 and the third letter corresponds to the leg 3 connections to the dc bus. For example, in the switching state (− + −):—represents the leg 1 lower switch (S_4) is turned on and connected to negative of dc bus, + indicates upper switch in leg 2 (S_3) is turned on and connected to the positive of dc bus,

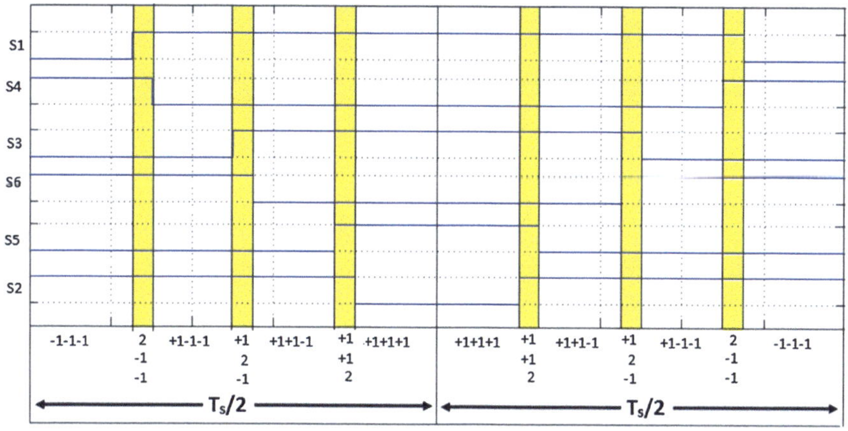

Fig. 2 Switching sequence for two level ZSI

and – represents lower switch in leg 3 (S_2) is turned on and connected to the negative dc bus and the shoot through states are denoted as (2 * 2), first letter 2 represents two switches (S_1 and S_4) in leg 1 are turned on, second letter * indicates undefined for leg 2 and third letter indicates two switches (S_5 and S_2) in leg3 are turned on. During the shoot through state the output voltage will be zero. The phase and line voltage for different switching states are shown in Table 1.

Table 1 Phase and line voltage for different switching states

State	Switches On	Phase voltages			Line to line voltages		
		V_{rn}	V_{yn}	V_{bn}	V_{ry}	V_{yb}	V_{br}
V_1	+ – –	$2/3*V_{dc}$	$-1/3*V_{dc}$	$-1/3*V_{dc}$	V_{dc}	0	$-V_{dc}$
V_2	+ + –	$1/3*V_{dc}$	$1/3*V_{dc}$	$-2/3*V_{dc}$	0	$-V_{dc}$	$+V_{dc}$
V_3	– + –	$-1/3*V_{dc}$	$2/3*V_{dc}$	$-1/3*V_{dc}$	$-V_{dc}$	$+V_{dc}$	0
V_4	– + +	$-2/3*V_{dc}$	$1/3*V_{dc}$	$1/3*V_{dc}$	$-V_{dc}$	0	$+V_{dc}$
V_5	– – +	$-1/3*V_{dc}$	$-1/3*V_{dc}$	$2/3*V_{dc}$	0	$-V_{dc}$	$+V_{dc}$
V_6	+ – +	$1/3*V_{dc}$	$-2/3*V_{dc}$	$1/3*V_{dc}$	$+V_{dc}$	$-V_{dc}$	0
V_7	– – –	Load terminals are short circuited by turning on all lower switches					
V_8	+ + +	Load terminals are SC by turning on all upper switches					
V'_1	2 * *	Source terminals are SC by turning on all switches in I leg					
V'_2	* 2 *	Source terminals are SC by turning on all switches in II leg					
V'_3	* * 2	Source terminals are SC by turning on all switches in III leg					
V'_{12}	2 2 *	Source terminals are SC by turning on all switches in I and II leg					
V'_{23}	* 2 2	Source terminals are SC by turning on all switches in II and III leg					
V'_{31}	2 * 2	Source terminals are SC by turning on all switches in III and I leg					
V'_{123}	2 2 2	Source terminals are SC by turning on all switches in all three legs					

Shoot through period (T_{st}) = D_{sh}*T_f and Number of shoot through states (N_{st}) in a cycle for Sine PWM (SPWM) is given by $N_{st} = F_c/25$.

Where D_{sh} = Shoot through duty ratio, $1/T_f$ = output frequency or reference frequency and

$1/T_c$ = carrier frequency or switching frequency

In case of SPWM, if the carrier frequency is 10 kHz, 400 times the shoot through is applied within one complete cycle. This reults in additional loss and stress on the switches. This limitation can be overcome by using SVPWM. In case of SVPWM, the number of shoot through states can be controlled by selecting the proper switching sequence as explained in Sect. 4.

3 SVPWM for Three Level ZSI

In traditional diode clamped multilevel inverter (DCMLI) the switches S_1 and S_3, S_2 and S_4 are operated in complementary. Due to the three possible connections to the dc bus for each of the three phases, the three-level inverter can produce a total $(3)^3 = 27$ switching states. In multilevel inverter the instantaneous error between the desired voltage vector and actual voltage vector is very less, this leads to lower harmonic distortion. Three level DC ZS MLI is represented in Fig. 3. The switch numbers are represented with the following notation 'S_{mn}' where m represents the leg and 'n' represents the switch number in that leg. Space vector diagram for three level is shown in Fig. 4.

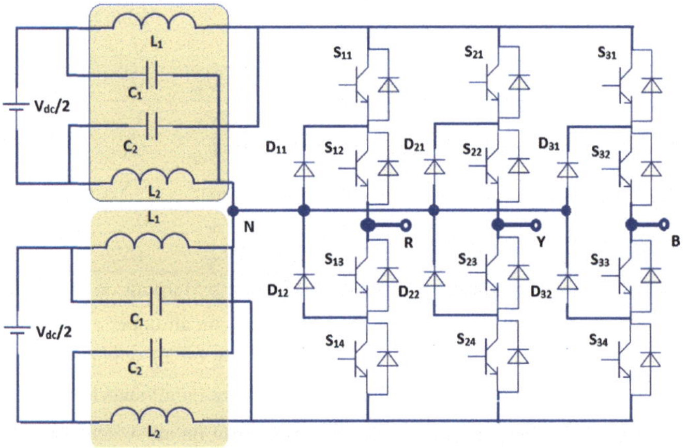

Fig. 3 Three level DC ZSI

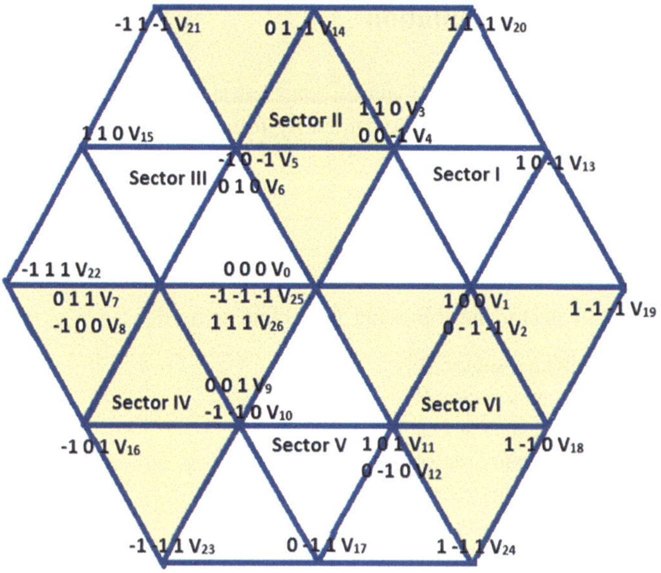

Fig. 4 SV diagram for three level inverter

Based on the magnitude, the total voltage vectors are classified as:

1. Zero vector or shoot through vector will have a length of zero. (V_0, V_{25} and V_{26})
2. Small vectors will have a length of $V_{dc}/2$ (V_1–V_{12})
3. Medium vector will have a length of $\sqrt{3}V_{dc}/3$ (V_{13}–V_{18})
4. Large vector will have a length of $2/3 * V_{dc}$ (V_{19}–V_{24})

In ZSI, the shoot through state has to be inserted during null states without disturbing the active states. For inserting shoot through state (all the switches in any one leg or any two legs or all the switches are turned on) in upper impedance circuit S_1, S_2, S_3 and D_2 should be conducting. Similarly to get shoot through state in lower impedance circuit S_2, S_3, S_4 and D_1 should be conducting. Both impedance networks should be boosted for the same duration of time to avoid voltage unbalance and complete shoot trough can be obtained by turning on all the switches in any one leg or in any two legs or in all three legs.

In three level DCMLI, three-level switching states are represented as (1 0 −1) 1 represents leg 1 is connected to positive of DC bus, 0 represents the leg 2 is connected to source neutral and −1 represents leg 3 is connected to negative of the DC bus. In ZSI, the state 2 represents all switches in any one leg are turned on (complete shoot through), $2'$ indicates first three switches in any one leg are turned on (upper shoot through) and $2''$ represents last three switches in a leg is turned on (lower shoot through).

4 MATLAB Implementation

As shown in Fig. 4, the three level space vector has six sectors. In each sector there are four triangles. SVPWM is implemented by following below steps.

4.1 Sector Identification

For identifying the sector the following flow chart is used

$0^\circ \leq \theta \leq 60^\circ$ indicates sector I
$0^\circ < \theta \leq 120^\circ$ indicates sector II
$120^\circ < \theta \leq 180^\circ$ indicates sector III
$180^\circ < \theta \leq 240^\circ$ indicates sector IV
$240^\circ < \theta \leq 300^\circ$ indicates sector V
$300^\circ < \theta \leq 360^\circ$ indicates sector VI

4.2 Triangle Identification

Each sector consists of four triangles as shown in Fig. 5 for sector 1. In this section triangle identification is explained for sector—I and it remains same for rest of the sectors.

The expressions for m_1 and m_2 [4] is given by:

$$m_1 = V_{ref}\left(\cos\alpha - \frac{1}{\sqrt{3}}\sin\alpha\right) \quad (1)$$

Fig. 5 Triangles identification

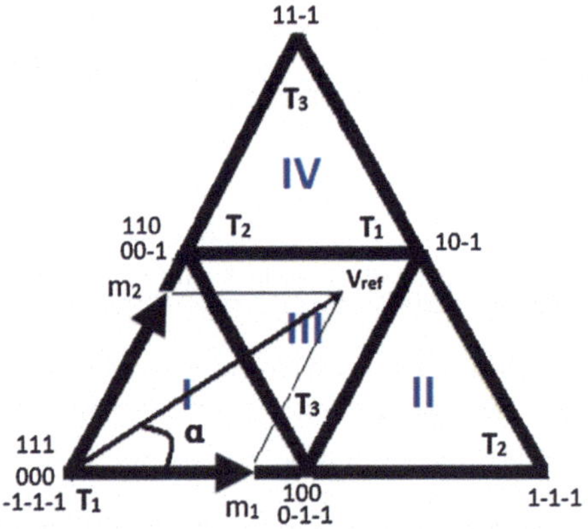

$$m_2 = V_{ref}\left(\frac{2}{\sqrt{3}} sin\,\alpha\right) \tag{2}$$

Reference vector lies in triangle I if m_1, m_2 and $m_1 + m_2$ is less than 0.5
Or a reference vector lies in triangle II if $m_1 > 0.5$
Or a reference vector lies in triangle III if m_1, m_2 are less than 0.5 and $m_1 + m_2$ is greater than 0.5 or reference vector lies in triangle II if $m_2 > 0.5$.

4.3 *Calculating the Switching Times of T_1, T_2, T_3 and T_{st}*

Finding the switching times T_1, T_2 and T_3 [5] will remain same as conventional inverter. In additional to these times the shoot through time needs to be calculated. The shoot through time depends on the amount of voltage boost required and also on the type of boosting method adopted. This work deals with the simple boost method as it is easy for the implementation.

$$\text{DC link voltage} = 2V_L - V_{dc} \tag{3}$$

$$\text{Boost factor}(B) = V_c/V_{dc} \tag{4}$$

$$\text{Shoot through duty ratio}(D_0) = (B-1)/(2*B) \tag{5}$$

$$T_{st} = D_0 * T_s \tag{6}$$

Shoot through must be inserted during zero/null vectors or pivot vector time in order to avoid voltage distortion and voltage unbalance.

4.4 *Finding the Switching Sequence*

Switching sequence when the reference vector is in sector 1 is shown in Table 2 for all the four triangles. To avoid voltage unbalance positive and negative pivot are used for equal interval of times [6].

Table 2 Switching pattern for sector I and triangle

			Ts/2								
Sector I	Triangle I	MLI	$T_1/8$		$T_2/4$	$T_3/4$	$T_1/4$	$T_2/4$	$T_3/4$	$T_1/8$	
			−1−1−1		0−1−1	00−1	000	100	110	111	
		ZSMLI	T_1^*	$T_{sh}/4$	$T_2/4$	$T_3/4$	$T_1/4$	$T_2/4$	$T_3/4$	$T_{sh}/4$	T_1^*
			−1−1−1	−12″−1	0−1−1	00−1	000	100	110	12′0	111
			Ts								
	Triangle II	MLI	$T_3/4$		$T_2/2$	$T_1/2$	$T_3/2$	$T_2/2$	$T_1/2$	$T_3/4$	
			0−1−1		1−1−1	10-1	100	10−1	1−1−1	0−1−1	
		ZSMLI	T_3^*	$T_{sh}/4$	$T_2/2$	$T_1/2$	$T_3/2$	$T_2/2$	$T_1/2$	$T_{sh}/4$	T_3^*
			0−1−1	0−12″	1−1−1	10−1	100	10−1	1−1−1	0−12″	0−1−1
			Ts								
	Triangle III	MLI	$T_3/4$		$T_1/2$	$T_2/2$	$T_3/2$	$T_2/2$	$T_1/2$	$T_3/4$	
			100		10−1	00−1	0−1−1	110	10−1	100	
		ZSMLI	T_3^*	$T_{sh}/4$	$T_1/2$	$T_2/2$	$T_3/2$	$T_2/2$	$T_1/2$	$T_{sh}/4$	T_3^*
			100	2′00	10−1	00−1	0−1−1	110	10−1	2″00	100
			Ts								
	Triangle IV	MLI	$T_2/4$		$T_3/2$	$T_1/2$	$T_2/2$	$T_1/2$	$T_3/2$	$T_2/4$	
			00−1		11−1	10−1	110	10−1	11−1	00−1	
		ZSMLI	T_2^*	$T_{sh}/4$	$T_3/2$	$T_1/2$	$T_2/2$	$T_1/2$	$T_3/2$	$T_{sh}/4$	T_2^*
			100	2′00	11−1	10−1	110	10−1	11−1	2′00	100

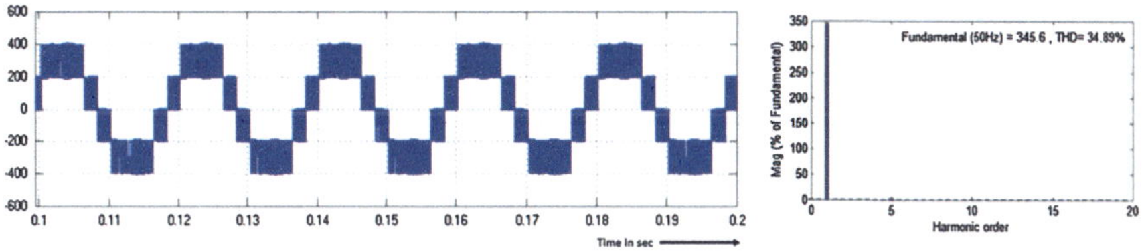

Fig. 6 Simulation results of three level SPWM output line to line voltage and its THD

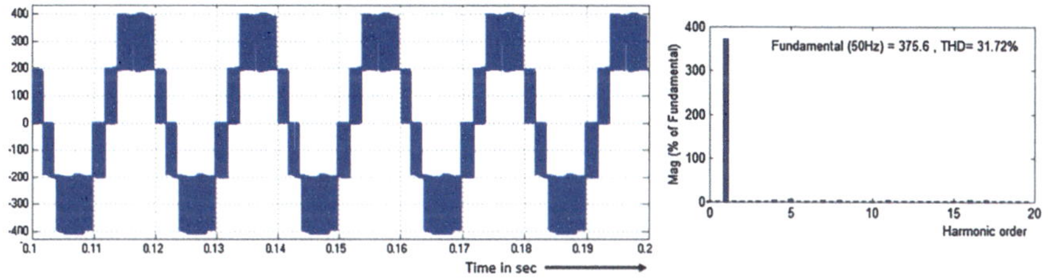

Fig. 7 Simulation results of three level SVPWM output line to line voltage and its THD

5 Simulation Results

The parameters considered for the simulation are as follows: the input dc voltage source is V_{in} = 400 V. Inductor $L_1 = L_2$ = 3.3 mH and capacitor $C_1 = C_2$ = 500 μF. The switching frequency is considered as 10 kHz. Figures 6 and 7 shows output line to line voltage and corresponding THD of ZSI with Sine PWM and SVPWM technique.

6 Conclusion

In this work the three level DCZSI is simulated with SVPWM and SPWM using MATLAB tool. The reference vector is identified directly without mapping procedure and number of times the shoot through pulses applied is also controlled. By using SVPWM there will be a freedom for selecting the switching sequence, use of positive or negative pivot vectors and shoot through vectors, which enables the control of neutral point voltage.

References

1. Peng FZ (2003) Z source inverter. IEEE Trans Ind Appl 39:504–510
2. Manjunatha BM, Ashok Kumar DV, Vijay Kumar M (2015) Advanced pulse width modulation techniques for Z source multi level inverter. World Acad Sci Eng Technol Int Sci Index 99 Int J Electrical Comput Energ Electron Commun Eng 9(3):337–342
3. Thangaprakash S, Krishnan A (2010) Comparative evaluation of modified pulse width modulation schemes of Z-source inverter for various applications & demands. Int J Eng Sci Technol 2(1):103–115
4. Kocalmiş A, Sünter S (2006) Modelling and simulation of a multilevel inverter using space vector modulation technique
5. Gebreel AAGM (2011) Simulation and implementation of two-level and three-level inverters by Matlab and RT-LAB. Dissertation, The Ohio State University
6. Jiao Yang, Lee Fred C, Sizhao Lu (2014) Space vector modulation for three-level NPC converter with neutral point voltage balance and switching loss reduction. IEEE Trans Power Electron 29(10):5579–5591

Impact of Distribution Generation on Losses of Distribution System

K. Kirubarani and A. Peer Fathima

Abstract The paper put forward a Load Flow algorithm for radial distribution system to evaluate the losses which is the basic requirement of many power system operation, planning and control problem. The Backward Forward Sweep method is utilized with amendment in the line data handling. The proposed method utilizes BIBC and BCBV matrices to estimate the line flow and voltage drop to reduce the memory requirements. The Distributed Generations (DGs) are incorporated into the Distribution load flow program to study the impact of penetrations of DG for Distribution system. A program is written in MATLAB R2012. This is used to test the efficiency of the proposed algorithm on IEEE 33 bus system with and without DG. The result demonstrates the capability and simplicity of the algorithm.

1 Introduction

In Distribution system Feeder distributes energy from the substation to the load points. In Radial Distribution System (RDS) the power losses can be reduced significantly by reducing the power flow through the branches and can be achieved by integrating the Distributed Generation (DG) into the network. Load flow technique is used to govern the power loss through each feeder or the network.

The load flow analysis of a power network provides a steady state solution through which various parameters like currents, voltages and power losses can be calculated. The load flow analysis is essential for the analysis of distribution system, and to investigate the matters related to planning, design and the operation and control. Traditional load flow methods like Gauss-Seidel, Newton Raphson are not applicable to the distribution system due to its characteristics like high R/X ratio and radial nature [1].

K. Kirubarani (✉) · A. Peer Fathima
SELECT, VIT, Chennai, India
e-mail: k.kirubarani2013@vit.ac.in

A. Peer Fathima
e-mail: peerfathima.a@vit.ac.in

K.R. Attele et al. (eds.), *Emerging Trends in Electrical, Communications and Information Technologies*, Lecture Notes in Electrical Engineering 394,
DOI 10.1007/978-981-10-1540-3_42

Though many solution techniques are discussed for Load flow analysis it should meet certain requirements in distribution system such as high convergence speed, less memory requirement simple and highly reliable [2]. Due to the non-applicability of transmission system power flow techniques many researchers developed the solution technique to Load Flow analysis for Distribution Systems. [3–6]. The details about the distribution system structure are required in these methods.

The significance of integrating DG into Distribution System is continuing to increase in many countries. Meanwhile the cost to serve residential and commercial customers is more than the cost to serve industrial customers. Therefore the integration of DG provides power to a single home, business or industrial facility will be indispensible and helpful. Additionally, if the planning and operation of DG can be incorporated into Distribution Automation (DA) efficiency and reliability of the distribution system could be improved [7, 8].

Usually, power in Distribution system flows from substations to the end of feeder. Conversely, the placement of DG leads to reverse power flow and affects the voltage profiles in the feeders. Hence the existing power flow methods have to be modified. A direct power flow technique using BIBC and BCBV matrices was proposed for solving the three-phase distribution system in [9] to obtain the Load flow solution.

This paper proposes a simple Forward Backward Sweep load flow method with the integration of the constant power factor DG model into the Distribution System. Only with the line and bus data requirement this method reduces the memory requirement and improves the convergence. The proposed method is applied and is verified on IEEE 33 bus system with DG. The results obtained show that the projected method can be used to analyze the impact of DG in Distribution System.

2 Mathematical Problem Formulation

A Single line diagram of a balanced three phase Radial Distribution Network (RDN) with a single source at one node and the other nodes are connected with loads shown in Fig. 1 is considered for the algorithm development.

Fig. 1 Single line diagram of a radial distribution network

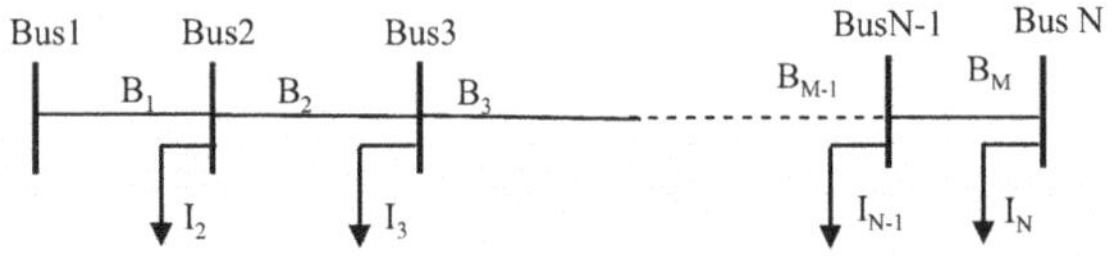

2.1 Calculation of Node Currents and Branch Currents

The load and the current injection in the kth iteration at the node i can be represented as follows, and thus the power injected at the node is converted to current injection.

$$S_i = P_i + Q_i \quad \text{where } i = 1, 2, 3, \ldots, N \tag{1}$$

$$I_i^k = \left(\frac{P_i + Q_i}{V_i^k}\right)^* \tag{2}$$

where I_i^k, V_i^k are the current and voltage magnitude at bus i.

For the distribution network shown in Fig. 1 the branch current in terms of node current can be represented in the matrix form using Kirchoff's current law. This can be simplified and gives the relationship between load current and branch current of the system as shown in Eq. (3). Elements of BIBC matrix are only 0's and 1's.

$$\begin{bmatrix} IB_1 \\ IB_2 \\ . \\ . \\ IB_M \end{bmatrix} = \begin{bmatrix} 1 & 1 & . & . & 1 \\ 0 & 1 & . & . & 1 \\ . & . & . & . & . \\ . & . & . & . & . \\ 0 & 0 & . & 0 & 1 \end{bmatrix} \begin{bmatrix} I_2 \\ I_3 \\ . \\ . \\ I_N \end{bmatrix}$$

$$[IB] = [BIBC][I] \tag{3}$$

Procedure for BIBC Matrix Formation

1. First Initialize a null matrix BIBC with the order (M × (N − 1)). Here M and N is the number of branches and the nodes of the system to be evaluated.
2. Consider a branch B_{ij} is located between node i and node j. If 'i' is the source node of the network then (j − 1, j − 1)th element of BIBC is replaced by +1.
3. If the branch B_{ij} is emanating from other than the source node then copy the column segments of (i − 1)th node to column segments of (j − 1)th node and replace (j − 1, j − 1)th element by +1.
4. By repeating steps 2 and 3 form the BIBC matrix.

2.2 Bus Voltage Calculation

Initially a nominal voltage is assumed to exist at all the nodes of the RDN and by using KVL the drop in voltage and the node voltages V_2, V_3, V_4, …, V_N are calculated using the branch currents and sending end voltages as follows.

$$V_i = (1 + j0)\ \text{p.u.} \quad i = 1, 2, 3, \ldots, N$$

$$\therefore \quad V_2 = V_1 - \Delta V = V_1 - I_{B1} * Z_1 \tag{4}$$

$$V_3 = V_2 - I_{B2} * Z_{23} \tag{5}$$

$$V_4 = V_3 - I_{B3} * Z_{34} \tag{6}$$

Substitute the Eqs. (4) and (5) in Eq. (6),

$$V_4 = V_1 - I_{B1} * Z_{12} - I_{B2} * Z_{23} - I_{B3} * Z_{34} \tag{7}$$

From the above equation it can be seen that the node voltage of the network can be expressed in terms of branch currents and line impedance. The BCBV matrix shows the relation between the node voltage and the branch current. In general it can be expressed as,

$$[\Delta V] = [BCBV]\ [I_B] \tag{8}$$

Procedure for BCBV Matrix Formation

1. First initialize a null matrix BCBV with the order ((N − 1) × M). Here M and N is the number of branches and the nodes of the system to be evaluated.
2. Consider a branch B_{ij} is located between node i and node j. If 'i' is the source node of the network then (j − 1, j − 1)th element of BIBC is interchanged by Z_{ij}.
3. If the branch B_{ij} is emanating from other than the source node then copy the row segments of (i − 1)th node to row segments of (j−1)th node and replace (j − 1, j − 1)th element by the impedance Z_{ij}.
4. By repeating steps 2 and 3 for all the branches form the BCBV matrix.

$$[BCBV] = \begin{bmatrix} 1 & 0 & . & 0 \\ . & . & . & . \\ . & . & . & . \\ 1 & 1 & . & 1 \end{bmatrix} \begin{bmatrix} Z_{12} & 0 & . & 0 \\ . & . & . & . \\ . & . & . & . \\ 0 & 0 & . & Z_{NN-1} \end{bmatrix}$$

$$[BCBV] = [BIBC]^T [ZD]$$

$$[\Delta V] = [BIBC]^T\ [ZD]\ [I_B]$$

The magnitude of the current flow through the RDN system can be calculated as follows for all the nodes.

$$I_i^k = I_i^r(V_i^k) + jI_i^r(V_i^k) \tag{9}$$

$$[\Delta V] = [BCBV]\ [I_B] \tag{10}$$

$$\mathrm{V}_{\mathrm{i}}^{\mathrm{k}+1} = \mathrm{V}_{\mathrm{i}}^{0} + [\Delta V]^{k+1} \tag{11}$$

Real and reactive power losses can be obtained using the branch currents and the total loss of the RDN can also be then calculated as follows.

$$\mathrm{P}_{\mathrm{loss}} = \mathrm{I}_{\mathrm{B(m)}}2 * \mathrm{R(m)} \qquad \text{for } \mathrm{m} = 1,\ 2,\ 3,\ \ldots,\ \mathrm{N} \tag{12}$$

$$\mathrm{Q}_{\mathrm{loss}} = \mathrm{I}_{\mathrm{B(m)}}2 * \mathrm{X(m)} \qquad \text{for } \mathrm{m} = 1,\ 2,\ 3,\ \ldots,\ \mathrm{N} \tag{13}$$

$$\text{Total Real Power Loss} = \sum_{k=1}^{N} P_{loss}(k) \tag{14}$$

$$\text{Total Reactive Power Loss} = \sum_{k=1}^{N} Q_{loss}(k) \tag{15}$$

Flowchart for the Load flow solution of the RDN is provided in Fig. 2.

3 Load Flow Solution with DG into the Distribution System

A DG unit is assumed to be placed at bus i in the single source RDN with $\mathrm{N_B}$ branches. The power flow through the branches being connected from i will vary with other branches connected in between the substation and bus i unaffected. At ith node the apparent power and node current are calculated without DG as

$$\mathrm{S} = \mathrm{S_{Di}} = \sum \mathrm{P_{Di}} + \mathrm{jQ_{Di}} \tag{16}$$

$$\mathrm{I_{Di}} = \mathrm{I}_{\mathrm{Di}}^{\text{without DG}} = \left(\frac{\mathrm{S_{Di}}}{V_i}\right)^{*} \tag{17}$$

The active power and reactive power at ith node has been altered. The DG reactive power may flow from source or it may be taken from the system. The apparent power and current injection at this node can be expressed as follows

$$\mathrm{S_{DG\_i}} = \sum \mathrm{P}_{\mathrm{Gi}}^{\mathrm{DG}} + \mathrm{jQ}_{\mathrm{Gi}}^{\mathrm{DG}} \qquad \mathrm{i} = 1,\ 2,\ 3,\ \ldots,\ \mathrm{N}$$

$$\text{Therefore,} \quad \mathrm{S} = \mathrm{S_{Di}} - \mathrm{S_{DG\_i}} \left(\frac{\mathrm{S_{Di}} - \mathrm{S_{DG\_i}}}{V_i}\right)^{*} \tag{18}$$

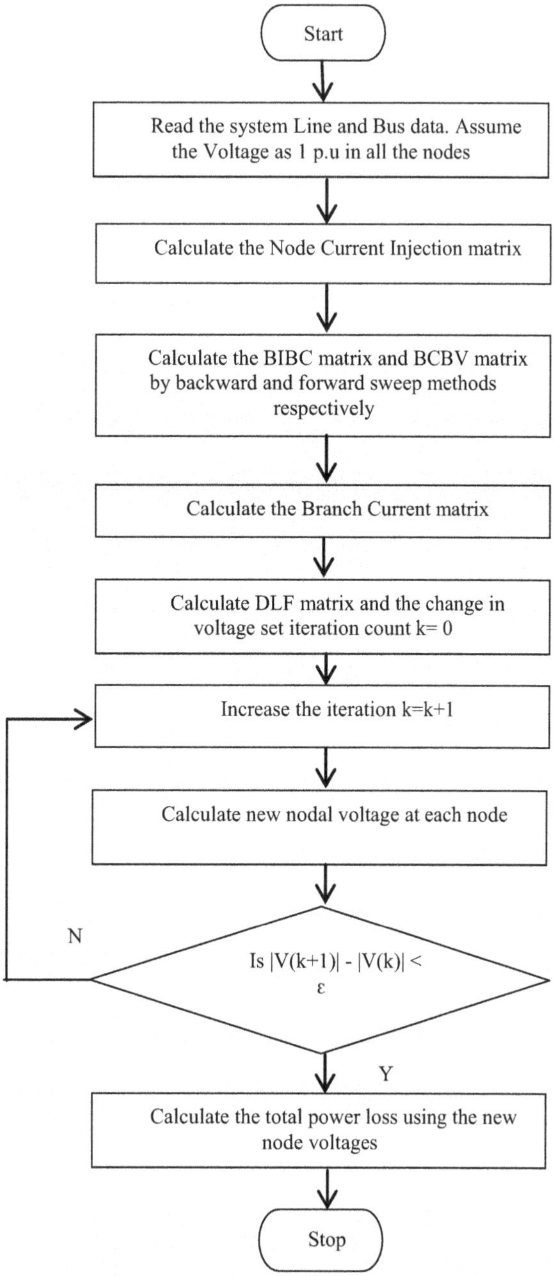

Fig. 2 Flow chart for load flow algorithm of radial distribution network

To study the influence of DG in a RDN, the DG is incorporated into the Load flow algorithm. DG supplies real power to the system and it may take small reactive power from the source. Due to modified injected power the voltage drop in the line reduces considerably which in turn reduces the current flow. The injected apparent power to the system is modified as shown in Eq. (18) and the load flow algorithm is applied to the system with DG. Hence the active power loss can be reduced significantly by placing DG in the RDN.

4 Results and Discussion

To validate the efficiency of the proposed method, a 33-bus, 12.66 kV radial distribution system is considered from [10]. At nominal voltage the total real and reactive power loads are 3715 kW and 2300 kVAr respectively. The load flow solution voltage results for 33-bus RDN have been given in Table 1. The real and reactive power losses of the system are 6.12 and 6.869 % of their respective roads. The minimum voltage is experienced at node 32 and is 0.9443 p.u. The voltage profile of the system at various nodes is given in Table 1. Figure 3 Shows the voltage profile of IEEE 33 bus RDN without DG. The proposed algorithm is implemented to the RDN with DG as described above. The results of the Load flow solution with DG are given in Table 2. The DG has placed at bus 3 with different capacity to implement the proposed algorithm. It is observed that the DG incorporation into the Distribution system reduces the losses significantly. The result shows the impact of DG in the RDN and also reduces its active power loss considerably.

Table 1 Load flow solution bus voltages of 33-bus RDN

Bus no.	Voltage in p.u.	Bus No.	Voltage in p.u.	Bus no.	Voltage in p.u.
1	1.0000	12	1.0804	23	1.0167
2	1.0285	13	1.0806	24	1.0119
3	1.0382	14	1.0837	25	1.0278
4	1.0469	15	1.0870	26	1.0248
5	1.0295	16	1.0877	27	0.9867
6	1.0113	17	1.0888	28	0.9577
7	1.0368	18	1.0077	29	0.9499
8	1.0493	19	0.9835	30	0.9468
9	1.0608	20	0.9710	31	0.9453
10	1.0640	21	0.9557	32	0.9443
11	1.0698	22	1.0264	33	

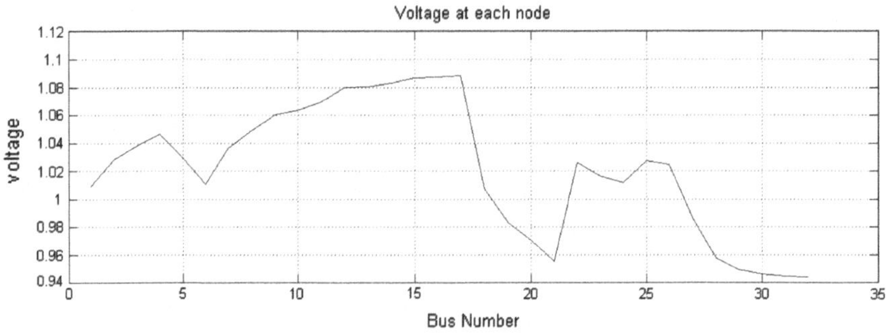

Fig. 3 Voltage profile of 33-bus RDN

Table 2 Load flow solution of 33-bus RDN with DG

DG size	Min voltage	Real power loss (KW)	Reactive power loss (kVAr)
Base case	0.9433	227.35	157.99
1 MVA	0.9680	211.42	147.56
2 MVA	0.9065	162.80	106.79

5 Conclusion

The proposed Backward Forward Sweep Load Flow algorithm to assess the active power losses demonstrates the impact of incorporation DG into radial distribution system. The program is written in MATLAB R2012 to test the efficiency of the proposed algorithm on IEEE 33 bus system with and without DG. The algorithm is applied to the case without DG and with DG with various capacity and location of DG. The results show that the integration of DG into the network reduces the active power losses of the network significantly. This algorithm is further can be useful to reconfiguration of the feeder, optimal DG placement and so on.

References

1. Scott B (1974) Review of load flow calculation methods. Proc IEEE 62(7):916–929
2. Kersting WH (1984) A method to teach the design and operation of a distribution system. IEEE Trans Power App Syst 103(7):1945–1952
3. Cespedes Renato G (1990) New method for the analysis of distribution networks. IEEE Trans Power Delivery 5(1):391–396
4. Jasmon GB, Lee LHC (1991) Distribution Network reduction for voltage stability analysis and load flow calculations. Elec Power Energy Syst 13(1):9–13
5. Goswami SK, Basu SK (1991) Direct solution of distribution system. IEE Proc. C 188 (1):78–88
6. Das D, Nagi HS, Kothari DP (1994) Novel method for solving radial distribution networks. IEE Proc C 141(4):291–298

7. Willis HL, Scott WG (2000) Distributed Power generation-planning and evaluation. Marcel Dekker Inc., New York
8. Rabinowitz M (2000) Power Systems of the future 1. IEE Power Eng. Rev. 20(1):5–16
9. Teng JH (2003) A direct approach for distribution system load flow solutions. IEEE Trans Power Delivery 18(3):882–887
10. Liu J, Salama RR, Mansour RR (2002) An efficient power flow algorithm for distribution systems with polynomial load. Int J Elec Eng Edu 39(4):372–386

Performance Analysis of Shunt and Hybrid Active Power Filter Using Different Control Strategies for Power Quality Improvement

S. Shamshul Haq, D. Lenine and S.V.N.L. Lalitha

Abstract This paper includes assay and correlation of three phase shunt active to extricate its reference currents with distinct control strategies to evaluate their dynamic performance under different load conditions for both Shunt Active Power Filter (SAPF) and Hybrid Shunt Active Power Filter (HSAPF). The instantaneous active and reactive power (p-q) theory and synchronous reference frame theory (SRF) are compared. Extensive simulations are carried out for p-q and d-q theory's for both shunt APF and hybrid shunt APF with PI controller. The compensation capability of Hybrid shunt APF is greater than SAPF for harmonics, reactive power compensation and dynamic performances capability. Performances comparison is analyzed using Matlab/Simulink. Simulation outcome is given to compare and validate the control techniques for SAPF and HSAPF during load changing condition.

Keywords Shunt active power filter (SAPF) · Hybrid shunt active power filter (HSAPF) · Synchronous reference frame (SRF) · Instantaneous active and reactive power (p-q)

1 Introduction

Immense employ of power electronic devices causes contamination of current at the source side which leads to poor power quality. The trouble of harmonics caused by the hike use of non-linear loads, like, thyristor based converters and power electronic based loads has unveiled numerous methods which have become indispensable.

S. Shamshul Haq (✉) · S.V.N.L.Lalitha
KL University, Vijayawada, India
e-mail: shaik.shamshul@gmail.com

D. Lenine
RGMCET, Nandyal, India

K.R. Attele et al. (eds.), *Emerging Trends in Electrical, Communications and Information Technologies*, Lecture Notes in Electrical Engineering 394,
DOI 10.1007/978-981-10-1540-3_43

The active power is delivered by maintaining the voltage with the use of reactive power. Merits like, low cost, easy maintenance and high efficiency are offered by passive power filters [1]. The demerits of PPF dominates its advantages because of their poor dynamic performances, resonances problems and impact of filtering characteristics due to small change of system parameters [2–5] etc. Though high initial cost is a drawback of APF, It eliminates current harmonics and compensates reactive power for linear/nonlinear loads [2, 3]. But, the dual quality of HAPF as shown in Fig. 1, enhances compensation features and lowers the current ratings which in turn minimizes the cost of APF [2, 3].

2 Instantaneous Active and Reactive Power Theory

It is based on instantaneous active and reactive power in time domine. Three phase voltage and currents are converted into αβ0 stationary reference frames. Zero sequence components exits only for three phase four wire system. In this paper three phase three wire system is designed so zero sequence components does not exist. Clarke and inverse Clarke transformation is as shown below in Eqs. (1) and (2).

$$\begin{bmatrix} V_\alpha \\ V_\beta \end{bmatrix} = \sqrt{\frac{2}{3}} \begin{bmatrix} 1 & -1/2 & -1/2 \\ 0 & \sqrt{3}/2 & -\sqrt{3}/2 \end{bmatrix} \begin{bmatrix} V_a \\ V_b \\ V_c \end{bmatrix} \tag{1}$$

$$\begin{bmatrix} i_{L\alpha} \\ i_{L\beta} \end{bmatrix} = \sqrt{\frac{2}{3}} \begin{bmatrix} 1 & -1/2 & -1/2 \\ 0 & \sqrt{3}/2 & -\sqrt{3}/2 \end{bmatrix} \begin{bmatrix} i_{La} \\ i_{Lb} \\ i_{Lc} \end{bmatrix} \tag{2}$$

Instantaneous active power and reactive power is given by Eq. (3).

$$\mathrm{P_L = V_\alpha i_\alpha + V_\beta i_\beta} \tag{3}$$

$$\mathrm{q_L = V_\alpha\, i_\beta - V_\beta i_\alpha} \tag{4}$$

Active and reactive power can be written as

$$\mathrm{P_L = \bar{P}_L + \tilde{P}_L} \tag{5}$$

$$\mathrm{q_L = \bar{q}_L + \tilde{q}_L} \tag{6}$$

where $\bar{P}_L$ and $\tilde{P}_L$ are the DC component and harmonic component of active power. Where $\bar{q}_L$ and $\tilde{q}_L$ are the DC component and harmonic component of reactive power.

AC components are obtained by using low—pass filters. Reference signals are obtained using below Eq. (7). The three phase reference quantities are given to hysteresis controller to generate gate pulses for VSI as shown in Fig. 2. This control strategy could not yield an adequate solution for non ideal source voltage [6].

$$\begin{bmatrix} i_{sa}^* \\ i_{sb}^* \\ i_{sc}^* \end{bmatrix} = \sqrt{\frac{2}{3}} \begin{bmatrix} 1 & 0 \\ \frac{1}{2} & \frac{\sqrt{3}}{2} \\ \frac{-1}{2} & \frac{-\sqrt{3}}{2} \end{bmatrix} \begin{bmatrix} V_\alpha & V_\beta \\ -V_\beta & V_\alpha \end{bmatrix}^{-1} \begin{bmatrix} P^* \\ q^* \end{bmatrix} \quad (7)$$

3 Synchronous Reference Frame Theory

SRF theory is another control algorithm for generation of triggering pulse is implemented in this paper. Three phase load current (i_{La}, i_{Lb}, i_{Lc}), voltages (V_{sa}, V_{sb}, V_{sc}) and DC voltage across the capacitor are sensed as feedback signals. By using parks transformation the load current are into the dq0 frame as given below

$$\begin{bmatrix} i_{Ld} \\ i_{Lq} \end{bmatrix} = \sqrt{\frac{2}{3}} \begin{bmatrix} \cos\theta & \cos(\theta - \frac{2\pi}{3}) & \cos(\theta + \frac{2\pi}{3}) \\ -\sin\theta & -\sin(\theta - \frac{2\pi}{3}) & -\sin(\theta + \frac{2\pi}{3}) \end{bmatrix} \begin{bmatrix} i_{La} \\ i_{Lb} \\ i_{Lc} \end{bmatrix} \quad (8)$$

A Three phase locked loop (PLL) is used to synchronize the current signal with the PCC voltage. The d-q current components are given to low pass filter to extract the DC component of i_{Ld} and i_{Lq}. The d-q axis currents consist of fundamental and harmonic components.

$$i_{Ld} = \bar{i}_{Ld} + \tilde{i}_{Ld} \quad (9)$$

$$i_{Lq} = \bar{i}_{Lq} + \tilde{i}_{Lq} \quad (10)$$

Using inverse clerk's transformation the harmonic components are converted back into three phase quantities. The three phase reference quantities are given to hysteresis controller to generate gate pulses for VSI [7] the complete block diagram of controller is shown in Fig. 3.

$$\begin{bmatrix} i_a^* \\ i_b^* \\ i_c^* \end{bmatrix} = \sqrt{\frac{2}{3}} \begin{bmatrix} \cos\theta & -\sin\theta \\ \cos(\theta - \frac{2\pi}{3}) & -\sin(\theta - \frac{2\pi}{3}) \\ \cos(\theta + \frac{2\pi}{3}) & -\sin(\theta + \frac{2\pi}{3}) \end{bmatrix} \begin{bmatrix} i_{Ld} \\ i_{Lq} \end{bmatrix} \quad (11)$$

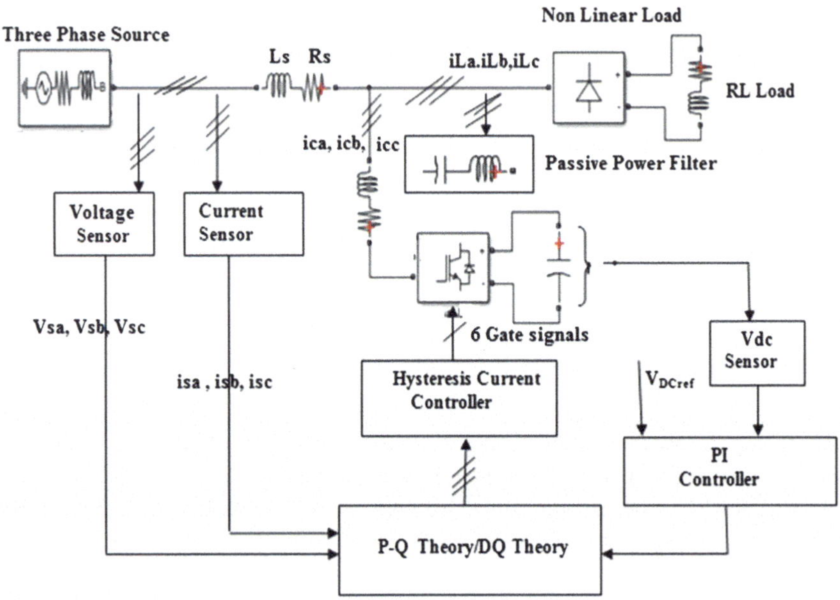

Fig. 1 Block diagram of Hybrid SAPF

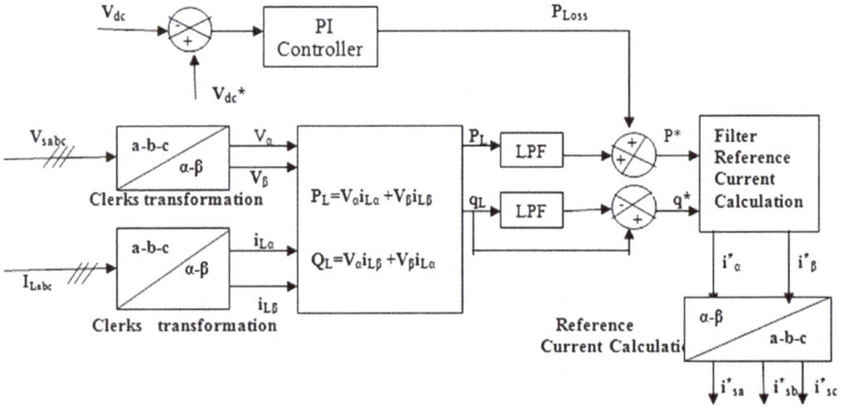

Fig. 2 Instantaneous p-q theory for reference current extraction

4 Analysis and Simulation Results

Table 1 shows system parameters to analyze the dynamic performance of SAPF and HSAPF varying load is considered in this paper, shunt active power filter is designed for 26 KVA. Up to time t = 0.3 s the active power load is 8.8 KW after t = 0.3 s it increases to 11 KW and at t = 0.6 s it decrease to 4.5 KW. Performances analysis of

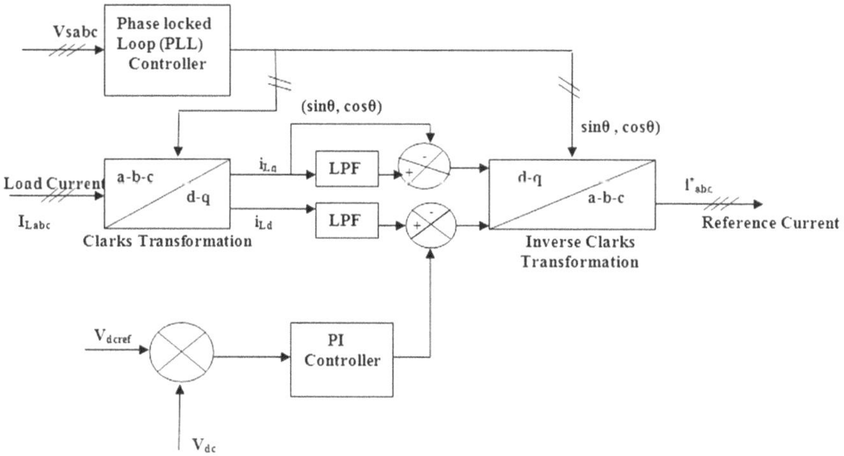

Fig. 3 d-q theory for reference current extraction

SAPF and HSAPF is verified using both p-q and d-q theory's. Performances analysis is carried out using Matlab/Simulink. A PI controller is used to obtain reference current from the distorted line current. Tables 2 and 3 shows the analysis of the result.

As depicted in Figs. 4 and 5 Source current THD without compensation is 23.78 % which is more than the IEEE standard. After compensation THD has reduced to 3.83 % by using SAPF using d-q theory and 1.11 % using HSAPF. In HSAPF passive power filters are designed to compensate 5th and 7th harmonics. Figures 6, 7, 8, 9, 10, 11 shows the waveforms with and without compensation.

Performances comparison of SAPF and HSAPF are given Table 4. It is seen that settling time of Vdc, overshoot of Vdc voltage and overshoot of active power is less in HSAPF when compared to SAPF. It observed that compensating current of three phase inverter is low in case the of HSAPF.

Table 1 System parameters

Sr. no	System parameters	Value of parameters
1.	Source (Vs)	$415V_{L\text{-}L}$ (rms), f = 50 Hz
2.	Interfacing inductance	L = 10 mH
3.	Passive filter	$L_5 = 0.73$ H, c5 = 5.54 μF, $L_7 = 0.52$ H, $c_7 = 3.54$ μF
4.	DC link capacitor	2200 μF
5.	DC link voltage	800 v
6.	Switching band range	±0.02

Table 2 Source current THD comparison

Active power (KW)	Source current % THD value				
	With out compensation	d-q theory		p-q theory	
		SAPF	HAPF	SAPF	HAPF
4.5	26.26	1.16	0.87	2.11	1.09
8.8	24.56	1.82	1.00	1.76	0.55
11	23.78	3.83	1.11	3.19	0.38

Table 3 Compensating current comparison

Active power (KW)	Shunt active power filter injecting current (A)			
	d-q theory		p-q theory	
	SAPF	HSAPF	SAPF	HSAPF
4.5	4	2	4	2
8.8	7	4	7.5	4
11	9	5.5	9.1	5.8

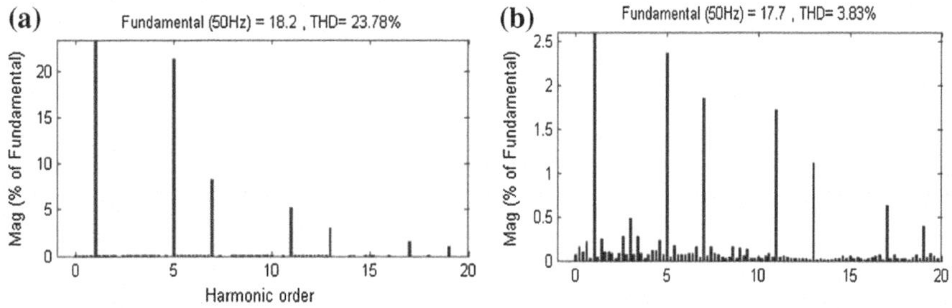

Fig. 4 Source current THD for phase A **a** Without compensation **b** SAPF using d-q theory

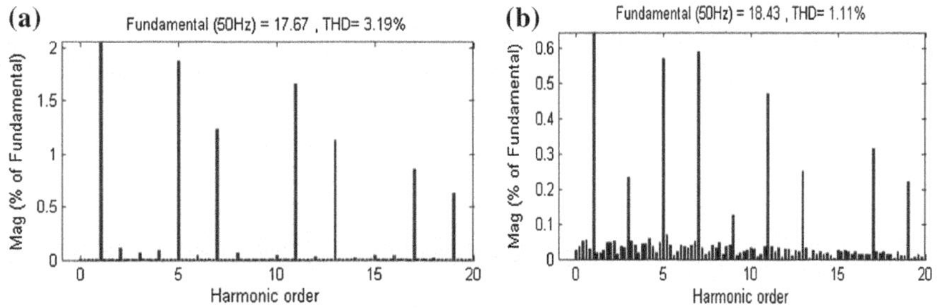

Fig. 5 Source current THD for phase A **a** With SAPF p-q theory **b** HSAPF using d-q theory

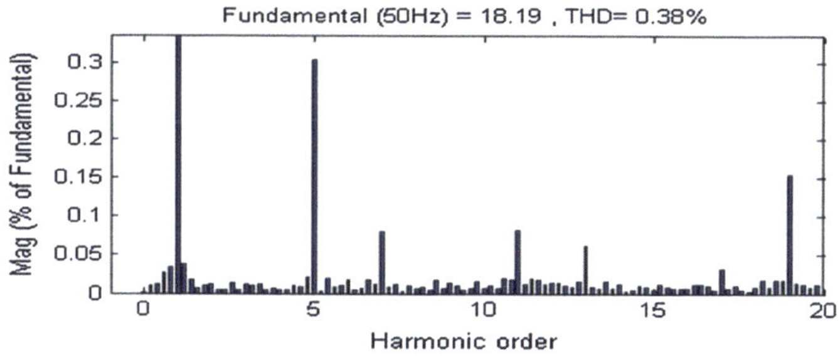

Fig. 6 Source current THD for phase A with HSAPF using p-q theory

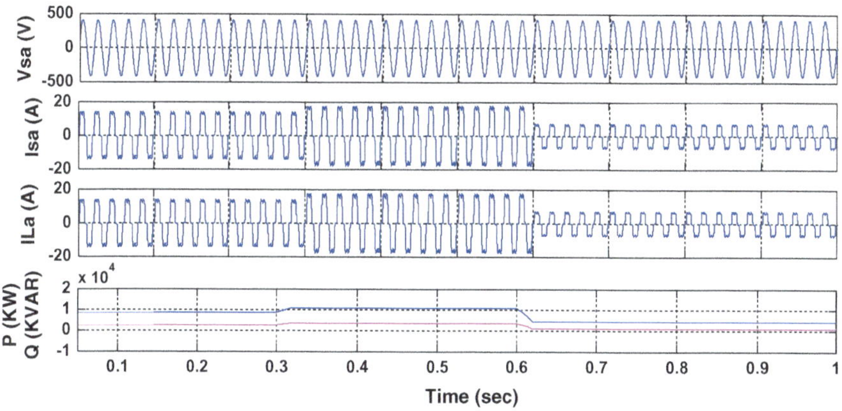

Fig. 7 Simulation results waveform without compensation

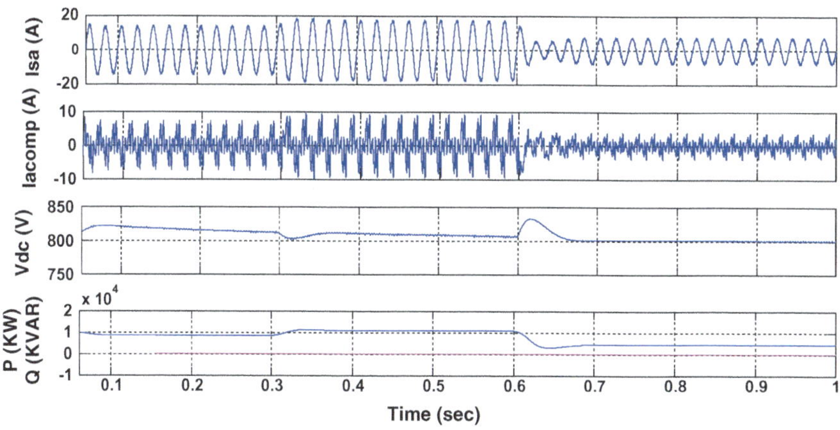

Fig. 8 Simulation results waveforms with SAPF using d-q theory

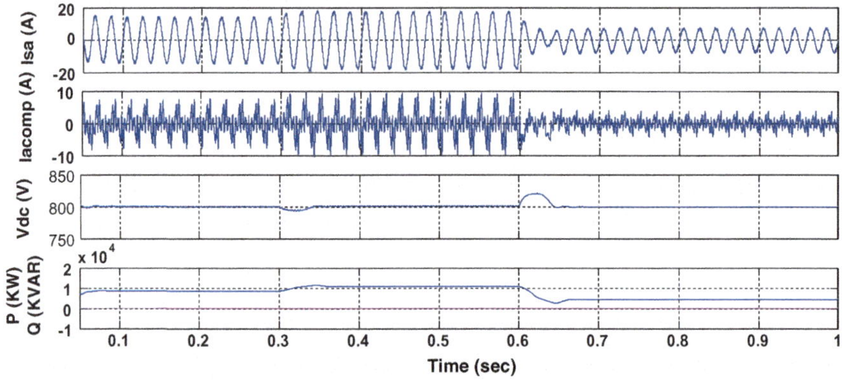

Fig. 9 Simulation results waveforms with SAPF using p-q theory

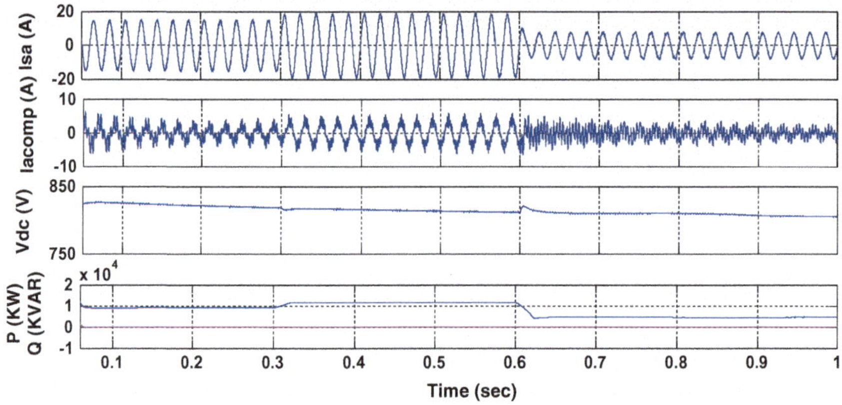

Fig. 10 Simulation results waveforms with HSAPF using d-q theory

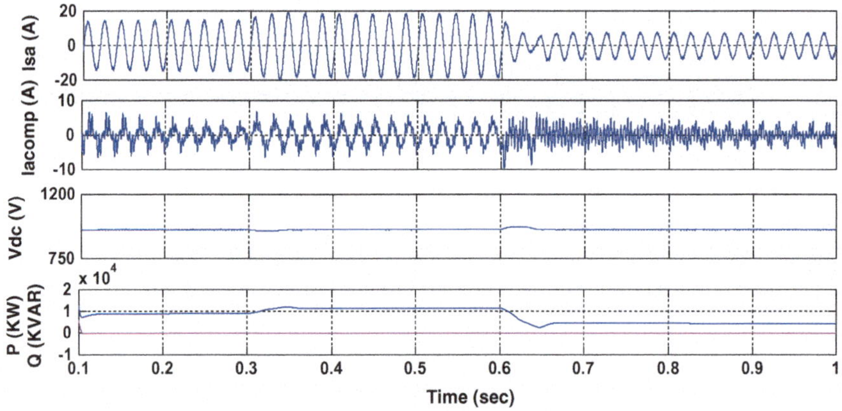

Fig. 11 Simulation results waveforms with HSAPF using p-q theory

Table 4 Dynamic performances comparison of SAPF and HSAPF

Performance parameters	d-q theory		p-q theory	
	SAPF	HSAPF	SAPF	HSAPF
Settling time for Vdc (s)	0.06	0.025	0.05	0.05
Max overshoot for Vdc (V)	11	2	6	6
Max undershoot for Vdc (V)	28	9	19	20
Settling time of active power (s)	0.06	0.02	0.07	0.061
Max overshoot of active power (W)	500	200	625	520
Max undershoot of active power (W)	1500	550	1780	1890

5 Conclusion

This paper comprises of implementation of SAPF and HSAPF with p-q and d-q theory's using PI controller for power quality improvement at distribution system. These control theories are effective to compensate harmonics of source current, enhance power factor for steady state and dynamic load change conditions. Thus, HSAPF results are more efficient than SAPF. Finer harmonic compensation is acquired from p-q theory with HSAPF. And the dynamic performances are greater with d-q theory than p-q theory. Thus HSAPF is more appropriate than SAPF to improve power quality in distribution system.

References

1. Moran L, Rivas D, Espinoza JR, Dixon, JW (2003) Improving passive filter compensation performance with active techniques. IEEE Trans Ind Electron 50(1):161–170
2. Lee W-C (2015) Cost-effective APF/UPS system with seamless mode transfer. J Electr Eng Technol 10(1):195–204
3. Jung Y-G (2013) Graphical representation of the instantaneous compensation power flow for single-phase active power filters. J Electr Eng Technol 8(6):1380– 1388
4. Al-Haddad K, Hamadi A, Rahmani S, Mendalek N A new control technique for three-phase shunt hybrid power filter. IEEE Trans Ind Electron
5. Trinh QN, Lee HH (2013) An advanced current control strategy for three-phase shunt active power filters. IEEE Trans Ind Electron 60(12):5400–5410
6. Mikkili S, Panda AK (2011) Instantaneous active and reactive power and current strategies for current harmonics cancellation in 3-ph 4-Wire SHAF with both PI and fuzzy controllers. Energy Power Eng (3):285–298
7. Sebasthi Rani K, Porkumaran K (2011) Performance evaluation of PI and fuzzy controller based Shunt active power filter. Eur J Sci Res 61(3). ISSN 1450–216X

A Unique Tuning of PID Controller Using Particle Swarm Optimizer for Modern Multilevel Inverter Fed Micro-grid System

P. Hemachandu, V.C. Veera Reddy, N. Kusuma, D. Mohan Reddy and P. Divya Prasad

Abstract Elimination of harmonics in high level inverter is an optimized problem which is interprets by using particle swarm optimization (PSO) technique. The derived objective function is used for the tuning of PID controller to minimize the harmonic content in output voltage. The classical tuning of PID controller usually produces a high overshoot, imperative heuristics approaches are employed to improve the capability of classical techniques. This controller predicts the switching angles and optimum modulation index essentials for an improved output voltage and prevents the sudden variations of the asymmetrical 15-level modern multilevel inverter with fever switches. Here, this model has several inputs such as grid voltage, difference voltage, controlled target voltage. By means of these parameters, this proposed controller makes the better tuning values for enhanced quality voltage at grid, fast steady state response. In this process, the proposed methodology provides a pure sinusoidal current is in-phase with the grid voltage, then interfacing to the grid by PSO-PID controller. A simulink model is designed to validate the performance evaluation of this proposed work using Matlab/Simulink platform and results are conferred.

Keywords Asymmetrical modern multilevel inverter • Fuel stacks • PV arrays • Particle swarm optimization (PSO) • PID controller • Total harmonic distortion (THD)

P. Hemachandu (✉) · V.C. Veera Reddy · P. Divya Prasad
Department of Electrical & Electronics Engineering, SV University, Tirupati, India
e-mail: chandusvuphd@gmail.com

V.C. Veera Reddy
e-mail: veerareddy_vc@yahoo.com

N. Kusuma · D. Mohan Reddy
Department of Electronics & communication Engineering, SVTM, Madanapalle, India
e-mail: kusuma.nt@gmail.com

K.R. Attele et al. (eds.), *Emerging Trends in Electrical, Communications and Information Technologies*, Lecture Notes in Electrical Engineering 394,
DOI 10.1007/978-981-10-1540-3_44

1 Introduction

The classical proportional, integral, derivative (PID) controllers have been used widely for generating the switching states for proposed MLI. To improve the capabilities of classical PID parameter tuning techniques, multiple intelligent techniques have been suggested to improve the PID tuning for grid connected modern MLI. The sustained employ of fossil fuels has concluded in the global warming of green-house exertions. However, as the afford of fossil fuels are depleted in future days and become more expensive that's why explores the superior technologies in recognition of renewable energy sources, which hold the outstanding advantages [1] may ability to handle the micro grid system with facile connectivity [2]. For splendid scalability and attainable flexibility for excellence of safe, clean, eco-friendly specifications, the photo-voltaic (PV) and fuel cell (FC) are imperatively used as chief power generators to form as a co-generation system The eminent power density is attained by intermittent formation of PV/FC have good dynamic stability, the fast step load changes will be minimized by support of co-generation system [1–4]. An excellent, power generation schemes are widely used in many micro-grid/residential applications in very near future as interconnected system via advanced power conditioning units. There are two flavors such as; (i) high step up DC/DC boost converter, (ii) DC-AC inverter module, DC/DC boost converter are used to strengthen voltage at DC bus and DC-AC inverter act as the barrier between the DC bus and grid system.

Enhancement of voltage quality, amelioration of efficiency and low stress on switches by using multi-level inverters [5, 6]. This technology has been used over many years ago and getting superior advantages like as exalt voltage quality, low THD, harmonic eradication, incredible efficiency and low dv/dt stress by high voltage levels to governing the superior topologies. The basic characteristics of classical multilevel inverter topologies are neutral point clamped module [7], flying capacitor module [8] cascade H-bridge module [9] are regular topologies utilized by many researchers, but these are useless due to several disadvantages. NPC and FC topologies needs extra switching devices and difficult to generate asymmetrical voltages. Asymmetrical voltage technique is mostly used in cascaded H-bridge module, but it requires more devices. Several inverter structures have been explored in [5–9].

In this paper, novel asymmetrical modern multi-level inverter topology has been investigated by using advanced modulation schemes for micro grid applications with optimum tuning of PID gains by PSO algorithm. Attaining the optimal THD resolution, without adding any complex circuit and low number of switches and less gate circuits are demanded. The main characteristic of PSO algorithm is a stochastic nature based on natural search and selection principle. There are several evidences of intelligence for the well posed domains in animals, plants and normal living things. The optimum parameters are designed for closed loop control action of proposed 15-level inverter to eradicate the harmonic content, improve the transient stability, acquiring the qualitated RMS voltage. Several comparisons are made for different control objectives with classical over proposed intelligent controllers. Finally, the validation of the proposed module for micro grid system with PID-PSO controller is evaluated by using Matlab/Simulink platform and results are conferred.

2 Proposed Co-generation Scheme

Figure 1 depicts the overall structure of proposed co-generation system with effective PID-PSO control objective for micro grid system via modern asymmetrical 15-level inverter. This co-generation system comprises of a combination of fuel cell (FC) stacks, photo-voltaic (PV) arrays are required for attractive power production attitude due to high reliability incredible efficiency, tidiness. While, the stiff power are attained by polymer electrolytic membrane fuel cell (PEMFC) as a primary candidate under low contrasted temperatures.

The PV/FC outcome power is converted into unique self-reliant voltage sources with myriad relationships which are interfaced to the power inverter module via high voltage gain DC-DC converter. This DC link acts as barrier in between these converter modules. The inverter module consists of two different modules; primary is level generation can be formed as a sub-multiple modules. This module consists of two switches per module and holds the DC link voltage with several voltage values like Vdc, 2 Vdc, 4 Vdc, generates the 7-level DC voltage using 6 switches The secondary circuit as a full bridge converter, it converts the 7-level DC voltage into 15 levels AC voltage as a polarity generation concept. In this process, the proposed system generates a sinusoidal outcome of both voltage/currents, these are

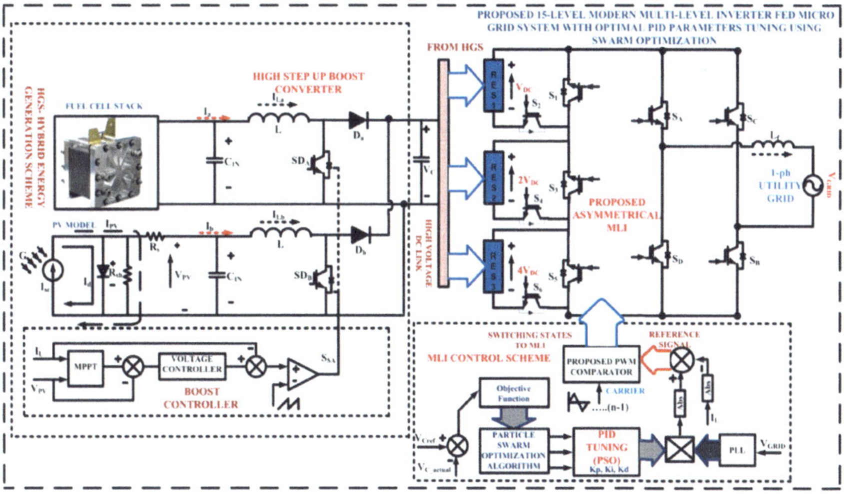

Fig. 1 Overall structure of proposed co-generation system with PID-PSO controller for micro grid system via modern asymmetrical 15-level inverter

in-phase with each other and related to grid voltage, then interfaced to utility grid via inductive filters. This converter requires only 10 switches for generating 15-levels, compare to basic inverters and have more opportunities. This inverter is powered by effective controller by recognizing the grid voltage by a detection circuit and fed to PLL circuit so as to produce a sinusoidal signal with unified amplitude. The capacitor voltage is recognized and then compared over the referred voltage, and that outcome coming from that comparison is drive to the PID controller. The PLL circuit output and the controller output are multiplied to produce a reference signal, while the current of the 15-level inverter is recognized by a current detection circuit. The reference signals coming from the comparison are proceed to integrate with proposed PWM generation circuits for production of switching states for 15-level inverter topology according to characteristic table.

3 Proposed Modulation Scheme

Plentiful modulation approaches are attained for multi-level inverters with efficient techniques are used in many applications can be classified as phase shifted multi carrier switching scheme is proposed by Bahr and Rama Rao in [10], based on these schemes author implement a unique modulation scheme, which have better results compared to formal methods.

The classical type variable switching frequency is proposed by Mao et al. in [11] and compare to proposed PWM technique, this technique mainly regards to different carrier frequencies ranges with respect to multiplication factor. The carrier waves with carrier frequencies of 3050, 5050, and 7050 Hz are compared with the reference signal as depicted in Fig. 2. The proposed multi-carrier modulation scheme is nothing but merging of both phase and level shifted PWM techniques to trounce the problem it regards to switching action of level shift technique and

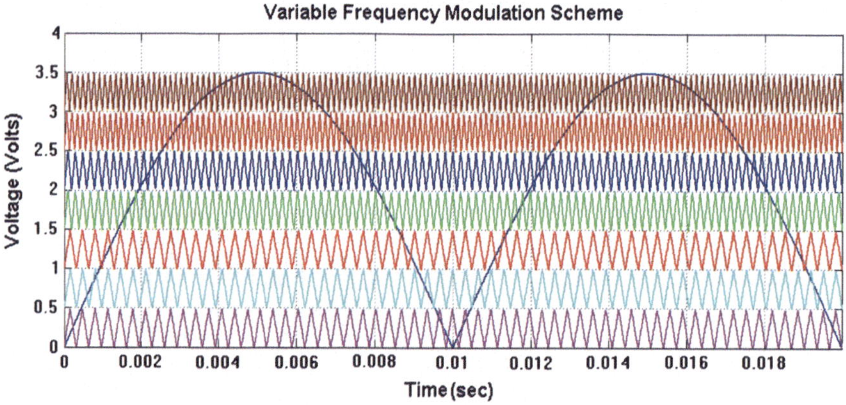

Fig. 2 Switching strategy for variable frequency modulation scheme

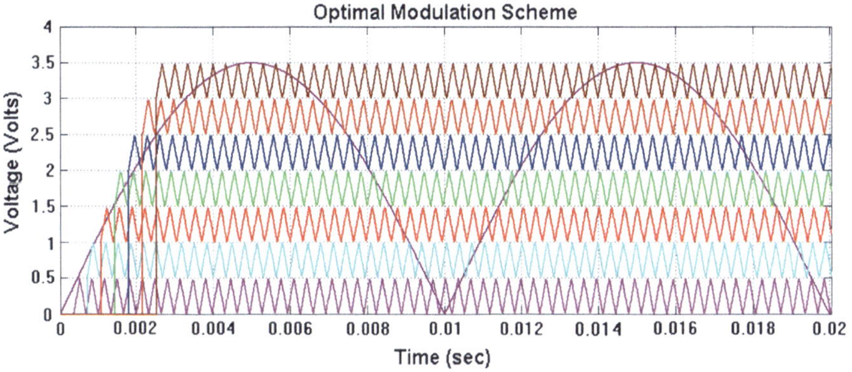

Fig. 3 Switching strategy for optimal modulation scheme

overcome the phase imbalance condition. In this scheme all carriers required equal frequency with difference of peak amplitude is equal, which are vertically disposed. The optimal modulation scheme for 15-level proposed multilevel inverter as depicted in Fig. 3.

$$MI_a = \frac{V_{ref}}{V_{cr(q-1)}} \tag{1}$$

$$\varnothing_{sh} = \frac{360^{\circ}}{4(n-1)} \tag{2}$$

For this optimal scheme, (n − 1) carriers are required and are defined based on requirement and essentially disposed vertically by measurable angle. Where V_{ref} constitutes the reference signal coming from PID-PSO controller, V_{cr} constitutes the carrier signal, MI_a constitutes the modulation index and $\varnothing_{sh}$ constitutes the disposed phase angle.

As, Table 1 shows the switching pulses for 15 Level MLI for production of DC voltage levels and Table 2 shows the switching pulses for 15 Level MLI for production of polarities, in that "L" specifies the switch is at OFF state and "H" specifies the switch is at ON state.

Table 1 Switching pattern level selection scheme for proposed inverter module

Vo	S_1	S_2	S_3	S_4	S_5	S_6
7 Vs	L	H	L	H	L	H
6 Vs	H	L	L	H	L	H
5 Vs	L	H	H	L	L	H
4 Vs	H	L	H	L	L	H
3 Vs	L	H	L	H	H	L
2 Vs	H	L	L	H	H	L
Vs	L	H	H	L	H	L

Table 2 Switching pattern level selection scheme for proposed inverter module

Vo	S_A	S_B	S_C	S_D
Zero state	L	H	L	H
Positive state	H	H	L	L
Negative state	L	L	H	H

4 PID-PSO Controller

The classical PID controller includes of Proportional; Integral; Derivative gains in feedback systems. The transfer function of specified plant for this controller is given as;

$$C(s) = K_p + \frac{K_i}{s} + K_d s \tag{3}$$

where, K_p, K_i, K_d, are the respective gains of the classical PID controllers that are tuning imperatively by classical trail and error method and proposed PSO method. PSO stands for Particle Swarm Optimization (PSO) is an optimized algorithm based on soft-evolutionary computational method. The general PSO is implemented from the research on many swarm like as bird flocking and fish schooling [12] techniques. A novel parameter called inertia weight is added to original PSO where the inertia weight is decreased linearly during iteration with addition to another common scheme of PSO is proposed by Clerc [13, 14]. Several particles are represented in a PSO by potentiality solution to the problem; every particle adjusts its flying capacity following to its possessed flying experience and its associate. Every particle is handled as a D-dimensional space point. The *i*th particle is constituted as $X_1 = (xi_1, xi_1, ..., xi_D)$. The best minimum fitness values of any particle is constituted and registered as $P_1 = (pi_1, pi_1,...., pi_D)$, is called as the *pbest*. The index function of the best particle over all the particles in the specified population is constituted by the symbol *g*, called as the *gbest*. The velocity for every particle of *i* is constituted as $V_1 = (vi_1, vi_1, ..., vi_D)$. The obtained particles are renovated respect to following equations:

$$v_{id}^{n+1} = w.v_{id}^{n} + c_1.rand().(p_{id}^{n} - x_{id}^{n}) + c_2.rand().\left(p_{gd}^{n} - x_{gd}^{n}\right) \tag{4}$$

$$x_{id}^{n+1} = x_{id}^{n} + v_{id}^{n+1} \tag{5}$$

where the c1, c2 = 1.2 are two attracted positive constants, *n* constitutes the number of iterations, rand () constitutes the numbers taken by randomly in between 0 and 1. Calculation of each particle new velocities by Eq. 4 and it is followed by its antecedent velocity and the distance of its present position from its best own

Table 3 Operating parameters for proposed 15-level inverter for grid system

S. no	Operating parameters	Values
1	PV source	50–200 V
2	Fuel cell stacks	50–150 V
3	Boost inductor (L)	1 mH
4	Capacitors (C)	780 μF
5	Filter inductor (Lf)	1.5 mH
6	Switching frequency	3050 Hz

experience and the group best experiences. Then the particles move toward a new position followed by Eq. 5. The appearance of each particle is recorded according to a pre-established fitness function, which is allied to the problem to be decoded. Inertia weight w is accompanied into the followed equation to balance the local and global search capability [12]. It could be a positive linear/non-linear function of time. A highly assured convergence of PSO is proposed and set as approximate of 0.9. Swarm size of 10–60, here generally taken as 40. This impressed algorithm can be used for optimum tuning of PID gains to ensure best control objective at standard operating conditions. PSO is commissioned to tune the K_p, K_i, K_d gains of the PID controller by offline model. A unique set of PID gain parameters can permits a acceptable system response and minimization of error index as specified in result (Table 3).

Figure 4 depicts the Several Simulations of Proposed 15-Level MLI topology for Grid Integrated System via PID-PSO Controller, in that (a) DC Link Voltage of proposed Controller over formal controller, the proposed has fast steady state response under step changes in inputs; (b) 15-Level Stair Case Output Voltage; (c) Grid Voltage and Current, both are in pure sinusoidal and in-phase each other specifies the unity power factor; (d) Total Harmonic Distortions (THD) of Proposed 15-Level MLI topology for Grid Integrated System via proposed Controller is 4.69 %, it is in IEEE standards limit without any additional filter circuit.

5 Conclusion

Attained simulation results, the optimum tuning of PID controller gains by PSO optimized tool have low steady state response, less peak overshoot, low THD content over classical controllers. Moreover, the classical tuning method is virtue for providing the erupted point of what the PID gains are setting. Consequently the benefit of utilizing smart optimization approach is constituted as a supplement solution to enhance the implementation of the PID controller. The several conclusions are updated from the evaluation of proposed 15-level MLI has fewer switching devices, low space requirement, low cost, high comfort. THD

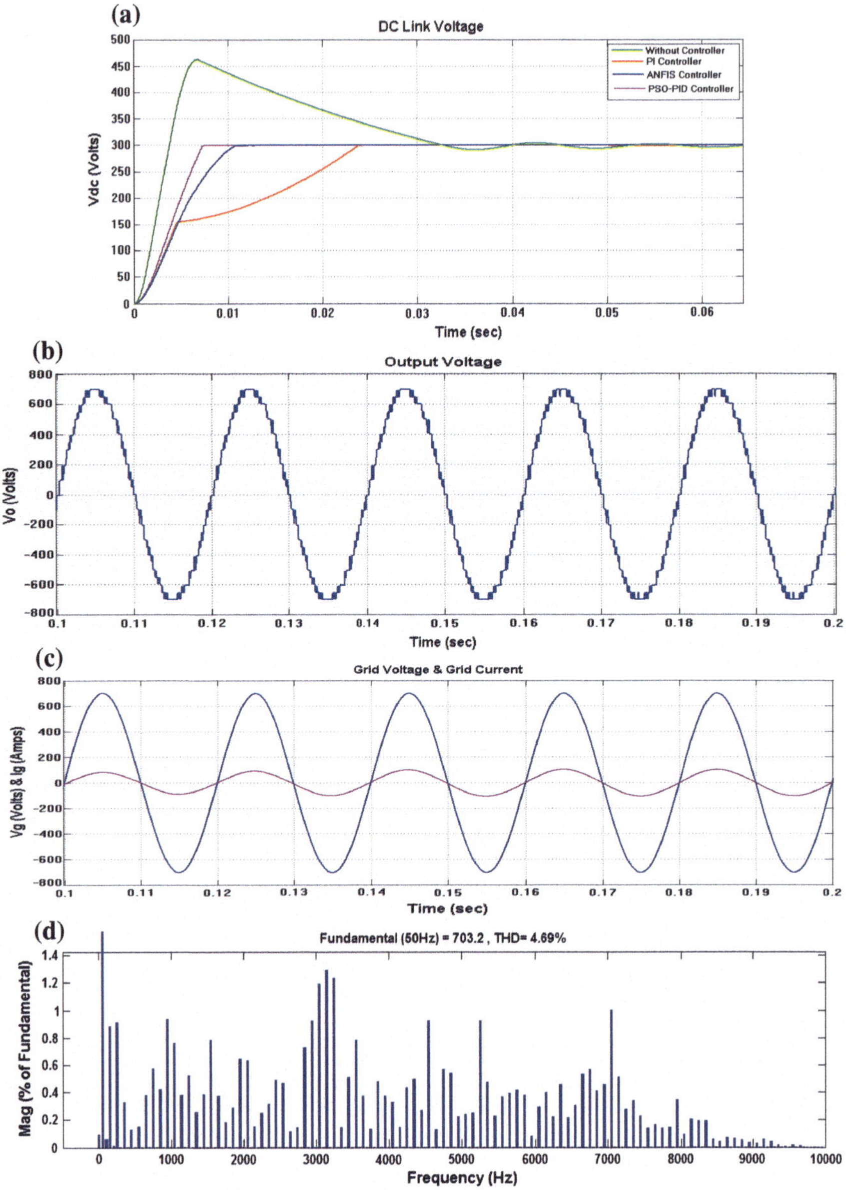

Fig. 4 Several simulations of proposed 15-level MLI topology for grid integrated system via PID-PSO controller **a** DC link voltage, **b** 15-level output voltage, **c** grid voltage and current, **d** THD of 15-level output voltage

Table 4 FFT analysis of output voltage with proposed controller over classical controllers

Type of controller	Regular PWM			Optimal PWM scheme			Variable frequency PWM scheme		
	PD (%)	POD (%)	APOD (%)	PD (%)	POD (%)	APOD (%)	PD (%)	POD (%)	APOD (%)
PI controller	7.75	7.76	7.76	7.65	7.67	7.67	7.68	7.60	7.71
Fuzzy controller	7.22	7.35	7.36	7.39	7.43	7.44	7.29	7.32	7.30
ANFIS controller	6.65	6.76	6.76	6.85	6.88	6.90	6.28	6.33	6.33
PID-PSO controller	5.46	5.56	5.67	5.67	5.70	5.72	4.69	4.70	4.70

comparison is takes place in Table 4, in that optimal modulation with proposed controller has better THD %reduction over classical schemes. The simulation results validates the proposed concept is very attractive for micro-grid system and attains qualitated power, meticulously manages the co-energy.

References

1. Mastromauro RA, Liserre M, Dell A (2012) Control issues in single stage PV systems: MPPT current and voltage control. IEEE Trans Ind Inform 8(2):241–254
2. Thounthong P, Pierfederici S, Martin JP, Hinaje M, Davat B (2010) Modeling and control of fuel cell/super capacitor hybrid source based on differential flatness control. IEEE Trans Veh Technol 59(6):2700–2710
3. Liserre M, Sauter T, Hung JY (2010) Future energy systems: integrating renewable energy sources into the smart power grid through industrial electronics. IEEE Ind Electron Mag 4 (1):18–37
4. Seme S, Stumberger G, Vorsic J (2010) Operational properties of a photovoltaic system with three single phase inverters. In: International conference on renewable energies and power quality, pp. 1–4
5. Hasegawa K, Akagi H (2012) Low-modulation index operation of a five-level diode-clamped pwm inverter with a dc-voltage-balancing circuit for a motor drive. IEEE Trans Power Electron 27(8):3495–3505
6. Pouresmaeil E, Montesinos-Miracle D, GomisBellmunt O (2012) Control scheme of three-level NPC inverter for integration of renewable energy resources into AC grid. IEEE Syst J 6(2):242–253
7. Barros JD, Silva JFA, Jesus EGA (2013) Fast-predictive optimal control of NPC multilevel converters. IEEE Trans Ind Electron 60(2):619–627
8. Choi S, Saeedifard M (2012) Capacitor voltage balancing of flying capacitor multilevel converters by space vector PWM. IEEE Trans Power Delivery 27(3):1154–1161
9. Maharjan L, Yamagishi T, Akagi H (2012) Active-power control of individual converter cells for a battery energy storage system based on a multilevel cascade pwm converter. IEEE Trans Power Electron 27(3):1099–1107
10. Bahr SME, Rama Rao KS (2011) A new multi carrier based PWM for multilevel converter. In: 2011 IEEE Applied Power Electronics Colloquium (IAPEC) IEEE 2011, pp. 63–68

11. Mao X, Ayyanar R, Krishnamurthy HK (2009) Optimal variable switching frequency scheme for reducing switching loss in single-phase inverters based on time-domain ripple analysis. IEEE Trans Power Electron 24(4):991–1001
12. Shi YH, Eberhart RC (1998) A modified particle swarm optimizer. In: IEEE International Conference on Evolutionary Computation, Anchorage, Alaska
13. Shi YH, Eberhart RC (2000) Comparing inertia weights and constriction factors in particle swarm optimization. Proceedings of The 2000 Congress on Evolutionary Computation, vol. 1, pp. 84–88
14. Clerc M (1999) The Swarm and the queen: towards a deterministic and adaptive particle swarm optimization. In: Proceedings of the Conference on Evolutionary Computation, pp. 1951–1957

Optimal Operation of an Integrated Power Distribution System Fed with Renewable Energy Sources, Diesel Generation and Battery Storage

D. Ravi Kumar, K.C. Archana and G.S. Raju

Abstract Recent developments in the electric utility industry is to encourage the entry of renewable energy sources for power generation in the distribution system. Wind energy systems, Photovoltaic systems, diesel engines and gas turbines are considered as Distributed Generators (DG). An Integrated Power Distribution System (IPDS) is a distribution system independent of grid that includes diesel generators, Wind Park, Solar Photovoltaic and batteries etc. This IPDS is considered for voltage stability analysis in the literature with Solar and Wind energy sources. However, it appears that no attempt has been made to reduce the losses using with integration of renewable energy sources. Further, optimal scheduling for integration of Distributed Generators (DG) is not considered in the literature so far. The main objective of this paper is to reduce the diesel consumption and losses by utilizing the energy from renewable energy sources and battery storage. The optimal scheduling of the distributed generators in IPDS is also considered. IEEE 14 Bus system is analyzed with Wind, Solar, Battery, Diesel generation and Static Var Compensator (SVC) using Power System Analysis Toolbox in MATLAB.

Keywords Integrated power distribution system · Distributed generation · Optimal scheduling · PSAT

1 Introduction

In recent years, the interest in the distributed energy sources is increasing due to various technical, economical and environmental factors. By this point of view, studies on renewable energies focus more and more attention on their optimal

D. Ravi Kumar (✉) · G.S. Raju
VNR VJIET, Secunderabad, India
e-mail: ravikumar_d@vnrvjiet.in

K.C. Archana
VCE, Hyderabad, India

K.R. Attele et al. (eds.), *Emerging Trends in Electrical, Communications and Information Technologies*, Lecture Notes in Electrical Engineering 394,
DOI 10.1007/978-981-10-1540-3_45

utilization. Wind energy and Solar energy are the two renewable energy sources most common in use. Wind energy has become the least expensive of renewable energy technologies in existence. Photovoltaic Cells (PVC) convert the energy from sunlight into DC power [1]. PVC offer added advantages over other renewable energy sources and require practically no maintenance. Hybridization of wind and solar power sources provide a realistic form of power generation.

Renewable energy generation offers clean, abundant energy gathered from self-renewing resources such as wind and the sun etc. When the power demand increases, generation also should increase. So, renewable energy sources can be used to supply constant loads. The distributed generations have several advantages such as reducing the real power loss, required transmission capacity and the traditional generation expansion, etc. [2].

The output of wind and solar generators is determined by the climate and weather conditions [3]. The effect of wind power on the transient fault behaviour is considered by replacing the power generated with two wind turbines, increasing gradually the rate of wind power penetration and changing the placement of wind resources [4]. The algorithm using Particle Swarm Optimization [5] identifies the optimal location of Solar PV with minimum active power losses. However, it appears integration of various renewable energy sources have not been considered.

In this paper, the optimal scheduling of the distributed generators in IPDS with Analytical method using Current injection is considered. Load scheduling is applied to IEEE 14 Bus test system with Wind, Solar, Battery, Diesel generation and SVC. The simulation of test system is carried out using MATLAB Power System Analysis Toolbox (PSAT) [6].

2 Load Scheduling

2.1 Integrated Power Distribution System

An IPDS of capacity 30 MW is considered for the simulation process, with Wind power maximum capacity of 10 MW, Solar maximum capacity of 5 MW, Diesel maximum capacity of 10 MW. Maximum power rating of Battery is 8 MW. Initially a Load scheduling is prepared for a particular day which is repeated daily. In the next stage IEEE 14 bus system is considered for the Load Flow study. Wind, Solar, Diesel and battery are connected to the IEEE 14 bus system and load flow is performed and Load scheduling table is simulated on this system. SVC is connected to one of the buses to reduce the losses.

2.2 Load Scheduling Table

Steps involved in Preparation of Table:

- Load Scheduling in an IPDS with renewable energy sources, Diesel generator, non conventional energy sources and battery storage is performed by dividing the 24 h of day into 12 intervals of time duration 2 h during which loads are assumed to be constant. The table consists of Load demand, Wind Power, Solar Power, Diesel Power, Battery and Battery Capacity.
- Wind Power is divided in two columns one is Available and the other is Used. 'Available' means which is available for that particular time period and 'Used' means which is used for the load supply.
- Excess power from the Wind generation is used to charge the Battery. This mode of battery is called Charging state of Battery and is indicated as negative sign in the table. Scheduling is done considering the battery condition i.e. battery has to be charged to give output of 2 MW at the end of the each day. This is to ensure long life for the battery.
- Power can be used from battery which is known as discharging state of battery mode. This is indicated as positive sign in the Table 1.

2.3 Constraints

- Minimum power available from battery at the beginning and end of day is 2 MW.
- Diesel cost has to be reduced by utilizing the maximum power available from wind/solar.
- Solar power limits : 0–5 MW
- Wind power limits : 0–10 MW
- Diesel power limits : 2–10 MW

2.4 Power Balance Equation

- Diesel Plant power = (Load demand at that point) − (Available generation of renewable energy sources and battery discharge)

$P_d\,(\min) \leq P_d \leq P_d\,(\max)$
$P_w \leq P_w\,(\max)$
$P_s \leq P_s\,(\max)$
$P_d = P_l - P_s - P_w - P_b$
where P_d = Diesel Power; P_l = Load Power; P_s = Solar Power
P_w = Wind Power; P_b = Battery Power.
where, Available Generation implies that:
Total generating capacity—Losses—Capacity of the units not working.

Table 1 Optimal generation schedule

S. no.	Time	Load/ demand	Wind power (MW)		Solar power (MW)		Diesel power (MW)	Battery (MW) charging/ discharging	Battery capacity (MW) (Initially battery charged to 2 MW)
			Available	Used	Available	Used			
1	0–2 AM	6	8	4	–	0	2	–4	6
2	2 AM–4 AM	6	6	4	–	0	2	–2	8
3	4 AM–6 AM	8	4	4	–	0	2	2	6
4	6 AM–8 AM	12	6	6	1	1	2	3	3
5	8 AM–10 AM	14	8	8	1	1	4	1	2
6	10 AM–12 PM	20	10	10	5	5	5 + 2	–2	4
7	12 PM–2 PM	17	4	4	5	5	8		4
8	2 PM–4 PM	15	4	4	5	5	4	2	2
9	4 PM–6 PM	14	6	6	1	1	6	1	1
10	6 PM–8 PM	12	6	6	–	0	6		1
11	8 PM–10 PM	10	8	8	–	0	2		1
12	10 PM–12 AM	11	10	9	–	0	2	–1	2

2.5 Daily Load Scheduling Table

Daily load scheduling table of the different schedule for generation available for this system, the one which gives minimum diesel consumption is given in Table 1.

3 Simulation and Results

3.1 Implementation to IEEE 14 Bus System

PSAT 2.1.8 version is used for load flow study in this paper. The load scheduling given in Table 1 is applied for IPDS of maximum capacity 30 MW. IEEE 14 bus system is taken for the load flow analysis. IEEE 14 bus system implemented with PSAT in MATLAB is shown in Fig. 1. Diesel Generator bus is considered as Slack Bus. Wind Energy Generator is connected at Bus 3, solar photovoltaic generator is connected at Bus 6 and Battery is connected at Bus 2. Newton-Raphson Method is used for Power Flow Computation. Power flow is conducted for every interval of Time period mentioned in the Load scheduling table and is analyzed. Power flow results are given in Table 2. Diesel generator is used to supply the losses in the system as it has been used as a Slack bus.

From Table 1, Interval 4 is considered for load flow analysis. Data from Load scheduling Table is: Load = 12 MW; Wind Energy = 6 MW; Solar Energy = 1 MW; Diesel energy = 2 MW; Battery Power = 3 MW.

3.2 Optimal Placement of Static Var Compensator

For the reduction of losses in IPDS, SVC is used to provide necessary reactive power.

IEEE 14 bus system with optimal placement of SVC and Diesel generator is shown in Fig. 2.

1. SVC is connected at different buses and its location for minimum losses is obtained for each time interval and losses are given in Table 2.
2. Q_{max} and Q_{min} ranges for SVC are chosen as 0.15–0.06 p.u.
3. SVC position is changed at different buses and found that the placement at BUS 7, the losses are less compared to the other places. From Table 2, it is observed that the losses are reduced by using the SVC. For example, for interval 4 the losses are reduced to 0.17375 MW from 0.17805 MW. In a similar way, for other intervals also losses are reduced. Further reduction in losses is also proposed by optimal location of Diesel Generator.

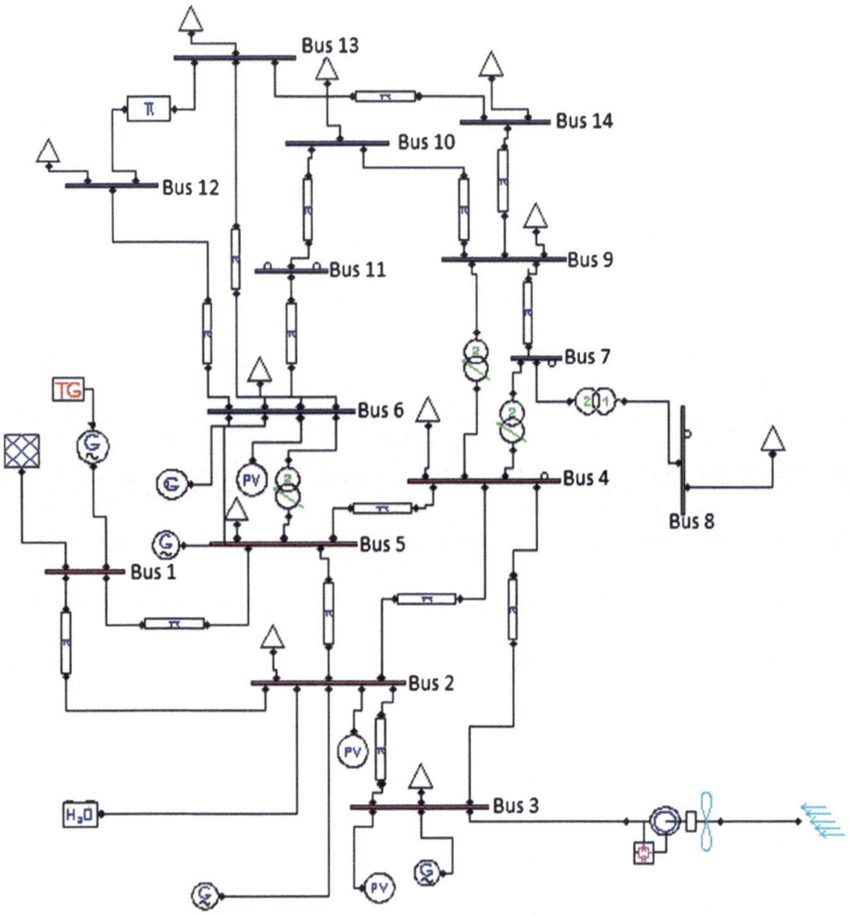

Fig. 1 IEEE 14 bus system implemented in MATLAB with PSAT

3.3 Optimal Placement of Diesel Generator

1. The energy from Renewable Energy Sources is uncertain and the considered IPDS is independent of Grid, so, to compensate losses Diesel Generator is used as backup for the IPDS.
2. Diesel Generator connected bus is considered as Slack bus.
3. Diesel generator is placed at different buses to obtain its best location for minimum losses. As the position of the Wind, Solar and Battery Bank cannot be changed, Diesel generator is changed and system is analyzed.
4. BUS 8 Position gives the minimum losses as compared to the other locations.

Table 2 Comparisons of losses

INT	Time	Load/ demand (MW)	Real power losses without SVC (MW)	Real power losses with SVC (MW)	Reactive power injected at Bus 7 *QGen (Mvar)*
1	0–2 AM	6	0.10798	0.10479	0.94743
2	2 AM–4 AM	6	0.10798	0.10479	0.94743
3	4 AM–6 AM	8	0.12482	0.1213	1.7588
4	6 AM–8 AM	12	0.17805	0.17375	2.0203
5	8 AM–10 AM	14	0.22114	0.21632	2.181
6	10 AM–12 PM	20	0.33047	0.32372	2.5015
7	12 PM–2 PM	17	0.23679	0.23096	2.2105
8	2 PM–4 PM	15	0.19452	0.18941	2.0705
9	4 PM–6 PM	14	0.20633	0.20144	2.1584
10	6 PM–8 PM	12	0.18205	0.17766	2.0496
11	8 PM–10 PM	10	0.17623	0.17239	1.9334
12	10 PM–12 AM	11	0.19985	0.19579	2.0145

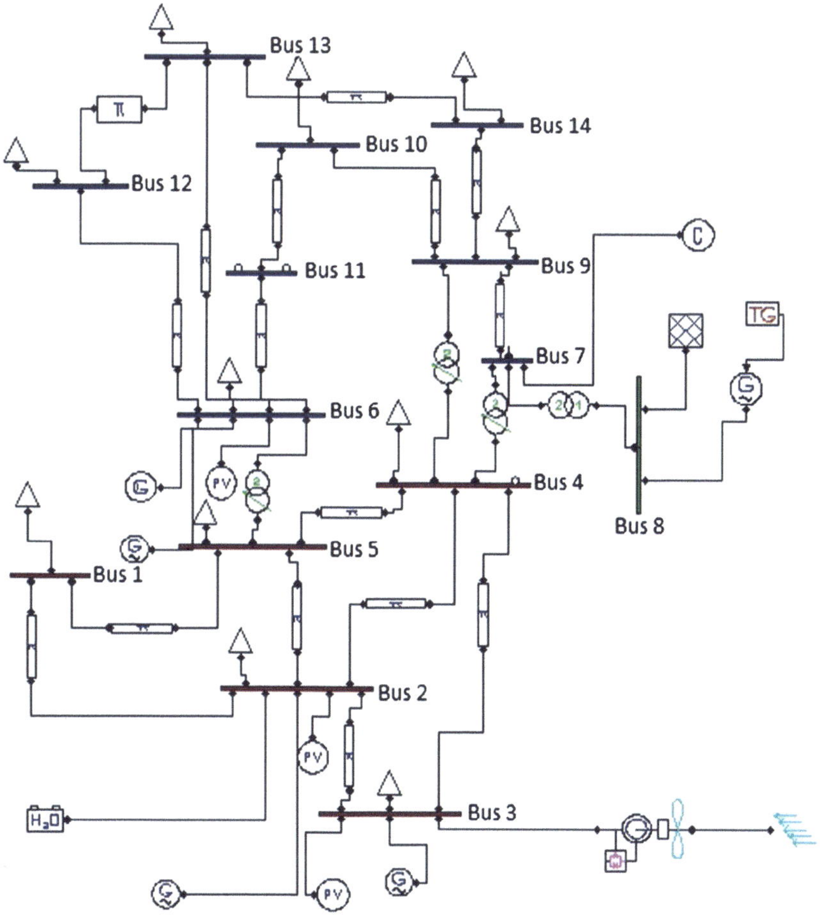

Fig. 2 IEEE 14 bus system with optimal placement of SVC and diesel generator

4 Conclusions

Daily Load scheduling table is prepared and IEEE 14 Bus system is analyzed by considering IPDS with wind, solar, Diesel generation and Battery storage. Losses without reactive power injection and with reactive power injection are also analyzed. Bus 7 is found to be the best location for reactive power injection with SVC for minimum Real power losses. Bus 8 is found to be the best position for the diesel generator.

Acknowledgments The authors thank Dr. M. Ramamoorty, Former Director General, CPRI, Bangalore, Distinguished Professor, VNRVJIET for his valuable guidance and also thank the management of VNRVJIET for the support and encouragement.

References

1. Rai GD Non conventional Energy sources. Khanna Publishers
2. Hong Y-Y, Chiu C-S, Li C-T (2007) KW scheduling in an autonomous system. In: Power Tech, IEEE Lausanne, pp. 1730–1735
3. Ani Vincent Anayochukwu (2013) Simulation of photovoltaic, diesel hybrid power generation system with energy storage and supervisory control. Int J Renew Energy Res 3(3):605–614
4. Warsono, King DJ, Ozeveren CS, Bradely DA (2007) Economic load dispatch optimization of renewable energy in power system using genetic algorithm. IEEE Trans. Power Syst 2174–2179
5. Naimi D, Bouktir T (2008) Impact of wind power on the angular stability of a power system. Leonardo Electron J Pract Technol 12:83–94
6. Milano F (2005) An open source power system analysis toolbox. IEEE Trans Power Syst 20(3):1199–1206

Mitigation of Power Quality Problems in Distribution System Using D-STATCOM

N. Visali, Kamarthi Sridevi and N. Sreenivasulu

Abstract Now-a-days, the most important discussing topic in the world of power systems is maintenance of power quality. After generating voltage, the engineers in the substations are struggling for transmitting as well as distributing of power to the receiving end, since different loads at the ends of distribution are very sensitive to the fluctuations in the voltage, interruptions of voltage and harmonics. This paper shows the improvement of Voltage Sag and THD using LCL Passive Filter along with the Distribution Static Compensator (D-STATCOM) which works with the principle of Voltage Source Converter (VSC).

Keywords D-STATCOM · Voltage Source Converter (VSC) · Controller · LCL passive filter · Harmonic distortion · FACTS

1 Introduction

Usually, the generation and transmission of power is a very difficult task, in which so many components within the system are to be worked at a time for effective maximization of output. Within those components, the reactive power plays a significant role. The active power has to be delivered through the transmission lines by maintaining the voltage.

N. Visali · K. Sridevi (✉)
Department of Electrical and Electronics Engineering,
JNTUACEP, Pulivendula 516390, AP, India
e-mail: kamarthi.sridevi@gmail.com

N. Visali
e-mail: nvisali@gmail.com

N. Sreenivasulu
Department of Electrical and Electronics Engineering,
SRIT, Ananthapuramu 515701, AP, India

K.R. Attele et al. (eds.), *Emerging Trends in Electrical, Communications and Information Technologies*, Lecture Notes in Electrical Engineering 394,
DOI 10.1007/978-981-10-1540-3_46

The most important source for all voltage sags is the current increment within the small interval. Mainly the contributions of these voltage sags are starting of motor, energizing of transformer and faults.

In transmission and distribution systems, for enhancing the problems of power quality, the various approaches are existed. Among those, more effectively used device is D-STATCOM. Due to the presence of switching pattern in this device, the harmonics are caused while the harmonics in the currents can cause distortion of harmonics in waveform, reduction in power factor and also losses in the equipment, which can be improved by using the LCL passive filter.

Chen and Hsu [1] proposed the approach for harmonic analysis of STATCOM for minimizing the voltage and current harmonics. Bollen [2, 5] discussed about the voltage sag and its solutions in three phase systems and also characterized them in distribution systems of industries. Mienski et al. [3] discussed the modelling of D-STATCOM for improving power quality by shunt compensation. Jenkins [4] discussed the power electronics based FACTS devices in the distribution system for improving power quality. Almeida [6] discussed the problems of power quality and related solutions for them. Haque [7] introduced the DVR and D-STATCOM devices for Voltage sag compensation in distribution system.

In this paper, the method has been proposed to improve the performance of power quality by using D-STATCOM with LCL Passive filter and also the simulation results will be observed.

2 Distribution Static Compensator (D-STATCOM)

A STATCOM is simply a three phase inverter which is connected to grid through reactor and a coupling transformer. The voltage difference across the reactance causes the transfer of active and reactive power between the power system and the STATCOM. Generally, a STATCOM utilized in the distribution system is said to be D-STATCOM.

D-STATCOM comprises of a

(1) Coupling transformer
(2) Two-level Voltage Source Converter,
(3) Energy storage device and
(4) Controller.
(5) LCL Passive filter

Coupling transformer is three winding transformer with 230/11/11 kV where the primary side is connected to transmission system, secondary side to load and tertiary side to D-STATCOM (Figs. 1, 2, 3, 4, and 5).

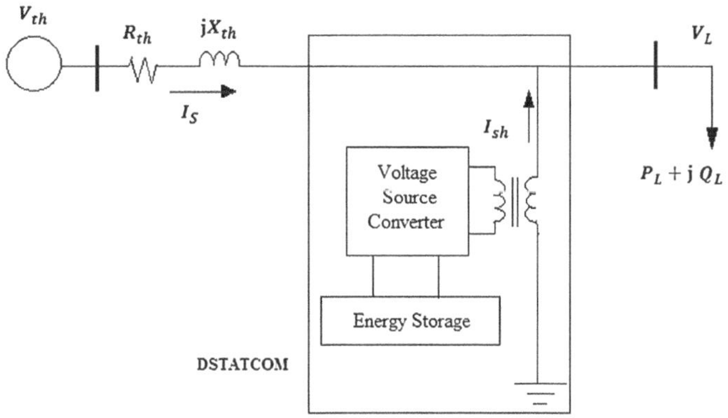

Fig. 1 Schematic diagram of D-STATCOM

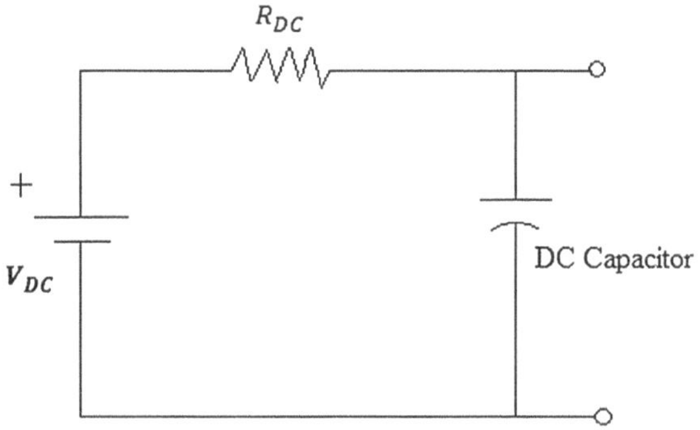

Fig. 2 Circuit diagram of DC storage

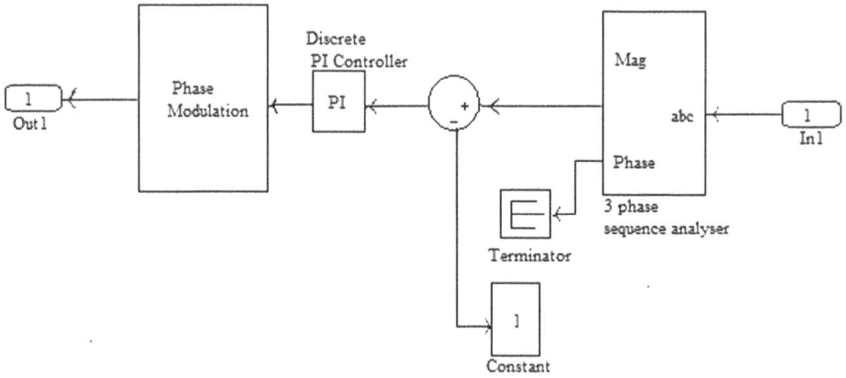

Fig. 3 Simulink model of D-STATCOM controller

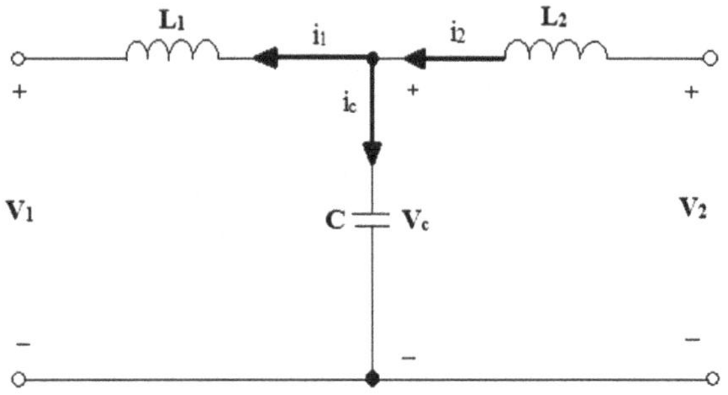

Fig. 4 LCL passive filter

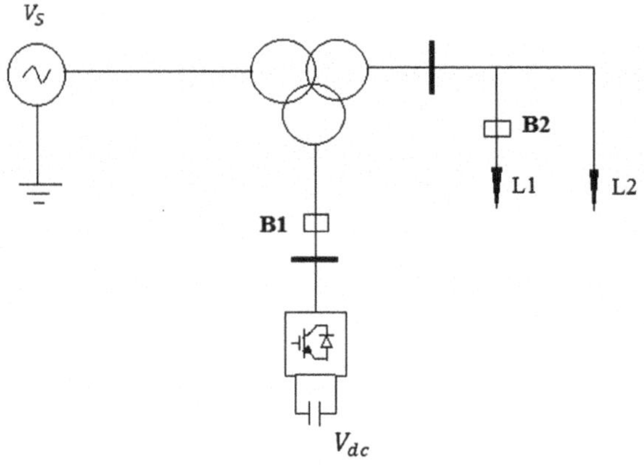

Fig. 5 Single line diagram of test system

The Voltage Source Converter will be supplied with a DC voltage by energy storage device then the technique of sinusoidal PWM is used to control the switching of VSC to maintain the required output voltage at the distribution system. The converter which is connected in parallel with the ac system has so many functions:

1. Regulation and reactive power compensation
2. Rectification of Power factor and
3. Removal of current harmonics.

The main reactive energy storage element is DC capacitor, which is connected in shunt with DC source, can carry the converter's input ripple currents.

The input of Proportional Integral controller (PI) is an error signal which is obtained from the difference of reference voltage and the measured rms value of terminal voltage. The error signal is processed by this controller and gives angle 'δ' as the output. Finally the 'δ' is given to the PWM signal generator for producing the signals for valves of VSC.

Earlier, passive filters and active power line filters were used for enhancement of Power Quality by filtering harmonics in the current waveform. For filtering the harmonics, both parallel and series LC circuit were used but parallel circuit (rejection circuit) are used only in some cases but normally the series LC circuit connection (accepter circuit) was used. At resonance frequency, if both the magnitudes of their reactances are equal then they subtract to zero. So, at this particular frequency the circuit will be short circuited.

The series tuned circuit will be used for compensating the harmonic currents which are produced by equipment i.e. the harmonic current which is flowing towards the supply source can be avoided by dividing the path to flow through the impedance in the acceptor circuit results distortion in the supply voltage. The main duty of this filter is reducing the harmonic current magnitude that flows towards supply network.

In active power filters, as it is an electronic converter necessarily generates and injects harmonic currents into the system for canceling the load current harmonics. For load compensation in ac system, it is installed in the point of coupling (PCC), if once installed the circulation of harmonic currents is limited. Even it can be used for compensation of harmonic currents, reactive power and negative sequence currents. So, it is named as active power line conditioners (APLC).

From fig 4. where V_1 = Grid side voltage i_1 = Grid side current
V_2 = Converter output voltage i_2 = Current through L_2
V_C = Voltage across filter capacitor i_c = Capacitor current

If the harmonic currents in the output voltage from the system are not handled properly, resonance oscillations and problems due to instability can be occurred. If only L-filter is used in the system, it introduces a resonance frequency within the system, so passive damping circuit along with L-filter is used for reducing the resonance.

Passive damping circuit which includes either purely resistive or a combination of resistors, capacitors and inductors is to be adopted. But the active damping in which the converter output voltage damping out the resonance oscillations is better option which results effective reduction of harmonics that are caused by the faults in distribution system.

3 Test System

The proposed test system consists of thevenin's equivalent circuit with 50 Hz transmission system connected to 230 kV primary side of three winding transformer, variable load to 11 kV secondary side and D-STATCOM to 11 kV tertiary side which is provided with 750 μF as energy storage device. Breakers 1 and 2 are used for controlling the operation period of D-STATCOM and load 1 connection to the system respectively.

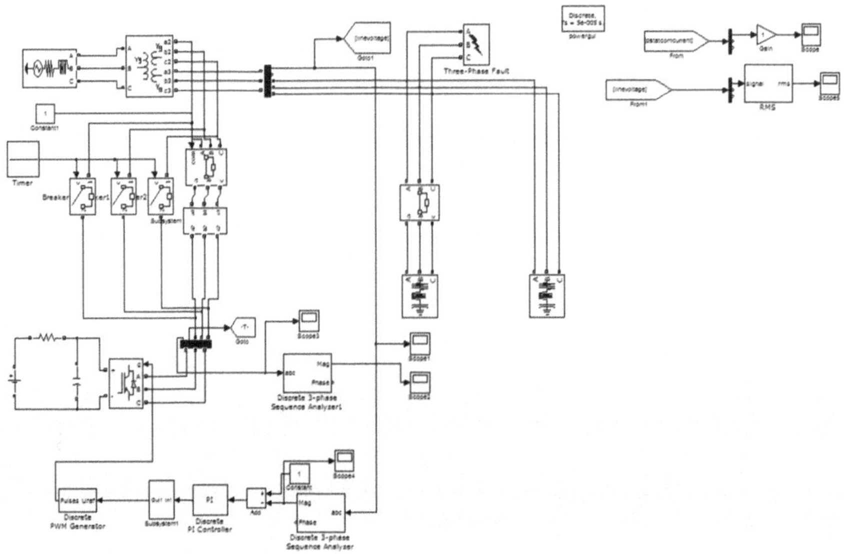

Fig. 6 Simulink model of proposed model

4 Simulink Model for the Test System

See Fig. 6.

5 Results

5.1 Simulation Results

In this proposed method, the results are observed for healthy system, system with fault without D-STATCOM and system with both fault and D-STATCOM. Here the fault is injected between 0.1–0.2 s with fault resistance 0.86 Ω.

5.1.1 For Healthy System

See Figs. 7 and 8.

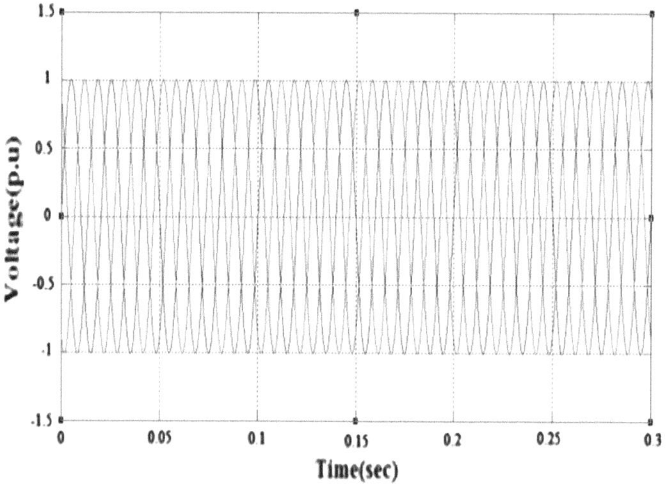

Fig. 7 Three phase voltage without fault

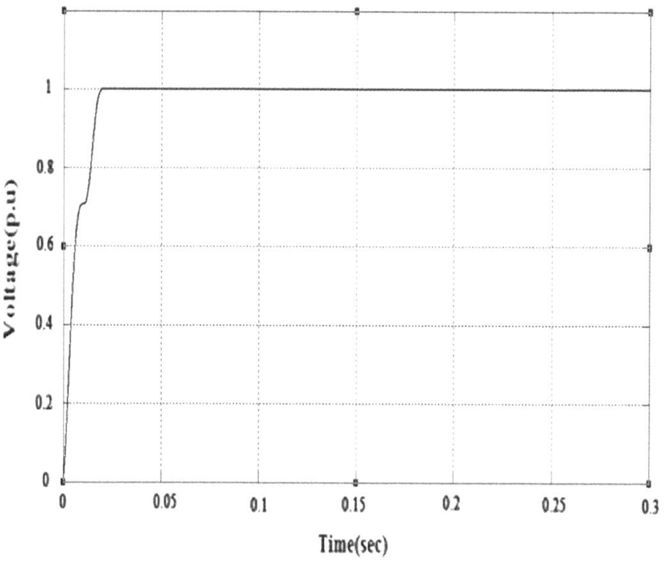

Fig. 8 Voltage without fault

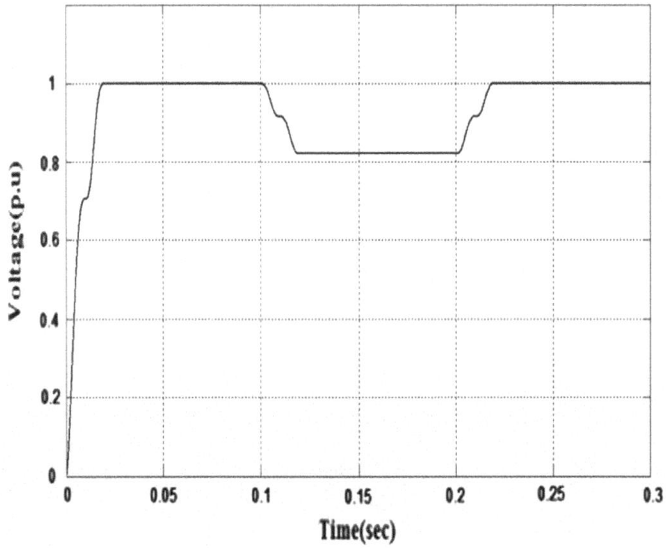

Fig. 9 Voltage sag with LL fault is 0.8210

5.1.2 System with LL Fault Without D-STATCOM

See Fig. 9.

5.1.3 System with LL Fault and D-STATCOM

See Fig. 10.

5.2 Simulation Results of Harmonic Distortion with D-STATCOM Without LCL Passive Filter

The simulation results are observed for percentage of Harmonic Distortion using FFT analysis before and after adding the LCL Passive filter to the proposed system (Figs. 11 and 12).

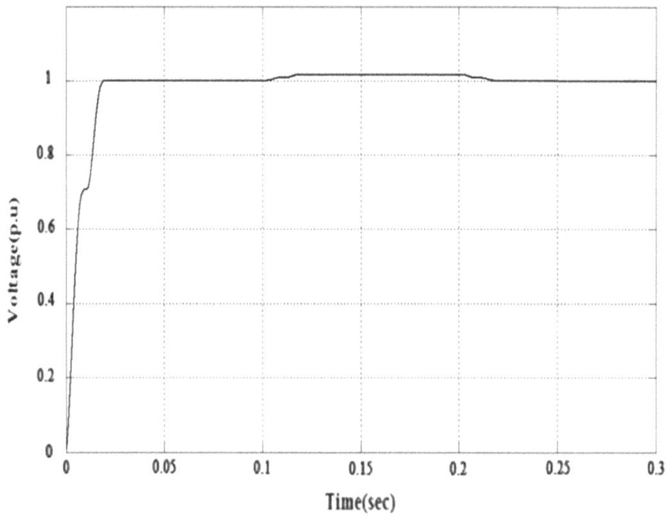

Fig. 10 Voltage sag with LL fault is 1.0152

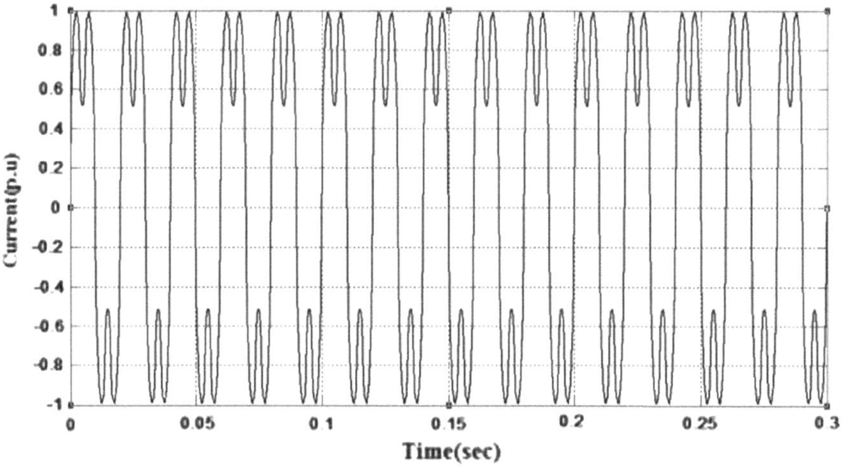

Fig. 11 DSTATCOM current

5.3 Simulation Results of Harmonic Distortion with D-STATCOM with LCL Passive Filter

By adding the LCL Passive filter to the system, there is improvement in dstatcom current as well as in the THD i.e. less than 5 % which is observed in Figs. 13 and 14.

The observation of THD values for other faults are tabulated (Table 1).

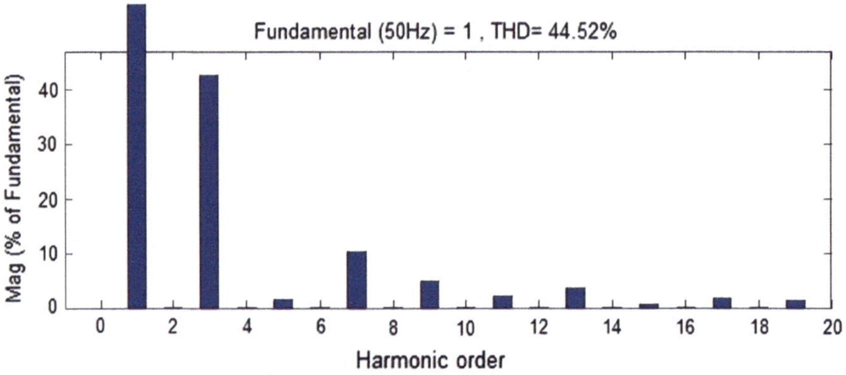

Fig. 12 Harmonic spectrum of output current for LL fault with THD 44.52 %

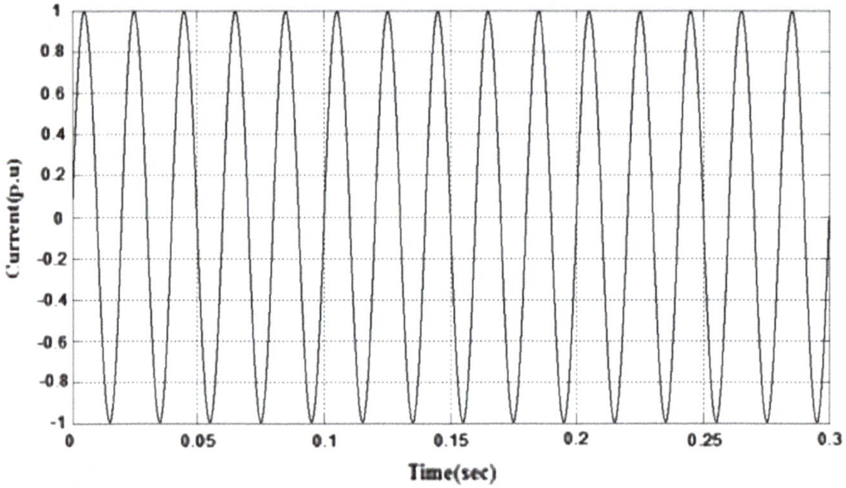

Fig. 13 DSTATCOM current

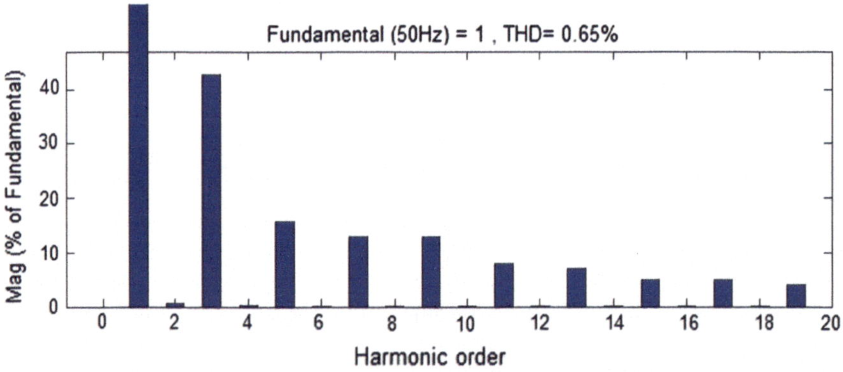

Fig. 14 Harmonic spectrum of output current for LL fault with THD 0.65 %

Table 1 Results of total harmonic distortion for other faults

Types of Fault	Without LCL passive filter (%)	With LCL passive filter (%)
SLG	48.26	1.14
DLG	88.67	1.11
TPG	63.63	1.11
LL	44.52	0.65

6 Conclusion

It is concluded that the simulation results of proposed system for Voltage sag without D-STATCOM and Total Harmonic Distortion without adding LCL Passive Filter to D-STATCOM are highly affected by the faults in the system. But the results of Voltage sag with D-STATCOM and Total Harmonic Distortion with LCL Passive filter are improved within the IEEE standards and additionally power factor is improved to unity which enhances the performance of the proposed system.

References

1. Chen B-S, Hsu Y-Y (2007) An analytical approach to harmonic analysis and controller design of STATCOM. In: Conference on IEEE Transactions on Power Delivery, vol. 22
2. Bollen MHJ (2001) Voltage sags in three-phase systems. Power Eng. Rev. IEEE 21 (9):8–11, 15
3. Mienski R, Pawelek R, Wasiak I (2004) Shunt compensation for power quality improvement using a STATCOM controller: modelling and simulation. IEEE Proc. 151(2)
4. Jenkins N (1998) Power electronics applied to distribution systems. In: IEE Colloquium, Flexible AC Transmission Systems, Ref. No. 19981500, pp. 311–317
5. Yaleinkaya G, Bollen MHG, Crossley P (1999) Characterization of voltage sags in industrial distribution systems. IEEE Trans Ind Appl 34(4):682–688
6. De Almeida A, Moreira L, Delgado (2003) Power quality problems and new solutions. In: International conference on renewable energies and power quality '03, Vigo, Spain, vol. 1
7. Haque MH (2001) Compensation of distribution systems voltage sags by DVR and DSTATCOM. In: Power Tech Proceedings, 2001 IEEE Porto, vol. 1, 10–13 Sept 2001

Location of IPFC Under Contingency Condition in Power System

B.V. Rami Reddy, P. Sujatha and Y.V. Siva Reddy

Abstract The interline power flow controller (IPFC) is the latest generation of flexible AC transmission systems (FACTS) device specifically used for the control of power flows in multi transmission system. If the load on the power system is heavily increased, then the system is at high risk because of line outages and consequent voltage instability problem. The power loss and voltage drop are reliable indicators of voltage security of power networks. Here we analyze the voltages, line apparent power flows and total power losses in the system. This paper also proposes an algorithm for optimal location of the IPFC so as to enhance voltage stability and to maintain the line flows within the limit under the over loaded line outage contingency in a power system network. The over loaded lines (outages) are ranked based on Severity Index. The effectiveness of the proposed method is tested for IEEE-30 bus system with the help of MATLab software.

Keywords FACTS devices • IPFC • Line flows • Line losses • Contingency condition • Severity index • Voltage stability improvement

1 Introduction

The introduction of the FACTS devices into the power system offered great opportunities for the power engineer in the area of operation and control of modern power systems. For example, FACTS devices are often planned for power flow regulation in the steady state thus enhancing the power transfer capability of existing transmission lines. Various types of FACTS devices and their location at

B.V. Rami Reddy (✉)
Department of EEE, R.C.E.W, Kurnool, Andhra Pradesh, India
e-mail: ramireddybv@gmail.com

P. Sujatha
Department of EEE, JNTUCE, Ananthapuramu, Andhra Pradesh, India

Y.V. Siva Reddy
Department of EEE, GPREC, Kurnool, Andhra Pradesh, India

K.R. Attele et al. (eds.), *Emerging Trends in Electrical, Communications and Information Technologies*, Lecture Notes in Electrical Engineering 394,
DOI 10.1007/978-981-10-1540-3_47

different places have varying advantages [1, 2]. The FACTS devices like Static VAR Compensator (SVC), Thyristor Controlled Series Capacitor (TCSC), Static Synchronous Series Compensator (SSSC) and Static Synchronous Compensator (STATCOM) are the first generation FACTS devices available in the literature for control of power flow in transmission systems [3].

The introductory FACTS devices were able to regulate either the flow of active or reactive power along a single transmission line. A breakthrough was made with the introduction of the UPFC [4], which is one of the most versatile FACTS devices and also capable of simultaneously controlling the flow of both active and reactive power in the transmission line. Another newly developed FACTS device, namely IPFC, further extends the capability of independently influencing the active and reactive power flows to simultaneous compensation of multiple transmission lines. These significant functions are made possible by the combination of multiple compensators coupled via a common dc link. Thus, both the UPFC and IPFC are defined as the combined compensators [5].

The IPFC is an advanced FACTS device aimed at controlling the power flow in multiline systems in a substation [6]. IPFC employs Voltage Source Inverter (VSI) as basic building block. Generally, it composes of two VSIs which are capable of transferring real power from one line to any other line and thereby facilitating transfer of real power among the lines, and also achieving independent control of series compensation of each individual line.

IPFC is presented as a power injection model and is implemented to study the effect of IPFC parameters on bus voltages, active and reactive power flows in the lines [7]. The applications of IPFC to improve damping of the system have been reported by few researches and they have applied IPFC to improve transient stability of power system [8]. It can also be utilized to compensate against reactive voltage drops and the corresponding reactive line power and thereby increase the effectiveness of the compensating system against dynamic disturbances [9]. The minimization of generation cost, transmission losses and maximization of the loadability of the transmission system can be achieved by optimally placing IPFC. Different operating conditions of the power system must be considered while determining the optimal size and location of the power flow controller.

Contingency analysis deals with the study of the impact and performance of the system during the outage of the power system components such as transmission lines, transformers and generators. Among these contingencies referring to major disturbances like loss of a transmission line or a generator may create sudden and large changes in both the configuration and the operating state of the power system. Contingencies sometimes may also result in severe violations of the operating constraints. Consequently, to have a secure operating evaluation and planning for contingencies forms an important aspect. [10, 11].

This paper proposes an algorithm for optimal location of the IPFC to improve voltage stability under the over loaded line outage contingency in a power system network. This paper also analyses the performance of the IPFC for various combinations of voltage magnitudes and angles at best IPFC location.

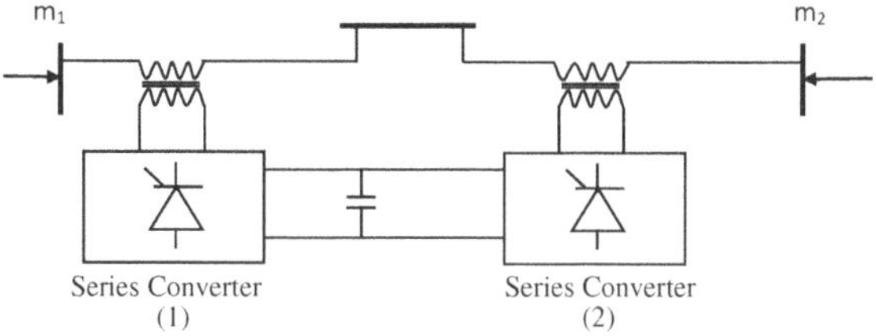

Fig. 1 Simple model of IPFC

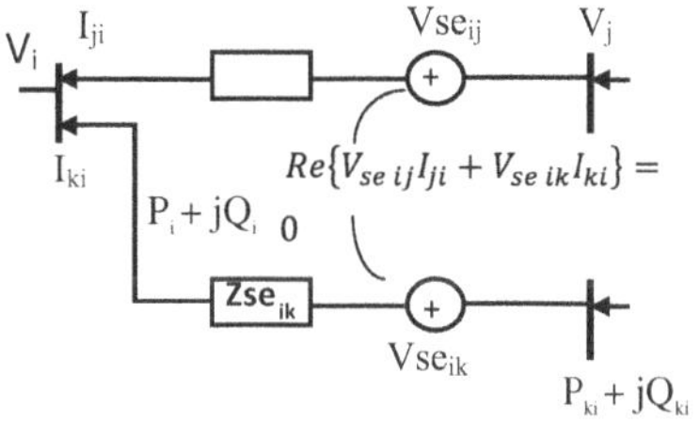

Fig. 2 Equivalent circuit of IPFC

2 Interline Power Flow Controller (IPFC)

In general the IPFC utilizes a number of DC to AC converters each providing a dedicated series compensation for a given line as shown in Fig. 1 and equivalent circuit shown in Fig. 2. The series compensation is achieved by employing two or more independently controllable static synchronous series compensators (SSSC) which are solid state voltage source converters (VSC). By maintaining DC link voltage at the desired level the combination of the series connected VSC can inject a voltage at fundamental frequency with controllable magnitude and phase angle. In practice the DC link is represented as a bidirectional link for exchange of active power among the converters. SSSC is employed for increasing real power transfer on a given line by directly compensating for the voltage drop due to inductive loading of a transmission network. In addition, active power can also be exchanged through these series converters via the common DC link in IPFC. It is noted that the sum of the active powers resulting from VSCs to transmission lines should be zero when the losses in the converter circuits are ignored.

3 Modeling of IPFC

The modeling for IPFC which will be referred to as power injection model is presented here. This model is helpful in understanding the impact of the IPFC on the power system in the steady state. Furthermore, the IPFC model can easily be incorporated in the power flow model. For steady state analysis of power systems the normal practice is to represent VSC as a synchronous voltage source injecting an almost sinusoidal voltage with controllable magnitude and angle. On this basis, the equivalent circuit of IPFC has been modified and is represented as shown in Fig. 3.

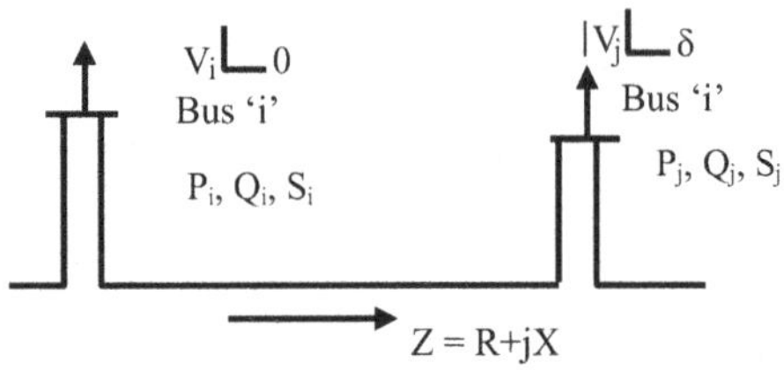

Fig. 3 Equivalent circuit of IPFC

In Fig. 3, V_i, V_j and V_k are the bus voltages at the buses i, j and k respectively, $V_x = V_x\angle\theta$ (x = i, j and k). In V_{se} it is the controllable voltage source injected by connecting in series, $V_{se} = V_{se}\angle\theta_{se}$ (n = j, k) and in Z_{se} (n = j, k) is the transformer impedance. The complex power injected into any bus can be determined by modeling IPFC as a current source. The line and the series coupling transformer's resistances are neglected for making the calculations simpler. The injected power at buses are summarized and The Power flow equations for IPFC can written as below,

$$\begin{aligned} P_i = V_i^2 g_{ii} - & \sum_{j=1,j\neq i}^{n} V_i V_j \left(g_{ij}\cos\left(\theta_j-\theta_i\right) + b_{ij}\sin\left(\theta_j-\theta_i\right)\right) \\ & - \sum_{j=1,j\neq i}^{n} V_i Vse_{ij}\left(\left(g_{ij}\cos\left(\theta_i-\theta se_{ij}\right) + b_{ij}\sin\left(\theta_i-\theta se_{ij}\right)\right)\right) \end{aligned} \tag{1}$$

$$\begin{aligned} Q_i = V_i^2 b_{ii} - & \sum_{j=1,j\neq i}^{n} V_i V_j (g_{ij}\sin\left(\theta_j-\theta_i\right) + b_{ij}\sin\left(\theta_j-\theta_i\right)) \\ & - \sum_{j=1,j\neq i}^{n} V_i Vse_{ij}\left(g_{ij}\sin\left(\theta_i-\theta se_{ij}\right) + b_{ij}\sin\left(\theta_i-\theta se_{ij}\right)\right) \end{aligned} \tag{2}$$

where: V = Bus voltage magnitude, θ = Voltage angle, V_{se} = magnitude of injected voltage, θ_{se} = Angle of injected voltage.

4 Voltage Stability Index Formulation

In this study the Voltage Stability Index [12] abbreviated by "L_{ij}" and referred to a line is formulated as the measuring unit in predicting the voltage stability condition in the system. The mathematical formulation presented here is very simple and also achieves faster computation. By using the second order linear voltage equation at the receiving bus on a two bus system the Lij is obtained (Figs. 4, 5 and 6).

From Fig. 3, the voltage quadratic equation at the receiving bus is written as

$$\lfloor V_j^2 - \left(\frac{R}{X}\sin\delta + \cos\delta\right)V_i V_j + \left(X + \frac{R^2}{X}\right)Q_j = 0\rfloor \tag{3}$$

Setting the discriminate of the equation to be greater than or equal to zero:

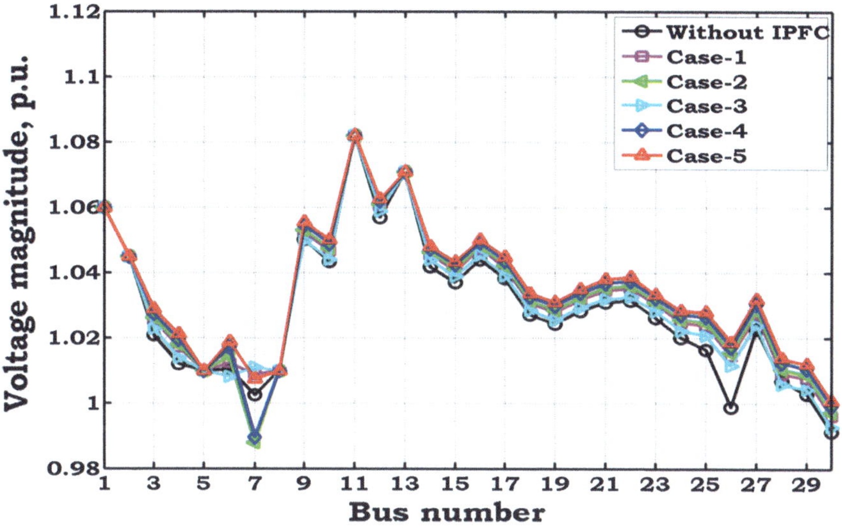

Fig. 4 Plot between bus number and voltage mangitude witout IPFC and with IPFC during various conditions

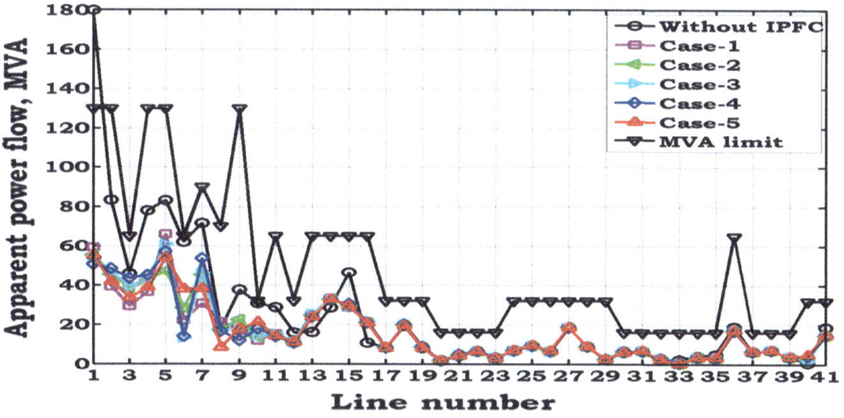

Fig. 5 Plot between line number and apparent power witout IPFC and with IPFC during various conditions

Fig. 6 Plot of power losses witout IPFC and with IPFC during various conditions

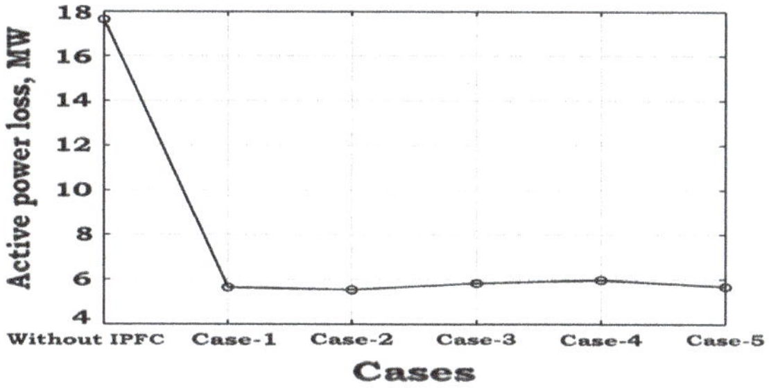

$$\lfloor\left(\frac{R}{X}\sin\delta+\cos\delta\right)V_i^2\rfloor-4\left(X+\frac{R^2}{X}\right)Q_j\geq 0 \tag{4}$$

Rearranging above equation, Voltage Stability Index "L_{ij}" is

$$L_{ij}=\frac{4Z^2Q_jX}{V_i^2(Rsin\delta+Xcos\delta)^2} \tag{5}$$

where: Z and X are line impedance and reactance respectively,

Q_j and is the reactive power at the receiving end, V_i and V_j are sending end and receiving end voltages.

5 Contingency Analysis

Contingency analysis aims at studying the effect of the outage of components of power system like transmission lines, transformers and generators on the power system network. Contingencies referring to disturbances such as transmission line outages or generator outages may cause large amount of load may stay connected or removed and thus resulting a change in either the state or configuration of the power system. Contingencies may result in severe changes of the operating parameters. Consequently, planning for contingencies forms an important aspect of secure operation of the power system network. Contingency analysis helps the power system engineer at many stages like network design, programmed maintenance, network expansion and also in the identification of network weaknesses and thus serve as an important tool for estimating security of the power system during operation and planning. Contingency analysis allows the power system to be operated defensively. Majority of the faults occurring in the power system network can cause serious troubles within a small time if the operator could not take fast remedial action. Keeping in view of these, modern computers are equipped with contingency analysis programs which model the power system network and are used to know outage events and give alert to the operators of potential overloads and voltage violations.

The most difficult practical problem to manage within contingency analysis is the correctness of the method and the speed of solution of the model used. The operator should have an idea of the performance of the existing network which is instable condition and also he should possess the knowledge of the effect of a particularly contingency like outage of a particular generator or transmission line.

Recently, due to the problems such as the congestion management, the minimization of the operational cost and the overall generating cost, the additional degree of freedom possessed by the FACTS devices have aroused great interest in the application of the FACTS devices, especially the UPFC, the IPFC and the generalized Unified Power Flow Controller (GUPFC), in the OPF control. However, very few publications have focused on the comparison between the performance of the UPFC and the IPFC in the OPF control. This paper proposes an algorithm for optimal location of the IPFC to improve voltage stability under the over loaded line outage contingency in a power system network.

6 Performance Index

The contingency analysis process gives an idea about the effect of individual contingency cases, hence the above process take large time to evaluate the contingency in the power system network. The contingency analysis is selected by calculating a kind of severity indices known as Performance Indices PI [13]. These indices values are calculated using the conventional power flow algorithms for individual contingencies. Based on the line flow limit in overloaded lines, contingencies are ranked in a manner where the highest value of PI is ranked first. This will continues till the no severe contingencies are found.

There are two kinds of performance indices used in power system networks, one is active power performance index (PI_P) and other one is reactive power performance index (PI_V). The active power performance index (PI_P) reflects the violation of line active power flows and is given as

$$PI_P = \sum_{i=1}^{L} \left(\frac{P_i}{P_{i^{max}}}\right)^{2n} \tag{6}$$

where: P_i = active power flow in line I, P_i^{max} = maximum active power flow in line i

n = specified exponent, L = number of transmission lines in the power system

The maximum power flow in each line will be calculated as

$$P_i^{max} = \frac{V_i * V_j}{X} \tag{7}$$

And other performance index parameter which is used in reactive performance index corresponding to the bus voltage magnitude violations. The value can be evaluated as below

$$PI_V = \sum_{i=1}^{N_{pq}} \left[\frac{2(V_i - V_{inon})}{V_{imax} - V_{imin}}\right]^2 \tag{8}$$

where V_i = voltage at bus I, V_{imax} and V_{imin} max. and min. values voltage limits, V_{inon} = average value of V_{imax} and V_{imin}, N_{pq} = total number of voltage buses.

7 Results and Conclusions

The proposed method is implemented in MATLAB working platform. The performance of proposed method is tested with IEEE 30 bus system. Initially severity indices known as Performance Indices are calculated and are ranked in a manner where the highest value of PI is ranked first. Based on the line flows (MVA) outage lines (lines which are overloaded) and contingency rank have been determined and are indicated in Tables 1 and 2 respectively. Also line flows under rank 1 contingency criterion are provided in Table 3.

Table 1 Over loaded lines of IEEE-30 bus system during contingency analysis

S. no.	Outage line	Overloaded lines	Line flow (MVA)	Line limit (MVA)	PI
1	1–2	1–3	312.783	130	2.406
		2–4	66.2477	65	1.0192
		3–4	281.6728	130	2.1667
		4–6	175.1941	90	1.9466
		6–8	40.2057	32	1.2564
2	1–3	1–2	274.0404	130	2.108
		2–4	86.1364	65	1.3252
		2–6	92.7082	65	1.4263
		6–8	35.5403	32	1.1106
3	2–4	1–2	163.1902	130	1.2553
		2–6	82.8385	65	1.2744
		6–8	34.3948	32	1.0748
4	3–4	1–2	271.089	130	2.0853
		2–4	84.8975	65	1.3061
		2–6	91.7550	65	1.4116
		6–8	35.2239	32	1.1007
5	2–5	1–2	171.3989	130	1.3185
		2–4	77.6706	65	1.1949
		2–6	105.4337	65	1.6221
		4–6	121.4176	90	1.3491
		5–7	110.1903	70	1.5741
		6–8	35.8277	32	1.1196
6	2–6	1–2	163.1085	130	1.2547
		2–4	74.6436	65	1.1484
		4–6	114.4738	90	1.2719
		6–8	36.3001	32	1.1344

(continued)

Table 1 (continued)

S. no.	Outage line	Overloaded lines	Line flow (MVA)	Line limit (MVA)	PI
7	4–6	1–2	203.7972	130	1.5677
		2–6	98.9857	65	1.5229
		4–12	66.9868	65	1.0306
8	5–7	1–2	183.331	130	1.4102
9	6–7	1–2	189.9598	130	1.4612
10	6–8	1–2	180.4949	130	1.3884
		6–28	48.2618	32	1.5082
11	9–10	1–2	179.4047	130	1.38
12	12–14	1–2	180.0452	130	1.385
13	12–15	1–2	180.6197	130	1.3894
14	12–16	1–2	180.1352	130	1.3857
15	14–15	1–2	179.8402	130	1.3834
16	16–17	1–2	179.9547	130	1.3843
17	15–18	1–2	180.042	130	1.3849
18	18–19	1–2	179.897	130	1.3838
19	19–20	1–2	179.8863	130	1.3837
20	10–20	1–2	179.9755	130	1.3844
		15–18	16.3239	16	1.0202
21	10–17	1–2	179.8006	130	1.3831
22	10–21	1–2	180.0833	130	1.3853
23	10–22	1–2	179.8806	130	1.3837
24	21–22	1–2	179.833	130	1.3833
25	15–23	1–2	180.0283	130	1.3848
26	22–24	1–2	179.8826	130	1.3837
27	23–24	1–2	179.8725	130	1.3836
28	24–25	1–2	179.7933	130	1.383
29	25–27	1–2	179.8072	130	1.3831
30	27–29	1–2	180.1101	130	1.3855
31	27–30	1–2	180.2028	130	1.3862
32	29–30	1–2	179.9245	130	1.384
33	8–28	1–2	179.8769	130	1.3837
34	6–28	1–2	179.9431	130	1.3842
		6–8	46.3583	32	1.4487

Table 2 Contingency ranking

S. no.	Outage line	Severity index	Rank	S. no.	Outage line	Severity index	Rank
1	**1–2**	**8.7949**	**1**	18	18–19	1.3838	23
2	1–3	5.9701	3	19	19–20	1.3837	24
3	2–4	3.6045	7	20	10–20	2.4046	10
4	3–4	5.9037	4	21	10–17	1.3831	31
5	2–5	8.1783	2	22	10–21	1.3853	17
6	2–6	4.8094	5	23	10–22	1.3837	25
7	4–6	4.1212	6	24	21–22	1.3833	30
8	5–7	1.4102	12	25	15–23	1.3848	20
9	6–7	1.4612	11	26	22–24	1.3837	26
10	6–8	2.8966	8	27	23–24	1.3836	28
11	9–10	1.38	34	28	24–25	1.383	33
12	12–14	1.385	18	29	25–27	1.3831	32
13	12–15	1.3894	13	30	27–29	1.3855	16
14	12–16	1.3857	15	31	27–30	1.3862	14
15	14–15	1.3834	29	32	29–30	1.384	22
16	16–17	1.3843	21	33	8–28	1.3837	27
17	15–18	1.3849	19	34	6–28	2.8329	9

Table 3 Line flows under rank-1 contingency

From bus	To bus	S flow	MVA limit	Margin limit	From bus	To bus	S flow	MVA limit	Margin limit
1	3	312.783	130	−182.783	15	18	7.8404	16	8.1596
2	4	66.2477	65	−1.2477	18	19	4.4918	16	11.5082
3	4	281.6728	130	−151.673	19	20	6.0022	32	25.9978
2	5	56.1075	130	73.8925	10	20	8.3951	32	23.6049
2	**6**	**29.0091**	**65**	**35.9909**	10	17	5.7973	32	26.2027
4	6	175.1941	90	−85.1941	10	21	18.5089	32	13.4911
5	7	52.5367	70	17.4633	10	22	8.7761	32	23.2239
6	**7**	**67.9993**	**130**	**62.0007**	21	22	2.503	32	29.497
6	8	40.2057	32	−8.2057	15	23	7.3842	16	8.6158
6	9	26.9166	65	38.0834	22	24	6.1744	16	9.8256
6	10	13.8073	32	18.1927	23	24	3.7824	16	12.2176
9	11	22.8785	65	42.1215	24	25	1.7065	16	14.2935
9	10	26.1654	65	38.8346	25	26	4.2654	16	11.7346
4	12	51.2579	65	13.7421	25	27	3.5429	16	12.4571
12	13	21.8916	65	43.1084	28	27	17.5846	65	47.4154
12	14	8.8638	32	23.1362	27	29	6.4196	16	9.5804
12	15	21.8485	32	10.1515	27	30	7.2948	16	8.7052
12	16	10.7842	32	21.2158	29	30	3.7553	16	12.2447
14	15	2.3373	16	13.6627	8	28	2.9555	32	29.0445
16	17	6.8978	16	9.1022	6	28	17.8011	32	14.1989

Table 4 Voltage magnitudes

Bus no.	Without	Case-1	Case-2	Case-3	Case-4	Case-5
1	1.06	1.06	1.06	1.06	1.06	1.06
2	1.045	1.045	1.045	1.045	1.045	1.045
3	1.020837	1.025552	1.026262	1.022836	1.027562	1.028899
4	1.012045	1.017312	1.018126	1.013987	1.019689	1.021329
5	1.01	1.01	1.01	1.01	1.01	1.01
6	1.010358	1.012468	1.014425	1.008207	1.017257	1.018915
7	1.002434	1.008811	0.987926	1.011041	0.989551	1.007447
8	1.01	1.01	1.01	1.01	1.01	1.01
9	1.05003	1.052258	1.053166	1.049843	1.054598	1.055624
10	1.043259	1.046638	1.047412	1.043946	1.048805	1.050049
11	1.082	1.082	1.082	1.082	1.082	1.082
12	1.056788	1.060382	1.061079	1.058948	1.062047	1.062608
13	1.071	1.071	1.071	1.071	1.071	1.071
14	1.041741	1.045531	1.046298	1.043947	1.047366	1.047985
15	1.036974	1.040737	1.041437	1.038896	1.042522	1.043316
16	1.043858	1.047362	1.048051	1.04534	1.04918	1.05008
17	1.03811	1.041537	1.042275	1.039022	1.043587	1.044746
18	1.027036	1.030706	1.031431	1.028529	1.032634	1.033607
19	1.024284	1.027888	1.02863	1.025524	1.029902	1.030974
20	1.028245	1.031795	1.032542	1.029343	1.033842	1.034962
21	1.030907	1.034646	1.035438	1.031939	1.036849	1.038095
22	1.031462	1.035296	1.03609	1.032596	1.037502	1.038743
23	1.026122	1.030367	1.031134	1.02817	1.032377	1.033345
24	1.02006	1.024951	1.025818	1.022314	1.027272	1.028448
25	1.016386	1.023951	1.024958	1.020853	1.026661	1.028042
26	0.998691	1.014721	1.015737	1.011594	1.017456	1.01885
27	1.022661	1.027435	1.028538	1.024079	1.030396	1.031884
28	1.006813	1.008988	1.010474	1.005733	1.012628	1.013898
29	1.00281	1.007686	1.008812	1.004258	1.01071	1.012228
30	0.991328	0.996262	0.997402	0.992794	0.999323	1.00086

Base on the results of line flows it can be concluded that the best location of IPFC will be such that interline power flow takes place between lines 2–6 and 6–7. In short it is read as 2–6–7 (Tables 4, 5 and 6).

IPFC LOCATION: 6–2–7

Voltage values

Case-1 Vseij = 0.02; Thseij = 072; Vseik = 0.10; Thseik = 360
Case-2 Vseij = 0.04; Thseij = 144; Vseik = 0.08; Thseik = 288
Case-3 Vseij = 0.06; Thseij = 216; Vseik = 0.06; Thseik = 216

Table 5 Line apparent power flows

Line no	Without	Case-1	Case-2	Case-3	Case-4	Case-5	Line limit
1	179.8264	59.47208	55.5219	54.12143	50.8346	55.39428	130
2	83.26183	39.46966	45.68394	46.30381	48.67692	42.05647	130
3	45.95788	29.43964	38.35691	39.58179	43.71073	33.89528	65
4	77.92641	36.66613	42.43029	43.01274	45.24646	39.02199	130
5	83.11859	66.29842	47.44745	61.02156	57.29246	54.24907	130
6	61.95546	22.31943	28.5559	13.29143	14.05878	38.07222	65
7	71.65108	30.93591	45.40385	46.29661	53.76256	38.53168	90
8	18.11872	21.45945	17.58856	16.86571	16.74498	8.707027	70
9	37.54465	17.24524	22.74425	18.61146	11.93575	18.12826	130
10	30.60232	12.07611	13.68856	14.3966	18.1302	21.37357	32
11	28.76067	15.30788	14.55741	15.37277	13.9281	14.25033	65
12	15.84003	11.27801	10.96543	10.94006	10.80724	11.13735	32
13	16.13931	24.80973	24.54832	25.52792	24.14602	23.86569	65
14	28.55022	33.38389	32.85225	32.85906	32.55582	33.07232	65
15	46.48143	28.80205	29.78736	29.44849	30.43123	29.69254	65
16	10.72805	21.46567	21.28129	21.87821	21.0439	20.91633	65
17	8.249855	8.343178	8.421711	8.468506	8.452871	8.348217	32
18	19.23466	19.6677	20.01886	20.17301	20.17465	19.73776	32
19	8.00534	8.235117	8.604458	8.752462	8.779008	8.319786	32
20	1.750135	1.827968	1.901815	1.947566	1.932963	1.830703	16
21	3.976752	4.211036	4.604166	4.726869	4.806056	4.331541	16
22	6.281252	6.363855	6.580712	6.649777	6.68563	6.429352	16
23	2.896999	2.978454	3.195705	3.259423	3.306035	3.048629	16
24	7.208406	7.127738	6.94041	6.858115	6.864059	7.089788	32
25	9.680744	9.595946	9.400787	9.317149	9.319372	9.554902	32
26	6.88069	6.708984	6.510201	6.291977	6.482794	6.775726	32
27	18.62382	18.75347	18.69912	18.7068	18.66702	18.71564	32
28	8.855086	8.953484	8.916976	8.922268	8.895382	8.927963	32
29	2.391588	2.563762	2.581609	2.589027	2.590211	2.567308	32
30	5.900249	6.223063	6.412591	6.536692	6.488813	6.224871	16
31	6.400275	6.861728	6.754018	6.774447	6.688916	6.782285	16
32	2.291648	2.593784	2.805286	2.911114	2.905035	2.62386	16
33	2.26144	0.34264	0.468959	0.574455	0.539754	0.333524	16
34	4.262281	3.501428	3.501428	3.501428	3.501428	3.501428	16
35	4.848405	3.21335	3.065147	2.972533	3.013456	3.208899	16
36	18.85763	17.08748	16.96832	16.84874	16.93703	17.12148	65
37	6.411282	6.409487	6.409076	6.410746	6.408387	6.407838	16
38	7.284724	7.282545	7.282046	7.284073	7.28121	7.280544	16
39	3.753032	3.75254	3.752428	3.752885	3.752239	3.752089	16
40	0.666002	2.93711	3.508159	1.966458	4.445017	5.053087	32
41	18.71879	14.71428	14.69893	14.44456	14.88212	15.19993	32

Table 6 Total power losses

	Without	Case-1	Case-2	Case-3	Case-4	Case-5
P loss, MW	17.64796	5.634803	5.532813	5.820498	5.966696	5.671125
Q loss, MVAr	34.41422	−4.34473	−8.96334	−4.67502	−4.00816	−7.57269

Case-4 Vseij = 0.08; Thseij = 288; Vseik = 0.04; Thseik = 144
Case-5 Vseij = 0.10; Thseij = 360; Vseik = 0.02; Thseik = 072.

References

1. Hingorani NG, Gyugyi L (2000) *Understanding FACTS: concepts and technology of flexible AC transmission systems*. The Institute of Electrical and Electronics Engineers, New York, NY, p 297
2. Saxena DR (2013) Multi-line power flow control using interline power flow controller (ipfc) in power transmission system. Int J Eng Comput Sci 2(11):3089–3093
3. Althowibi FA, Mustafa MW (2013) Power system voltage stability: indications, allocations and voltage collapse predictions. Int J Adv Res Electr Electron Instrum Eng 2(7):3138–3152
4. Gyugyi L, Schauder CD, Williams SL, Rietman TR, Torgerson DR, Edris A (1995) The unified power flow controller: a new approach to power transmission control. IEEE Trans Power Deliv 10(2):1085−1097
5. Teerathana S, Yokoyama A, Nakachi Y, Yasumatsu M (2005) An optimal power flow control method of power system by interline power flow controller (IPFC). In: Proceedings of 7th International Conference on Power Engineering, Singapore, pp. 1–6
6. Sayyed AN, Gadge PM, Sheikh RU (2014) Contingency analysis and improvement of power system security by locating series FACTS devices "TCSC and TCPAR" at optimal location. IOSR-JEEE 19−27. ISSN: 2320–3331
7. Zhang J, Yokoyama A (2006) A comparison between the UPFC and the IPFC in optimal power flow control and power flow regulation. IEEE Trans 1-4244-0228-X/06, 339−345
8. Kumkratug Prechanon (2010) Application of interline power flow controller to increase transient stability of power system. J Comput Sci 6(12):1490–1493
9. Nimje AA, Panigrahi CK, Mohanty AK (2011) Interline power flow controller: review paper. Int Electr Eng J 2(3):550–554
10. Gasim Mohamed SE, Yousif Mohamed A, Abdelrahim, YH (2012) Power system contingency analysis to detect network weaknesses. In: ZEC Infrastructure 2012, 18–20 Jun 2012, Amman, Jordan, pp. I3-1–I3-11
11. Sayyed, ANL, Gadge PM, Sheikh RU (2014) Contingency analysis and improvement of power system security by locating series FACTS devices (TCSC and TCPAR) at optimal location. IOSR J Electr Electron Eng (IOSR-JEEE) e-ISSN: 2278−1676, p-ISSN: 2320-3331, 19−27
12. Naresh Babu AV, Sivanagaraju S, Padmanabharaju C, Ramana T (2010) Multi-line power flow control using Interline Power Flow Controller (IPFC) in power transmission systems. World Acad. Sci. Eng. Technol. 39:776–780
13. Contingency ranking and selection—NPTEL material, chapter 5.6, pp. 245–249

Author Biographies

Sri. B.V. Rami Reddy had completed his B.Tech and M.Tech in 1999 and 2005 with first class in the department of Electrical and Electronics Engineering from J.N.T. University Hyderabad. Now he is pursuing his PhD in J.N.T. University, Ananthapur. He is working as a Associate Professor in the Department of Electrical and Electronics Engineering. His scientific research is focusing on Power Systems and Distribution Systems.

Dr. P. Sujatha has completed her B.Tech and M.Tech in 1993 and 2003 in the department of Electrical and Electronics Engineering from J.N.T. University Hyderabad and her Doctorate degree from J.N.T. University, Ananthapur in 2011. Now she is working as a Professor and Head of the Department of Electrical and Electronics Engineering in J.N.T. University college of Engineering, Ananthapuramu. Her scientific research is focusing on Power Systems and Energy Management.

Dr. Y.V. Siva Reddy had completed his B.Tech and M.Tech in 1995 and 2000 respectively in the department of Electrical and Electronics Engineering from J.N.T. University Hyderabad and obtained doctoral degree from J.N.T. University, Ananthapur in 2010. He is working as a Professor in the Department of Electrical and Electronics Engineering in G.P.R Engineering College, Kurnool. His areas of research are Power Systems and Industrial Drives.

The manufacturer's authorised representative in the EU is Springer Nature Customer Service Centre GmbH, Europaplatz 3, 69115 Heidelberg, Germany. If you have any concerns regarding our products, please contact ProductSafety@springernature.com

Printed and bound by CPI Group (UK) Ltd, Croydon, CR0 4YY
30/09/2026
02225440-0014